Martin Philipp Rohde

In-situ H_2O removal via hydrophilic membranes during Fischer-Tropsch and other fuel-related synthesis reactions

In-situ H_2O removal via hydrophilic membranes during Fischer-Tropsch and other fuel-related synthesis reactions

by
Martin Philipp Rohde

KIT Scientific Publishing

Dissertation, Karlsruher Institut für Technologie
Fakultät für Chemieingenieurwesen und Verfahrenstechnik, 2010

Impressum

Karlsruher Institut für Technologie (KIT)
KIT Scientific Publishing
Straße am Forum 2
D-76131 Karlsruhe
www.ksp.kit.edu

KIT – Universität des Landes Baden-Württemberg und nationales
Forschungszentrum in der Helmholtz-Gemeinschaft

KIT Scientific Publishing 2011
Print on Demand

ISBN 978-3-86644-640-3

IN-SITU H_2O REMOVAL VIA HYDROPHILIC MEMBRANES DURING FISCHER-TROPSCH AND OTHER FUEL-RELATED SYNTHESIS REACTIONS

zur Erlangung des akademischen Grades eines
DOKTORS DER INGENIEURWISSENSCHAFTEN (Dr.-Ing.)

der Fakultät für Chemieingenieurwesen und Verfahrenstechnik des
Karlsruher Institut für Technologie (KIT)
genehmigte

DISSERTATION

von

Dipl.-Ing. Martin Philipp Rohde

aus Wilhelmshaven

Referent:	Prof. Dr.-Ing. Georg Schaub
Korreferent:	Prof. Dr.-Ing. Andreas Seidel-Morgenstern
Tag der mündlichen Prüfung:	16. Juli 2010

Für Francka

Danksagung

Die vorliegende Arbeit entstand im Zeitraum von September 2002 bis April 2007 während meiner Tätigkeit als wissenschaftlicher Angestellter am Engler-Bunte-Institut, Bereich Gas, Erdöl und Kohle der Universität Karlsruhe (TH).

In erster Linie möchte ich bei meinem Doktorvater Prof. Dr.-Ing. Georg Schaub für sein großes Vertrauen, die Unterstützung und Förderung innerhalb der letzten Jahre am Institut bedanken. Er ermöglichte mir neben der Durchführung eines Teiles dieser Arbeit im Rahmen eines dreijährigen EU-Projektes die Teilnahme an zahlreichen nationalen und internationalen Tagungen.

Bei Prof. Dr.-Ing. habil. A. Seidel-Morgenstern bedanke ich mich für die Übernahme des Korreferats und die angenehmen Gespräche und die Möglichkeit, die Ergebnisse am MPI Magdeburg präsentieren zu dürfen.

Desweiteren möchte ich den Professoren am Engler-Bunte-Institut Prof. Dr.-Ing. R. Reimert, Prof. Dr.-Ing. G. Schaub und Prof. H. Schulz für die angenehme Atmosphäre, die Offenheit und die Zusammenarbeit bedanken. Ich habe in den zahlreichen Diskussionen sehr viel über die Fischer-Tropsch Synthese im Speziellen und Prozeß- und Anlagentechnik fossiler und erneuerbarer Energien im Generellen gelernt. Die Tätigkeit am Institut hat meinen Werdegang geprägt und wird ihn weiterhin prägen.

Eine Arbeit zu Membranreaktoren hängt maßgeblich von der Qualität der verwendeten Membrane ab. Deshalb möchte ich mich hier bei ECN (Energy Research Centre of The Netherlands), insbesondere bei Jaap Vente und Henk van Veen, für die Unterstützung und die gute Zusammenarbeit erkenntlich zeigen.

Die Teilnahme am EU-Projekt LTCPO-GTL ermöglichte mir die Zusammenarbeit mit einer Reihe von Institutionen aus Forschung und Industrie. Hier möchte ich den offenen Austausch und Kooperation mit unseren Partnern ENI, IFP, ECN und dem Paul-Scherrer-Institut erwähnen. Dank geht an Roberto Zennaro und Gabriele Clerici von ENI, Francois Hugues und Reynald Bonneau von IFP und Frederik Vogel und Stefan Rabe vom PSI. Besonders Roberto Zennaro hat mich durch seine Persönlichkeit und seine klare Denk- und Arbeitsweise beeindruckt.

Weiterer Dank geht an Prof. M. Menendes, Universität Zaragoza, und Prof. F. Kapteijn, Universität Delft, für die unkomplizierte wissenschaftliche Zusammenarbeit.

Meinem jetzigen Arbeitgeber, insbesondere Dr. M. Senden, Dr. F. Schrauwen und Dr. U. Gantz, danke ich für die Unterstützung.

Zwei Menschen möchte ich hier hervorheben, die einen wichtigen Teil zum Gelingen dieser Arbeit beigetragen haben, nämlich meine Kollegen und Freunde Dominik Unruh und Alba Mena. Ich werde die Zusammenarbeit und die gegenseitige Unterstützung, die Diskussionen und die schöne Zeit immer in Erinnerung behalten. Auf eine lange Freundschaft.

Weiterhin möchte ich mich bei meinen ehemaligen Arbeitskollegen Ulrich Hennings, Markus Brune, Luan Nguyen, Sebastian Zürcher, Marius Hackel, Gelase Mbdinga, Vince van Buren, Frank Graf und Siegfried Bajohr sowie allen anderen Kollegen für die angenehme Zeit am und außerhalb des Instituts bedanken.

Hervorheben möchte ich zudem meine Diplom- und Seminararbeiter Johannes Ebert, Alberto Freijanes, Jens Buhmann, Christelle Perard und Manuel Mächtel. Simon Nonnenmacher und Johannes Baisch haben einen großen Beitrag geleistet, die Laborinfrastruktur aufzubauen. Die Durchführung der Arbeit wurde erst durch die Unterstützung aus den Werkstätten und dem Labor durch Frank Herter, Horst Haldenwang, Gustav Kliewer, Steffen Haury, Stefan Herbel und Sabina Hug ermöglicht.

Der grösste Dank geht natürlich an Francka für ihre Geduld, ihre uneingeschränkte Unterstützung und ihr Verständnis, und an meine Eltern und Geschwister, die mich während dieser Zeit immer wieder motiviert haben.

Contents

1 INTRODUCTION 1

1.1 H_2O removal during fuel-related synthesis reaction 4
1.2 Synthesis of fuels via CO_2 hydrogenation 7
1.3 Objectives and approach 9

2 LITERATURE REVIEW 13

2.1 Fischer-Tropsch synthesis 13
2.1.1 Reactions 15
2.1.2 Product selectivity 16
2.1.3 Kinetics 21
2.1.4 Catalyst stability 25
2.1.5 Fischer-Tropsch process 28
2.2 Membrane reactors and membranes 30
2.2.1 Membrane reactors 31
2.2.2 Membrane structure and materials 36
2.2.3 Mass transport through dense membranes 39
2.2.4 Mass transport in porous membranes 43
2.2.5 Permeation characteristics of hydrophilic membranes 47

3 MASS TRANSFER IN HYDROPHILIC MEMBRANES 53

3.1 Experimental 53
3.1.1 Applied membranes 53
3.1.2 Experimental set-up 55
3.1.3 Data analysis and definitions 57
3.2 Results and discussion 60
3.2.1 Permeation experiments with single gases 61
3.2.2 Permeation experiments with gas mixtures 63
3.3 Membrane assessment 65

4 KINETICS AND SELECTIVITY OF THE FISCHER-TROPSCH REACTION 69

4.1 Experimental 69
4.1.1 Applied catalysts 69
4.1.2 Experimental set-up 70
4.1.3 Data analysis and definitions 74
4.1.4 Fixed-bed reactor balance and kinetic analysis 76
4.2 Benchmarking of Fe- and Co-based catalysts 77
4.3 Kinetics and reference experiments 81
4.3.1 Fe-based catalyst Fe-GTLX 81
4.3.2 Co-based catalyst Co-GTL4 85
4.4 Product selectivity 86

5 *IN-SITU* H_2O REMOVAL DURING FISCHER-TROPSCH AND OTHER FUEL-RELATED SYNTHESIS REACTIONS 89

5.1 Experimental 89
5.1.1 Introduction and execution of *in-situ* H_2O removal experiments 90

5.1.2 Data analysis and definitions 94
5.1.3 Fixed-bed (membrane) reactor balance 96
5.2 *In-situ* H_2O removal during CO_2 hydrogenation by hydrophilic membranes 98
5.3 *In-situ* H_2O removal during DME/DEE synthesis by hydrophilic membranes 106
5.4 *In-situ* H_2O removal during FT synthesis by chemical reaction 109
5.5 Membrane stability 112

6 CASE STUDIES: POTENTIAL AND LIMITS OF IN-SITU H_2O REMOVAL 117

6.1 Da-Pe analysis of membrane reactors for *in-situ* H_2O removal 117
6.2 Case studies on *in-situ* H_2O removal during FT synthesis 120
6.2.1 Reference case 121
6.2.2 Protection of catalyst performance 123
6.2.3 Conversion of CO_2 containing synthesis gases 128
6.3 Application to other fuel related reactions 130

7 CONCLUSION AND OUTLOOK 137

8 SUMMARY 141

9 ZUSAMMENFASSUNG 151

10 REFERENCES 163

11 NOTATIONS 181

12 APPENDIX 189

12.1 Literature 189
12.2 Experimental plan and conditions 192
12.2.1 Mass transfer in hydrophilic membranes 192
12.2.2 Reaction kinetics 192
12.2.3 In-situ H_2O removal by membranes and chemical reaction 195
12.3 Analytical methods 197
12.3.1 Analysis of gas effluent streams by GC 197
12.3.2 H_2O measurement and oxygen balance 200
12.3.3 Carbon balance 202
12.4 Mass transfer in hydrophilic membranes 205
12.4.1 Mass transfer limitations 205
12.4.2 Membrane unit (permeation cell) performance 206
12.4.3 Dynamic permeation experiments 208
12.5 Kinetics and selectivity 211
12.5.1 Fischer-Tropsch synthesis 211
12.5.2 Low temperature CO/CO_2 shift reaction 223
12.5.3 DME/DEE synthesis 225
12.6 *In-situ* H_2O removal experiments 227
12.6.1 *In-situ* H_2O removal during CO_2 hydrogenation 227
12.6.2 *In-situ* H_2O removal during DME/DEE synthesis 229
12.6.3 *In-situ* H_2O removal during FT synthesis 231
12.7 Reactor and reactor models 232
12.7.1 (Membrane) reactor model – Additional calculations 232
12.7.2 Three-phase slurry bubble column reactor 242

1 Introduction

Meeting the growing energy demand will be one of the biggest challenges of this century. Economic, political, and environmental constraints have supported the development of new technologies in harvesting unconventional resources and providing new types of fuels:

- Prognoses indicate that the worldwide energy demand may increase significantly within the coming decades – e.g. up to 40% higher in 2030 compared to 2005 (ExxonMobil 2008), doubled primary energy consumption in 2050 compared to 2000 (Shell 2008) – due to continuing growth in world population and a strong growth in economies of Brazil, Russia, India and China. Fossil fuels will deliver the majority of primary energy, but the fraction of non-fossil energies will increase by 10-20% in 30 years. Liquid hydrocarbons will maintain their position in the overall scheme as no alternative concepts will be ready for the market. As the market launch of car-based fuel cells shifted, the car manufacturers want to bridge the gap to the 'hydrogen economy' or the age of 'electric mobility' by supporting the development of clean, sustainable synthetic hydrocarbons. In this context, the Volkswagen SunFuel® program promoting second generation biofuels (SunDiesel®, SunEthanol®) should be mentioned.
- The majority of conventional oil and gas reserves are concentrated within the 'strategic ellipse' in the Middle East and Russia. Political instabilities in the Middle East and diplomatic differences between Russia and Europe reveal the vulnerability of energy supply. Therefore, energy sources as coal, oil sands, oil shale and biomass have gained increasing public attention as they are distributed more equally around the globe. Recently, the United States and Sweden declared their desire to shift from oil dependent technologies to alternative resources, which is an example of the general change in public policy thinking about the energy challenge.
- Within the last seven years, the price of one barrel Brent oil has increased steadily from US$ 26 in 2001, until it reached a maximum of US$ 143 in July 2008. With oil prices above US$ 30-40 per barrel, the exploitation of non-conventional oil and gas reserves such as oil sands, oil shales, heavy crude oil, enhanced oil recovery, deep sea oil reserves, stranded (remote) and associated gas fields become feasible. Therefore, private oil companies are investing in bringing unconventional resources to market and thereby, establishing processes for future synthetic fuel production.

Fischer-Tropsch (FT), methanol and DME synthesis and methanol-to-hydrocarbon processes are different process routes based on synthesis gas from various competing feedstocks. Private oil companies use these new technologies as the key to market access. These technologies enable the exploitation of new raw materials and bring high-quality synthesized automotive fuels to the market. The Fischer-Tropsch synthesis gained worldwide attention when Sasol started-up their 34,000 bpd (Co-based slurry technology) and with Shell constructing its 140,000 bpd FT plant (Co-based multi-tubular fixed-bed technology) in Qatar.

- The Stern report (Stern 2006) and the recent series of IPCC reports on climate change (IPCC 2007) indicate that anthropogenic CO_2 emissions are responsible for global warming and that immediate action is needed to cut back energy consumption and CO_2 emissions. Today's mobility infrastructure is based on liquid hydrocarbons derived from crude oil and the transportation sector contributes about 20% of all global CO_2 emissions. Therefore, one can observe a growing production capacity of first generation biofuels as ethanol, FAME and palm oil and a move to the development of second generation biofuels as cellulose-based ethanol and butanol or biomass-based Fischer-Tropsch (BTL) synthesis. In particular, second generation biofuels are a step forward as they do not compete with food production and have a CO_2 reduction potential. The first BTL plant is currently being built by CHOREN in Germany, applying Shell's Fischer-Tropsch technology.

Synthetic fuels from synthesis gas have been produced on an industrial scale for several decades, known as coal-to-liquids (CTL) and gas-to-liquids (GTL). Fuels from biomass-derived synthesis gas (BTL) are a relatively new development. The most widespread synthesizing process is the Fischer-Tropsch (FT) synthesis, but methanol-to-hydrocarbon (MTHC) processes are also regarded as a promising route to convert coal, biomass and natural gas into high-octane gasoline and chemicals (Fougerit 1998, Fleisch & Sills 2004, Wurzel 2005). Furthermore, biomass can be converted to fuels in fermentation, extraction or thermo-chemical processes to obtain various fuel-suitable products as rapeseed oil, rapeseed methyl ester, ethanol, substitute natural gas (SNG) or bio crude oil (Specht 2003). These products can be used directly in engines, in combination with other fuels or as a source material for further applications (FNR 2005a). Table 1.1 gives an overview of synthetic fuels and biofuels discussed in literature as additives or substitutes for crude oil derived automotive fuels.

Table 1.1. Important physical properties of fuels and fuel components derived directly or indirectly (synthetic fuels) from fossil or renewable sources.

fuel	density[b] / kg/L	calorific value[c] / MJ/kg	cetane number /-	research octane number /-	boiling range[d] / °C	fuel equivalent / L	ref.
H_2	0.016	120	-	<88	-252	2.8	[4]
SNG	0.72	50	-	130	-161	1.4[e]	[4]
LPG	0.49	46	5	105-115	-42	0.86	[5]
gasoline	0.76	42.7	-	92	26-225	1	[4]
diesel	0.84	42.7	50	-	187-343	1	[4]
XTL[a]	0.76	43.9	>70	-	176-354	0.93	[2]
rapeseed oil	0.92	37.6	40	-	>300	0.96	[6]
FAME	0.87-0.90	37.1	56	-	180-330	0.91	[3]
methanol	0.79	19.7	3	>110	65	0.48	[4], [2]
ethanol	0.79	26.8	8	>100	78	0.65	[4]
butanol	0.81	33.05	25	96	118	-	[1]
MTBE	0.74	35.0	-	102	55.2	0.8	[4]
ETBE	0.74	36.4	-	102	73	0.83	[4]
DME	0.67	28.9	55-60	-	-26	0.59	[4], [5]
DMM_{3-8}			76		150-315		[8]
DEE	0.71	31	>125	-	34.6	-	[2]
DMC				104			[7]
DEC				125			[9]

[a] BTL, GTL, CTL, [b] at 15°C, [c] at 20°C, [d] at 1013 mbar, [e] in kg. SNG: substitute natural gas, LPG: liquefied petroleum gas, XTL: x-to-liquids, FAME: fatty acid methyl ester, MTBE: methyl tert-butyl ether, ETBE: ethyl tert-butyl ether, DME: dimethyl ether, DMM_{3-8}: poly-dimethoxy methane, DEE: diethyl ether, DMC: dimethyl carbonate, DEC: diethyl carbonate

[1] Asfar et al. (2003), [2] Bailey et al. (2006), [3] DIN E14214, [4] FNR (2005), [5] Ogawa et al. (2003), [6] FNR (2006), [7] Mills (1994), [8] Fleisch & Sills (2004), [9] Roh et al. (2003)

Some products obtained in the aforementioned processes can be converted into fuels with more specific properties. Well-known secondary products are the antiknock additives methyl or ethyl tert-butyl ether (MTBE/ ETBE). In recent years, dimethyl ether (DME) and butanol have attracted more interest due to several promising properties. Their efficiency is comparable to diesel but the exhaust gas is almost free of pollutant particles. In particular, DME has emerged as a clean multi-purpose fuel suitable for other applications, including LPG substitute and fuel for power generation (Fleisch, Sills & Briscoe 2002,

Fleisch & Sills 2004). DME is either produced through methanol dehydration or directly from synthesis gas. Recent studies have aimed to synthesize DME directly from biomass-originated synthesis gas, which is rich in CO_2 (Renk et al. 2006, Xu, Indala & Hertwig 2005, Hu 2005). DME can be distributed in the liquefied petroleum gas (LPG) network as an alternative to liquefied natural gas (LNG). TOTAL is also currently interested in developing DME as a clean commercial fuel to diversify and increase natural gas outlets, especially when a large investment in LNG is not economically justified (de Mestier du Bourg 2006). Due to their excellent fuel properties and their high oxygen content leading to a reduction in emissions of hydrocarbons, carbon monoxide, NO_x, and particulate matter, a whole range of oxygenates have been proposed as fuel additives as for e.g. polyether and acetals (Oppenlaender et al. 1980), dimethyl carbonate (DMC, Fleisch & Sills 2004), diethyl carbonate (DEC, Roh et al. 2003), poly-dimethoxy methane (DMM, Fleisch & Sills 2004) and diethyl ether (DEE, Bailey 2006), dialkyl polyformals from DME and formaldehyde (Sanfilippo, Patrini & Marchionna 2007).

1.1 H_2O removal during fuel-related synthesis reaction

H_2O is the main by-product of many fuel-related synthesis reactions. This is the case for the production of hydrocarbons from the synthesis gas, where 'excess' oxygen is removed as H_2O or CO_2. And this is also the case for many other synthesis reactions such as condensation reactions, oxidative dehydrogenation and carbonylation or acetal formation. High H_2O partial pressures or concentrations can affect reaction rates and the stability of catalytic systems negatively. Therefore, H_2O removal during the reaction may have several incentives as:

- *Accelerated reaction rates.* H_2O inhibits the reaction rate of various synthesis reactions due to strong competitive adsorption on the catalyst surface as reported for example for Fe-based FT synthesis or synthesis of dimethyl and diethyl ether over γ-Al_2O_3 (Bercic & Levic 1992, Butt, Bliss & Walker 1962). Selective H_2O removal during reaction could reduce the kinetic inhibition and leads furthermore to higher concentrations of the reactants, yielding higher reaction rates.
- *Improved catalyst lifetime.* Observations of reversible and irreversible catalyst deactivation induced by high H_2O partial pressure have been reported for various reactions. In the case of Co-based FT synthesis, various kinetic studies indicate catalyst deactivation at high H_2O/H_2 partial pressure ratios and high H_2O absolute pressures (e.g. Schanke et al. 1995, 1996, Dry 1990, van Berge et al. 2000, Bartholomew & Farrauto 2005, Saib

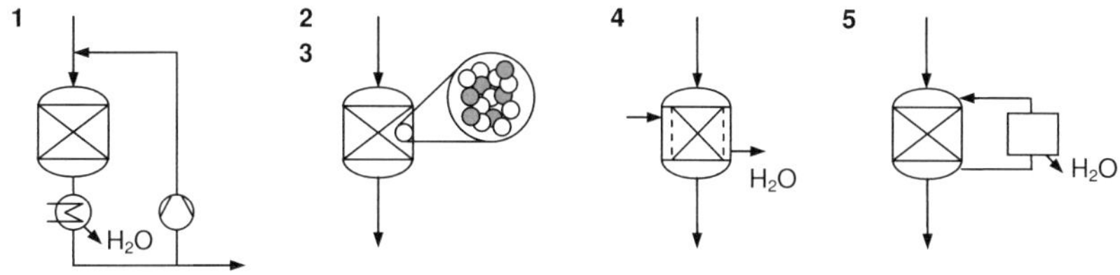

Figure 1.1. Concepts for H_2O removal, **1**: condensation in a recycle configuration, **2**: *in-situ* H_2O removal by reactive adsorption (catalyst/ adsorbent mixture), **3**: *in-situ* H_2O removal by H_2O consuming reaction (catalyst/ catalyst mixture), **4**: *in-situ* H_2O removal by reactive vapour permeation via selective, hydrophilic membrane, **5**: H_2O removal from liquid phase in external separation unit.

2006). In the case of Fe-based FT synthesis, water vapour leads to reoxidation of the catalyst (Dry 1990). Strong catalyst deactivation is also reported for dimethyl ether synthesis from methanol (Kim et al. 2006). In the methanol-to-gasoline (MTG) synthesis, the role of H_2O is two-fold: catalyst deactivation by coke deposition is slowed down at low H_2O concentrations, however, for H_2O contents above 50 wt%, irreversible deactivation by dealumination of the HZMS-5 zeolite occurs, restricting the operation of the MTG process (Aguayo et al. 2001). Another example is the production of dimethyl carbonate (DMC) from methanol and CO via the oxidative carbonylation (Di Muzio 1993), an environmentally safe and industrially favoured production route. Due to deactivation of the copper methoxy chloride catalyst, the H_2O concentration is limited to 10 wt%. Selective H_2O removal during reaction could diminish catalyst deactivation, and may allow operating at higher per pass conversions.

- *Conversion levels beyond equilibrium constraints.* Conversion levels of certain reaction are limited by equilibrium constraints. Examples are synthesis of methanol or hydrocarbons from CO_2 containing synthesis gases, dimethyl ether synthesis or CO_2/CO shift reaction. Selective H_2O removal during reaction can shift the equilibrium composition in favour of the desired product, resulting in conversion level exceeding the equilibrium constraints.

The *in-situ* removal of a specific component e.g. H_2O during reaction is possible if reaction and separation task are integrated into a single unit, which is one of the most important methods in the field of multifunctional reactors or process intensification (Agar & Ruppel 1988, Schmidt-Traub & Górak 2006). Figure 1.1 shows schematically different approaches for H_2O removal from the process. H_2O removal from a gas stream by condensation is the most straightforward concept. In the FT synthesis process, the reactor effluent is cooled down, and liquid H_2O and hydrocarbons are separated in a three-phase separator. The effluent cooler-separator arrangement for H_2O and product removal can be either integrated within a recycle around the reactor (Steynberg 2004) or between several reactor stages (Popp 1996). The resulting H_2O concentration at reactor outlet is determined by the per pass conversion and the per stage conversion, respectively. One advantage in FT synthesis is that the desired products (long-chain hydrocarbons) can be separated easily from the by-product water. In other processes (e.g. dimethyl ether synthesis, dimethyl carbonate synthesis), complex distillation line-ups are required to break up water-product-educt mixtures.

In an adsorptive reactor, the chemical reaction is coupled with adsorptive separation functionality (Reßler et al. 2006). H_2O removal by reactive adsorption was applied on lab-scale for the reverse water gas (CO_2/CO-) shift reaction at 250 °C (Carvill et al. 1996), for direct dimethyl ether synthesis from H_2-rich synthesis gases at 250 °C (Kim, Jung & Lee 2001, Reßler & Agar 2005) and for the Claus reaction at 250-350 °C (Elsner 2004) to overcome equilibrium limitations. The abovementioned studies show that reactive adsorption can be applied to a temperature range relevant for fuel-related synthesis reactions. A clear disadvantage is the dynamic, periodic operation, cycling between synthesis and adsorbent regeneration.

H_2O can be removed by a chemical reaction if an additional catalytic functionality is integrated. A proven example is the direct DME synthesis from H_2-deficient synthesis gas, where methanol synthesis, dehydration and CO/CO_2-shift functions are combined in a multifunctional catalyst system (Ogawa et al. 2003, de Mestier du Bourg 2006, Renk et al. 2006). H_2O is consumed by the shift reaction, which produces additional H_2 for the methanol synthesis. Fe-based FT catalysts are active for the shift reaction intrinsically, and the shift activity can be enhanced by promoters such as potassium. On lab-scale basis, non-shift active Co-based FT catalysts were combined with a low temperature shift catalyst (Post & Sie 1985, Chanenchuk, Yates & Satterfield 1991), whereas the focus was on *in-situ* H_2 production (internal shift) rather than on *in-situ* H_2O removal.

In reactive vapour permeation, H_2O is continuously removed via a highly selective hydrophilic membrane. Membrane reactor concepts for *in-situ* H_2O removal have been investigated for various synthesis reactions mainly with a focus on equilibrium displacement e.g. esterification reactions with zeolite membranes (Coronas & Santamaria 2004), gas-phase esterification reactions with (catalytic) active membranes (Bernal et al. 2002), gas-phase etherification from methanol and tert-butyl alcohol to MTBE (Salomon et al. 2000), and methanol synthesis from CO_2 with polymer and zeolite membranes (Struis &

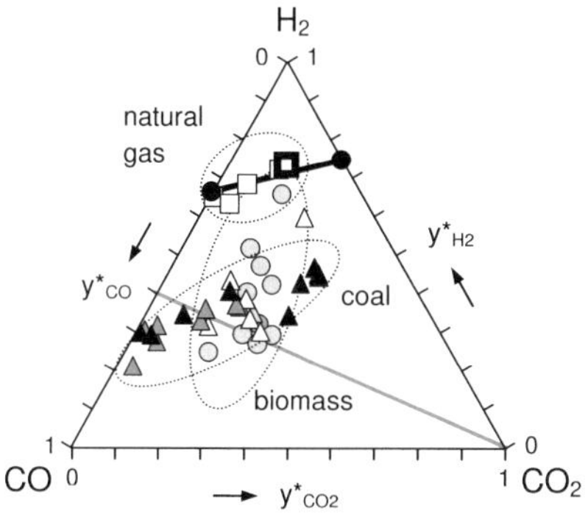

Figure 1.2. Composition of dry synthesis gases (inert components excluded), derived from different feedstocks and different gasification processes (data see Table 12.1).

Coal gasification: ▲: moving bed, △: fluidized bed, ▲: entrained flow gasifier

○: biomass gasification

□: natural gas gasification

●: $H_2/CO=2$ (FTS, stoichiometric)

●: $H_2/CO_2 = 3$ (CO_2 hydrogenation)

◘: LTCPO-GTL: natural gas low temperature catalytic partial oxidation (LTCPO-GTL 2005)

Stucki 2001, Galluci, Paturzo & Basile 2004). *In-situ* H_2O removal during FT synthesis and during DME synthesis from methanol was proposed by Espinoza et al. (2000) and by Lee, Youn & Bongkuk (2004, 2006), respectively, as H_2O acts as catalyst poison or inhibitor.

While the previous approaches deal with H_2O removal from the gas phase, the patent literatures describes concepts focusing on H_2O dissolved in the liquid phase. Here, the H_2O-rich liquid phase is routed from the reactor to an (external) H_2O removal unit and the nearly H_2O-free phase is recycled back. Zhang & Espinoza (2003) and Zhang et al. (2006) name a long list of different methods to remove H_2O from a hydrocarbon stream, such as degassing, partial condensation, flash vaporization, extraction, azeotropic distillation, absorption, stripping, liquid-liquid extraction, centrifugation, decantation. Battista, Lo & Tatterson (1999) proposed (for a different reaction type) to apply condensable stripping agents.

1.2 Synthesis of fuels via CO_2 hydrogenation

Figure 1.2 indicates that CO_2 is a potential constituent of synthesis gases from coal, biomass and natural gas (Schaub, Unruh & Rohde 2004). CO_2 content depends on the hydrogen/carbon ratio of the feedstock and on the gasification process. In particular coal- and biomass derived syngases contain significant amounts of CO_2 and are deficient in H_2 in contrast to natural gas derived syngases; however, the utilization of natural gas fields with high CO_2 content or the application of low temperature gasification processes may result in natural gas derived syngases with high CO_2 content.

The utilization of CO_2 removed from flue gas streams or recovered from the atmosphere has been considered as a feedstock for various concepts. The production of fuels and chemicals based on CO_2 as a carbon source will not have a major impact on short term greenhouse gas mitigation, but it could be an integral part of carbon management (Centi & Perathoner 2004). It may be of significant interest when potential climate change issues force human activities to strictly curb CO_2 emissions.

Various authors have addressed the conversion of CO_2 containing syngases or H_2/CO_2 mixtures to methanol (Ushikoshi et al. 1998, Struis & Stucki 2001, Galluci, Paturzo & Basile 2004, Joo, Jung & Jung 2004) and methanol and dimethyl ether using hybrid catalysts (Park, Jeon & Ihm 2004). Weimer et al. (1996, 1997) and Specht et al. (1998) discussed the concept of synthesis of zero emission methanol using CO_2 enriched from the atmosphere by adsorption and H_2 produced by high temperature electrolysis. Kaya (2004) proposed a CO_2 global recycling system for CO_2 fixation to mitigate global warming, in which the methanol synthesis from fossil CO_2 and H_2 from renewable sources plays a key role.

Fischer-Tropsch synthesis represents an alternative route for utilizing CO_2 or CO_2 containing syngases. Potassium promoted CO_2/CO-shift active Fe-based FT catalysts have been identified as the most promising catalyst systems, and various authors have concluded that CO_2 hydrogenation takes place in two steps: firstly, CO forms via the reverse water gas shift reaction, which is then subsequently converted to hydrocarbons in the FT reaction (Weatherbee & Bartholomew 1984, Lee, Lee & Chang 1990, Lee et al. 2000, Riedel et al. 2003, Rohde et al. 2004). This is contrary to methanol synthesis, where CO_2 acts directly as carbon source (Rozovskii & Lin 2003).

CO_2 is a highly oxidized and thermodynamically stable component. Therefore, only low conversion levels can be achieved for example in CO_2 hydrogenation to methanol, CO_2 hydrogenation to FT hydrocarbons or DMC synthesis from CO_2 and methanol due to thermodynamic constraints. In methanol synthesis, attempts are being made to overcome equilibrium limitations by *in-situ* removal of either H_2O or methanol via a selective membrane (Struis & Stucki 2001, Galluci, Paturzo & Basile 2004, Barbieri et al. 2002) or by removal of methanol via the consecutive reaction to dimethyl ether (DME) (Jun, Shen & Lee 1999).

CO_2 hydrogenation to FT hydrocarbons requires elevated temperatures (300-360 °C) favouring the CO_2/CO shift reaction, but resulting in a reduced yield of long-chain hydrocarbons. Unruh, Rohde & Schaub (2002) proposed applying hydrophilic membranes in an effort to keep operating temperatures low. By the selective removal of H_2O during the FT synthesis, the CO/CO_2 shift equilibrium is displaced in favour of CO and the conversion of CO_2 and hydrocarbon yield are enhanced, which was experimentally demonstrated in a lab-scale membrane reactor by Unruh at 225-250 °C (Rohde, Unruh & Schaub 2005).

A similar concept was applied in a project of the European Union on Low Temperature Catalytic Partial Oxidation and Gas-to-Liquids Conversion (LTCPO-GTL 2005) with the goal of converting CO_2 containing synthesis gases produced by low temperature gasification (see also Figure 1.2) in a FT reactor with integrated dehydration membrane.

1.3 Objectives and approach

The previous paragraphs outlined that H_2O can have a detrimental effect on various fuel-related reactions under certain conditions. The objective of this thesis is therefore to explore the potential of *in-situ* H_2O removal during fuel-related synthesis reactions.

A major part of this thesis was financed and carried out within a project of the European Union related to Low Temperature Catalytic Partial Oxidation and Gas-to-Liquids Conversion (LTCPO-GTL 2005). One work package focused on the development of a membrane reactor for conversion of CO_2 containing synthesis gases to FT products (CO_2 hydrogenation) using a novel hydrophilic membrane. Thus, this application forms a major part of the thesis; however, the scope should not be limited only to Fe-based FT synthesis and membranes. The thesis addresses the following questions:

- Is *in-situ* H_2O removal during Fe- and Co-based FT synthesis a feasible option for significantly enhancing reactor performance with regard to conversion and yield levels or catalyst lifetime? This question should be answered comprehensively based on experiments and mathematical case studies. With regard to H_2O removal by membranes, a definitive conclusion should be drawn concerning necessary membrane properties. Aspects of technical implementation and alternatives next to reactive vapour permeation should be addressed as well.
- Which other reactions beside the FT synthesis may benefit from *in-situ* H_2O removal? A more generalized description should be developed. The requirements on reaction rate and membrane properties, which have to be met for a viable concept of a multifunctional reactor, should be discussed in a broader sense.

The approach and the overall outline of this thesis are depicted in Figure 1.3. A comprehensive *literature study* summarizes permeability and permselectivity data for various types of hydrophilic membranes, which are suitable for selective H_2O removal at elevated temperatures. On the basis of this study, the potential, the limitations and the direction of future development of hydrophilic membranes should be outlined (Chapter 2). A *literature overview*

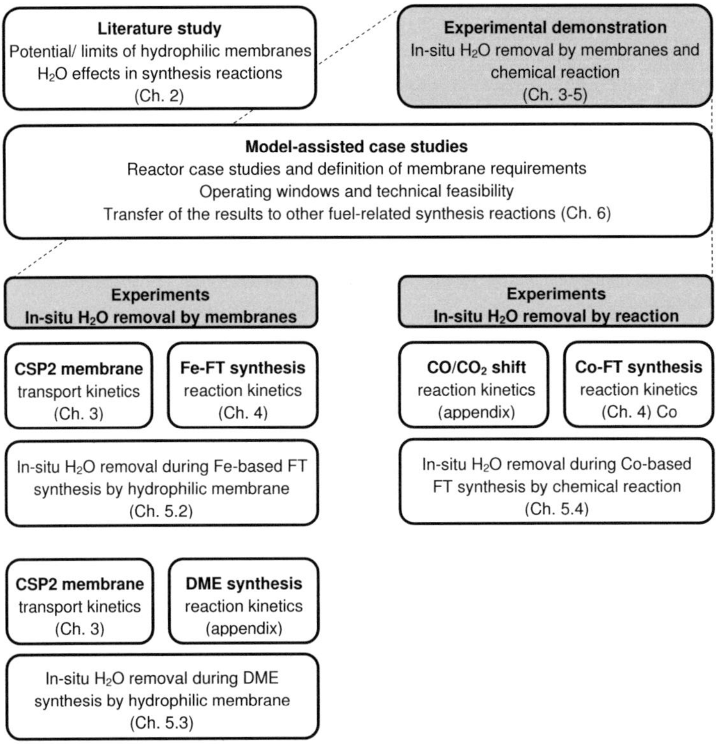

Figure 1.3. Scheme of the approach and outline of this thesis in chapters.

discusses H_2O effects on Fe- and Co-based FT catalysts with focus on H_2O induced deactivation (Chapter 2).

The literature overview is followed by the experimental part (Chapter 3-5) which covers two different approaches to remove H_2O during the reaction and three example reactions: (1) Fe-based FT synthesis applying a hydrophilic membrane, (2) dehydration of methanol to the dimethyl ether (DME) applying a hydrophilic membrane and (3) Co-based FT synthesis in a physical mixture with a low temperature CO/CO_2-shift catalyst.

The hydrophilic membranes were supplied within the LTCPO-GTL (2005) project by the Energy Research Centre of the Netherlands (ECN). The development and preparation of an own hydrophilic membrane was not the objective of this thesis. The permeation properties of the novel *ceramic supported polymer* (CSP) membrane were determined under non-reactive and reactive conditions by single gas experiments and permeation experiments with gas mixtures. The results should enable the assessment of the CSP membrane with respect to other types of membranes proposed *for in-situ* H_2O removal (Chapter 3).

In-situ H_2O removal experiments by means of the CSP membrane under reactive conditions were carried out in a lab-scale fixed bed membrane reactor. The discussion of the results is based on a suitable mathematical reactor model (model-assisted analysis). First, the CO_2 hydrogenation to long-chain hydrocarbons over a Fe-based FT catalyst was investigated as an example of a synthesis reaction with a parallel equilibrium reaction and H_2O inhibition (Ch. 5.2). Furthermore, the concept of reactive vapour permeation is applied to the dehydration of alcohols to ethers (DME/DEE synthesis), which represents a synthesis reaction with a known strong kinetic H_2O inhibition (Chapter 5.3). The operating parameters sweep gas composition, sweep ratio, sweep pressure and temperature were varied to demonstrate the effect of *in-situ* H_2O removal on conversion and yield. The membrane reactor performance was compared to a similar fixed bed reactor system without membrane.

As an alternative concept, the combination of a Co-based FT catalyst with a low temperature CO/CO_2-shift catalyst should be demonstrated experimentally (Chapter 5.3) as the Co-based FT synthesis is an example reaction where high H_2O partial pressures can lead to accelerated catalyst deactivation.

The thesis concludes with model-assisted case studies considering the technical (and economic) feasibility of the *in-situ* H_2O removal concept (Chapter 6). The potential and limits of *in-situ* H_2O removal in membrane reactors are discussed using Damköhler-Péclet analysis. A technical Co-based slurry reactor was applied as reference case and H_2O removal by membranes, by chemical reaction and from the liquid phase were assessed (Figure 1.1).

2 Literature review

This chapter gives the reader an introduction to the Fischer-Tropsch (FT) synthesis and to the principles of membrane separation and membrane reactors. The main objectives of the literature review are as follows: to discuss H_2O effects on Fischer-Tropsch product selectivity, kinetics and catalyst stability and to establish an overview on state-of-the-art hydrophilic membranes suitable for application at elevated temperatures.

2.1 Fischer-Tropsch synthesis

The synthesis of liquid hydrocarbons via Fischer-Tropsch (FT) route is a four step process consisting of (1) synthesis gas generation, (2) synthesis gas treating, (3) synthesis gas conversion and (4) product upgrading and separation (Figure 2.1). Natural gas, coal and biomass can be applied as feedstock. The feedstock is pre-treated and gasified in suitable gasifiers in the presence of a gasification medium such as O_2, air, CO_2 or steam. Depending on the feedstock and the type of gasification process, synthesis gases with varying H_2/CO partial pressure ratios, CO_2 and inert gas content are obtained (Figure 2.1). Higman & van der Burgt (2003) and Reimert & Schaub (2006) give a comprehensive overview on gas production from various feedstocks.

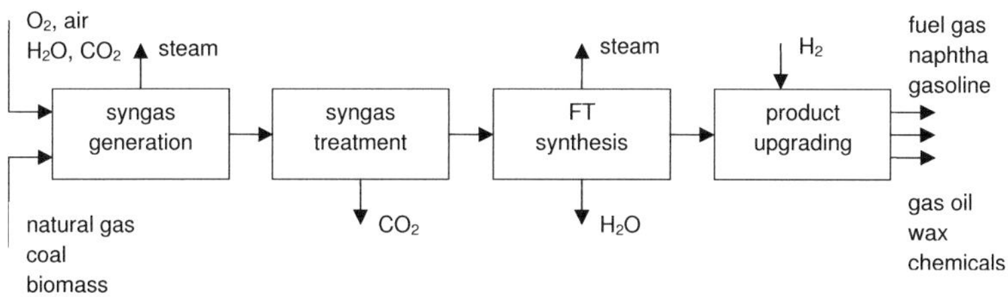

Figure 2.1. Simplified schematic of the production of liquid hydrocarbons from various feedstocks via the Fischer-Tropsch (FT) route.

In the subsequent cleaning and conditioning step, impurities and catalyst poisons as sulphur (H_2S, COS, CS_2) and nitrogen compounds (HCN, NH_3, NO_x), halogens, metal carbonyls, alkalis and tars need to be removed down to a ppb-level (Higman & van der Burgt 2003). Coal and biomass derived synthesis gases are deficient in H_2 and contain high fractions of CO_2. Under certain considerations it is desired to increase the H_2/CO ratio of synthesis gas before it is fed to the FT section. This step is referred to as carbon monoxide shift conversion and takes place in a catalytic shift reactor (Twigg 1989, Higman & Supp 2006, Higman & van der Burgt 2003). CO_2 and impurities are

removed from the syngas stream by physical or chemical washes (Kriebel 2006, Higman & van der Burgt 2003).
The treated and conditioned syngas is then converted in the FT synthesis section. Various types of Fischer-Tropsch reactors have been developed in the last decades, and Sie & Krishna (1999), Espinoza et al. (1999b) and Davis (2002) give a good overview. The new generation of FT plants applies Co-based FT catalysts, and thus operates in a low-temperature range (190-240 °C) with the aim of maximizing heavy hydrocarbon yield. Currently, two different reactor technologies have prevailed in commercial low-temperature synthesis:

- In three-phase slurry bubble column reactors, the synthesis gas is pushed upwards through the liquid product phase in which fine catalyst particles (50-500 μm) are suspended. Cooling tubes are immersed in the slurry and remove the reaction heat. The reactor is characterized by high mass and heat transfer rates due to extensive mixing in the heterogeneous hydrodynamic regime. Reactor up-scaling and the separation of catalyst and product are known key problems. The largest FT slurry bubble column reactor is operated with a nominal capacity of 17,000 bpd (Oryx GTL).
- In multi-tubular fixed-bed reactors, the catalyst is loaded into several thousand tubes. The synthesis gas flows downwards through tubes which are cooled by boiling water on the shell side. Multi-tubular fixed bed reactors are Shell's technology of choice. The reactor productivity of one multi-tubular reactor was increased from 3,500 bpd (SMDS Bintulu) to about 5,800 bpd (Pearl GTL). De Graaf & Schrauwen (2002) indicate a future reactor productivity potential of 10,000-15,000 bpd per reactor.

The heavy hydrocarbons from the synthesis section are upgraded by mild hydrocracking and separated by distillation into the respective product fractions such as naphtha, kerosene, gas oil or heavy waxes. The task of the mild trickle-flow hydrocracking process is the hydrogenation of olefins and oxygen compounds, hydroisomerization of the n-paraffins and hydrocracking of long-chain hydrocarbons to molecules with the desired chain length and desired boiling range. Typical operating conditions are 3-5 MPa and 300-350 °C (Eilers, Posthuma & Sie 1990).

In 2000, the installed coal-to-liquids (CTL) capacity was about 140,000 bpd, operated by the South African Synthetic Oil Limited (Sasol) applying low-temperature ARGE fixed bed reactors and high-temperature circulating fluidized bed (Synthol) and fixed fluidized bed (advanced Synthol) reactors (Fleisch, Sills & Briscoe 2002). The installed gas-to-liquids (GTL) capacity comprised of the Shell SMDS plant (14,700 bpd) in Malaysia and the PetroSA plant (former Mossgas, 25,000 bpd) in South Africa. The new generation of GTL plants is located in Qatar. Oryx GTL (SasolChevron/ QatarPetroleum) was started in 2007 and utilizes Co-based three-phase slurry FT technology (34,000 bpd). Pearl GTL (Shell/ QatarPetroleum) is currently under construction and is based on Shell's Co-based multitubular fixed-bed reactor technology (140,000 bpd). The predicted 'emergence of the GTL industry' (Fleisch, Sills, Briscoe 2002), however, was slowed down and eventually stopped around 2007 due to rapidly increasing costs.

A technical feasible biomass-to-liquids (BTL) concept is still in the development phase. One BTL plant (CHOREN Industries) of about 350 bpd is currently under construction in Freiberg, Germany. Thijmensen et al. (2002) and the DENA BTL feasibility study (DENA 2006) give an overview on the complexity and the potential and status of the BTL technology.

2.1.1 Reactions

The FT synthesis is an exothermic, heterogeneous catalyzed polymerization reaction. The key reaction is the hydrogenation of CO under typical FT conditions towards a wide range of linear and branched hydrocarbons and H_2O:

$$CO + 2\,H_2 \rightarrow (1/n)\,(C_nH_{2n}) + H_2O \qquad \Delta H_R{}^0{}_{250°C} = -158\ \text{kJ/mol CO} \qquad (2.1)$$

Most active metals for FT synthesis are Co, Fe, Ru and Ni. These metals have in common that they are active for hydrogenation reactions and susceptible to metal carbonyl formation (Pichler & Schulz 1970). Only Co- and Fe-based catalysts are applied in large scale industrial applications. Generally, Fe-based catalysts are produced by co-precipitation, whereas Co-based catalysts are generated by deposition of the active metal on a porous support material. Typical conditions for low-temperature FT synthesis are temperatures between 200-250 °C, pressures between 0.5 and 6 MPa and H_2/CO feed ratios between 1 and 2 (Dry 1990, Steynberg 2004, van Hardeveld 2007).

Figure 2.2 shows macroscopic reaction networks for FT synthesis over Co- and Fe-based catalysts, indicating that different reactions can take place simultaneously in addition to the FT reaction. Fe-based FT catalysts are active for the CO/CO_2 shift or water gas shift (WGS) reaction:

$$CO + H_2O \leftrightarrows CO_2 + H_2 \qquad \Delta H_R{}^0{}_{250°C} = -39.5\ \text{kJ/mol CO} \qquad (2.2)$$

Under typical FT conditions, the formation of CO_2 is favoured thermodynamically. The CO/CO_2 activity of Fe-based catalysts can be significantly increased by addition of promoters such as potassium (K). Co-based catalysts are generally inactive for the water gas shift reaction though

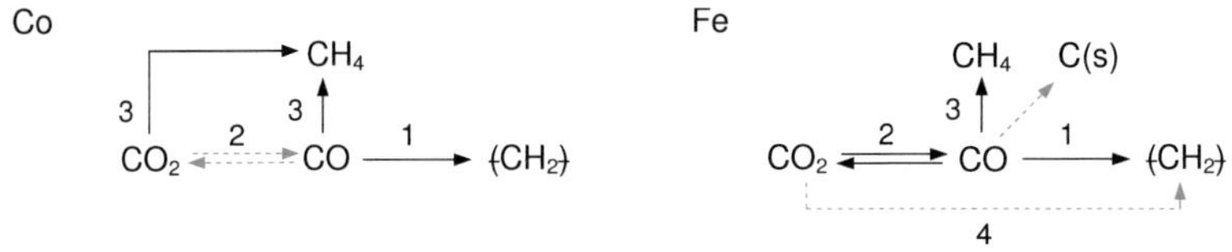

Figure 2.2. Macroscopic reaction network for Fischer-Tropsch (FT) synthesis on cobalt (left) and iron (right) catalysts. 1: FT synthesis, 2: water gas shift equilibrium, 3: methanation, 4: direct CO_2 hydrogenation, thick lines: main reaction, thin lines: side reaction, broken lines: possible (minor) reaction (Riedel 2003).

Keyser, Everson & Espinoza (2000) report that cobalt-manganese (CoMn) catalysts exhibit low CO/CO_2 activity. Due to the high CO/CO_2 shift activity, promoted Fe-based FT catalysts are able to convert CO_2 via the intermediate CO to hydrocarbons.

From a thermodynamic perspective, the formation of methane, short-chain hydrocarbons and carbon is highly favoured, but under typical Fischer-Tropsch conditions, the chain growing mechanism is not thermodynamically, but *kinetically controlled*. The reaction paths towards methane and carbon are kinetically not preferred, the reactions are 'frustrated' (Schulz 2003). However, under certain conditions such as high temperatures (>250 °C) and very high H_2/CO feed ratios, methane formation will be the predominant reaction for Co-based catalysts, and highly K-promoted Fe-based catalysts will show significant deposition of carbon:

$$CO + 3\,H_2 \rightarrow CH_4 + H_2O \qquad \Delta H_R^0{}_{250°C} = -213 \text{ kJ/mol CO} \qquad (2.3)$$

$$2\,CO \leftrightarrows C + CO_2 \qquad \Delta H_R^0{}_{250°C} = -175.6 \text{ kJ/mol CO} \qquad (2.4)$$

2.1.2 Product selectivity

The FT reaction is not selective towards a single reaction product – except for methane – or a narrow carbon range. Typical for a polymerization reaction, the product spectrum ranges from low molecular weight gaseous products to heavy waxes with a very high molecular weight. The formed hydrocarbon mixture is mainly comprised of linear paraffins and olefins (α- and internal olefins), minor amounts of branched paraffins and olefins and small amounts of oxygenates (alcohols, ketones and organic acids). The fraction of the minor products is significantly higher for Fe catalysts than for Co catalysts.

Schulz (2003) distinguishes according to the model of 'non-trivial surface polymerization' (Schulz, Beck & Erich 1988, Claeys 1997) between *major* and *minor reactions. Major reactions* are the methylene (CH_2) monomer formation, chain initiation, chain growth and chain termination. In the *ideal FT regime*,

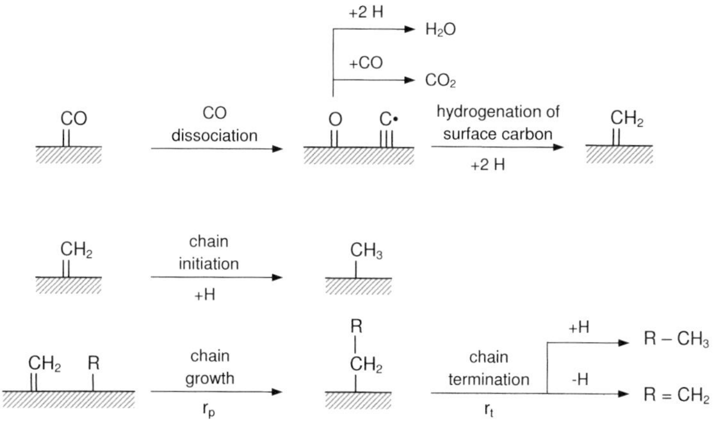

Figure 2.3. Schematic (top) of the carbene mechanism, adopted from Barthomolew & Farrauto (2005). r_p: rate of chain propagation, r_t: rate of chain termination.

linear α-olefins and linear paraffins are the primary products, whereas α-olefins would be the predominant species.

Several mechanistic schemes have been developed and they can be grouped into three principal types, which are based on different assumptions of CO activation, formation of the monomer species, and the addition of monomers to growing chains (Bartholomew & Farrauto 2005):

- The *carbene mechanism* (Figure 2.3) is based on CO adsorption and dissociation to adsorbed C (surface carbon) and O atoms, followed by a stepwise hydrogenation to CH_x monomers, chain initiation and insertion of CH_x monomers into metal-carbon bonds of an adsorbed alkyl chains (chain growth).
- According to the *enol* or *hydroxyl-carbene* mechanism, the adsorbed CO is then partially hydrogenated to an enol species –CH(OH). The condensation of two enol species yields an adsorbed –C(OH)-CH_3 species, which is hydrogenated to an olefin and H_2O.
- The *carbonyl insertion mechanism* suggests the insertion of adsorbed CO into metal-carbon bonds of adsorbed alkyl chains.

The *carbene mechanism* is a widely accepted reaction pathway for FT synthesis on Co, Fe, and Ru catalysts, though it cannot explain the formation of oxygenates. It is likely that these species are formed by the *enol* and *carbonyl insertion mechanism*.

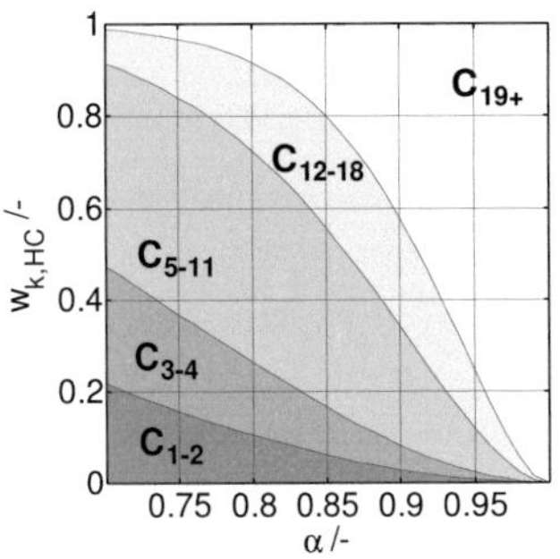

Figure 2.4. Weight fractions of different product fractions k as function of the chain growth probability α calculated with ASF-model.
C_{1-2}: off-gas, C_{3-4}: liquefied petroleum gas (LPG), C_{5-11}: gasoline, C_{12-18}: gas oil, C_{19+}: wax.

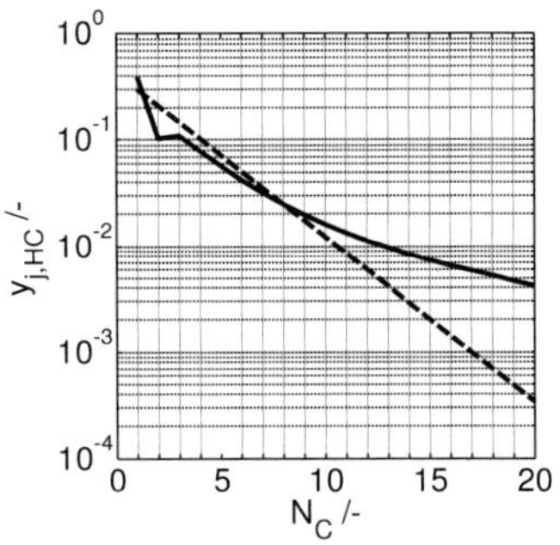

Figure 2.5. Molar product distribution (ASF plot): broken line: ideal ASF model ($\alpha = 0.7$), solid line: typical FT distribution of a Co-based catalyst.

The general product distribution is determined by the ratio of the chain propagation rate r_p and the chain termination rate r_t (Figure 2.3). A high chain propagation rate will lead to a high molecular weight product distribution, while high chain termination rate will yield to predominantly light products. For the *ideal FT regime*, the product distribution can be predicted by a simple statistical model with the chain growth propagation probability α as single parameter:

$$\alpha = \frac{r_p}{r_p + r_t} \tag{2.5}$$

This model is referred to as Anderson-Schulz-Flory (ASF) model and the ideal product distribution as ASF distribution. The molar fractions $y_{j,HC}$ and the weight fractions $w_{j,HC}$ (or carbon selectivity $S_{j,HC}$) of hydrocarbons containing j carbon atoms are calculated as follows:

$$y_{j,HC} = \alpha^{j-1}(1-\alpha) \quad \text{with } j = N_C \tag{2.6}$$

$$w_{j,HC} = S_{j,HC} = j\alpha^{j-1}(1-\alpha)^2 \quad \text{with } j = N_C \tag{2.7}$$

The value of α of a measured carbon distribution can be derived by plotting the logarithmic mole fractions versus the carbon number which yields in the ideal case a straight line with α as the slope (ASF plot, see also Figure 2.5). Figure 2.4 presents the weight fractions of different product fractions as function of the chain growth probability α, illustrating the broad spectrum of products and the accompanied selectivity limitations determined by the polymerization

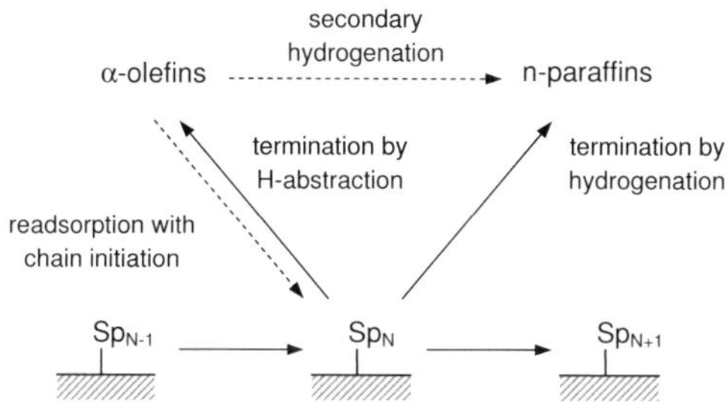

Figure 2.6. Schematic of chain growth, readsorption and secondary reactions of olefins. r_p: rate of chain propagation, r_t: rate of chain termination.

kinetics. The α value depends strongly on catalyst and process conditions. The chain growth probability increases with (1) decreasing temperature, (2) decreasing H_2/CO partial pressure ratio and (3) increasing pressure (Claeys 1997). The pressure effect is more pronounced for Co- than for Fe-based catalysts (at low temperatures), and the effect diminishes above 2 MPa. Typical α values for Co-based catalysts are 0.86-0.95, and for Fe-based 0.70-0.95 (Post et al. 1989, Bartholomew & Farrauto 2005).

The ASF model assumes that the rates of chain growth and termination are constant for all carbon numbers. However, analyses of real product distributions conclude that the chain growth probability varies significantly in the range C_{1-3} with carbon number and active metal, and the chain growth probability for C_{3+} generally increases with carbon number until it levels out at a carbon number range around C_{15-20}, resulting in a positive curvature in the ASF plot (Figure 2.5). Schulz (1993) explains these deviations from the *ideal FT regime* by *minor reactions*. These *minor reactions* can incorporate readsorption of primary α-olefins followed by secondary hydrogenation, isomerization or chain initiation and chain growth (Figure 2.6); CO insertion into olefins (hydroformylation) resulting in the formation of oxygenates; branching and methanation. The decrease of olefin fraction and the increase of the chain growth probability with increasing carbon number are explained by chain length dependent physisorption of alkyl species on the catalyst surface, chain length dependent solubility (Kuipers, Vinkenburg & Oosterbeck 1995, Kuipers et al. 1996, Schulz & Claeys 1999) or chain length dependent diffusivity (Iglesia 1997) of olefins in the wax filled catalyst pores. Multiple growth sites ('two-alpha phenomenon') or reaction paths are also discussed as a contribution to the observed deviations from the *ideal FT regime* (Patzlaff et al. 1992).

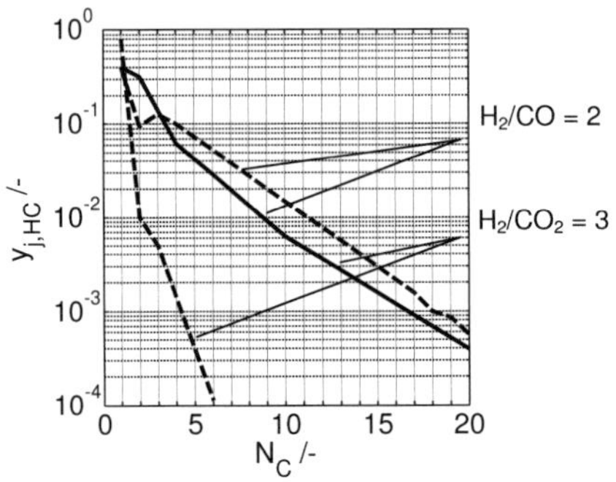

Figure 2.7. Comparison of molar product distributions of Co-based (broken line) and Fe-based (solid line) FT catalyst for synthesis gases with $H_2/CO = 2$ and $H_2/CO_2 = 3$ (Riedel et al. 2003).
* $Fe/Al_2O_3/Cu/K$: T = 250 °C, P = 1 MPa
* $Co/MnO/SiO_2/Pt$: T = 190 °C, P = 1 MPa

Matsumoto & Satterfield (1989) proposed an improved model which is able to describe the positive curvature in the ASF plot by applying to two chain growth probabilities and a transition carbon number ξ:

$$y_{j,HC} = \left(\alpha_1^{j-1} + \left(\frac{\alpha_1}{\alpha_2}\right)^{\xi-1} \alpha_2^{j-1} \right) \Big/ \left(\frac{1}{1-\alpha_1} + \left(\frac{\alpha_1}{\alpha_2}\right)^{\xi-1} \frac{1}{1-\alpha_2} \right) \quad \text{with } j = N_C \qquad (2.8)$$

Effect of synthesis gas composition on product selectivity

In general, one can conclude that higher H_2 partial pressures favour chain termination and higher CO partial pressures increase the chain propagation rate. Consequently, the chain growth probability and therefore the product distribution depend on the P_{H2}/P_{CO} ratio, and with increasing partial pressure ratio, the product distribution shifts to products with lower molecular weight. Co-based catalysts exposed to synthesis gases with a significant excess of H_2 mainly produce methane, while Fe-based catalysts are still able to maintain the FT regime. Fe-based catalysts are less invariant than Co-based catalysts against reaction conditions such as CO, CO_2 and H_2 partial pressures (Riedel et al. 1999, 2003). With H_2/CO and H_2/CO_2 syngases, the product distribution is merely different (Figure 2.7). Hence, CO_2 containing synthesis gases with a high P_{H2}/P_{CO} ratio can be converted on Fe-based catalysts to typical FT products, which is not possible with Co-based catalysts (Zhang et al. 2002, Riedel et al. 1999, 2003).

Effect of H_2O partial pressure on product selectivity

Various authors have investigated the effect of H_2O on product selectivity in co-feeding studies. In the case of a Co-based catalyst, the addition of H_2O (in the feed) or indigenous H_2O formed by the FT reaction increases the C_{5+} and decreases the methane selectivity (Claeys 1997, Claeys & van Steen 2002, Dalai et al. 2005, Storsaeter et al. 2005, 2005a). The positive effect depends on the support and increases as follows: Al_2O_3 < SiO_2 < TiO_2 (Storsaeter et al. 2005, 2005a). Significant increases in selective CO conversion by H_2O addition on Co-Ru/TiO_x, Co-Re/TiO_x or Co/TiO_X is claimed in the patent literature (Borodko & Somorjai 1999). H_2O inhibits secondary hydrogenation, hydroformylation, and oligomerization of α-olefins (van Steen 1993, Iglesia et al. 1993, Claeys 1997, Claeys & van Steen 2002). Similar observations were made for Fe-based catalysts. The addition of H_2O leads to a decrease in methane selectivity; however, the effect on C_{5+} selectivity is only minor (Satterfield et al. 1986, König et al. 1987). Satterfield et al. (1986) assume that H_2O inhibits hydrogenation and isomerization.

2.1.3 Kinetics

Many different kinetic rate equations for CO consumption have been proposed for Co- and Fe-based FT catalysts (Table 12.2 - Table 12.4). These macrokinetic models are either empirical models such as power law rate equations or based on different proposed surface reaction mechanisms such as Langmuir-Hinshelwood or Eley-Rideal type. For a comprehensive overview on macrokinetic models for Co- or Fe-based catalysts, the reader should refer to Zennaro, Tagliabue & Bartholomew (2000), Das et al. (2005), van der Laan (1999) or Bartholomew & Farrauto (2005).

Co-based catalysts

Table 12.2 gives an overview on representative macrokinetic models for CO consumption on Co-based FT catalysts. The rate law developed by Yates & Satterfield (1991) for Co-based FT catalysts represents an equation which is often employed to describe FT kinetics:

$$r_{FT} = k\frac{P_{CO}P_{H2}}{\left(1+aP_{CO}\right)^2} \tag{2.9}$$

The rate of CO consumption of Co-based catalysts increases with increasing P_{H2}/P_{CO} ratio, i.e. higher H_2 partial pressures accelerate the reaction, while CO supposedly strongly adsorbs to the catalyst surface and inhibits the reaction (Table 12.2). Generally, there is no effect of CO_2 on the intrinsic reaction rate and it acts as inert with the exception of a precipitated Co/SiO_2 catalyst, which deactivated in presence of CO_2 (Riedel & Schaub 2003).

The majority of rate laws do not consider H_2O partial pressure. However, various experimental studies have shown that the rate of CO consumption increases reversibly with H_2O partial pressure (van Steen & Schulz 1999, Claeys & van Steen 2002, Das et al. 2005, Dalai et al. 2005, Borg et al. 2006, Storsaeter et al. 2005, 2005a). According to Storsaeter et al. (2005, 2005a),

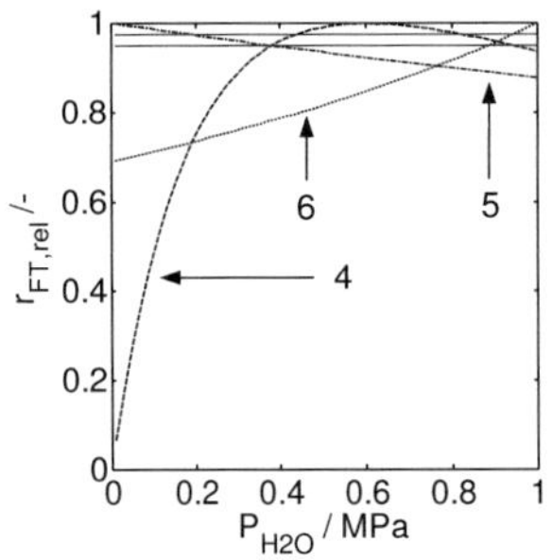

Figure 2.8. Relative reaction rates of the Fischer-Tropsch reaction on Co-based catalysts. Response of reaction rate to a change of H_2O partial pressure at constant $H_2/CO/CO_2$ partial pressures: $H_2/CO/CO_2/H_2O$ = 0.5/0.25/0/x MPa. Rate laws acc. to Table 12.2:
4: van Steen et al. (1999); 5: Withers et al. (1990); 6: Das et al. (2005)

the positive response on conversion depends on the support and increases as follows: Al_2O_3 < SiO_2 < TiO_2. On the other hand, it is observed that at high P_{H2O}/P_{H2} ratios, H_2O can also suppress the reaction rate and will expedite reversible and irreversible deactivation. In the last several years, a few rate laws were published incorporating a H_2O effect.Figure **2.8** shows the response of the reaction rate to a change of H_2O partial pressure for different rate laws (Table 12.2). Withers et al. (1990) observed a negative effect of H_2O on the reaction rate and proposed a rate similar to Huff & Satterfield (1984) for Fe catalysts. Van Steen & Schulz (1999) derived a rate law – which can be applied for Co-, Fe- and Ru-based catalysts – which shows a positive effect of H_2O at low and a negative effect at high conversion levels. They outline that H_2O partial pressure influences the inventory of surface carbon. On one hand, surface carbon is necessary for the formation of organic compounds; on the other hand, high concentrations can also inhibit the reaction. The rate is accelerated with increasing H_2O partial pressure as the concentration of surface carbon is reduced ('cleaning effect', Claeys & van Steen 2002). At high H_2O partial pressures, H_2O inhibits the rate as the concentration of surface carbon becomes depleted. This rate law can only be applied in back-mixed reactor systems.

Das et al. (2005) observed a positive effect on the reaction rate for Co/SiO_2 catalyst in H_2O co-feeding experiments. The increase in activity was reversible and interpreted as a kinetic effect. However, H_2O addition of more than 25 vol% (P_{H2O}/P_{H2} > 0.8) led to an irreversible loss of catalyst activity. Dalai et al. (2005) found a positive effect on large pore catalysts only, narrow pore catalysts showed little to no effect or a negative one. Iglesia (1997) and Borg et al. (2006) reaffirmed this observation. Iglesia (1997) explains the pore size dependent H_2O effect by the formation of H_2O rich phases that facilitate CO and H_2 transport within the catalyst structure. This capillary effect - observable in large pore catalysts - moderates the reactant transport resulting in a reduced CO deficiency in the catalyst core.

It can be concluded, that the effect of H_2O partial pressure on reaction rate of Co-based catalysts is very complex and not fully understood thus far. An additional difficulty is that at high H_2O partial pressures, the reversible kinetic effects are superimposed by H_2O induced catalyst deactivation.

Fe-based catalysts

Contrary to Co-based FT catalysts, Fe-based catalysts are active for the CO/CO_2 shift reaction. Van der Laan (1999) suggests from kinetic studies in a slurry reactor system that two different sites are present: iron carbides are active for the Fischer-Tropsch reaction, while magnetite (Fe_3O_4) seems to be the most active site for the CO/CO_2 shift reaction. The CO/CO_2 shift reaction determines the H_2O/CO_2 ratio in the gas phase and it can significantly alter the rate of the FT reaction (Satterfield et al. 1986). Conversion of H_2/CO synthesis gases on Fe-based catalysts with a low CO/CO_2 shift activity yields only small amounts of CO_2; H_2O is the main by-product. If the CO/CO_2 shift reaction is fast e.g. in the case of K-promoted Fe-based catalysts, a gas phase rich in CO_2 and H_2 is obtained already at medium CO conversion levels. The strong interaction between FT and CO/CO_2 shift reaction and the difficulties in deriving rate parameters of parallel reactions in integral reactors may be the reason for the wide variety of rate laws for the FT and CO/CO_2 shift reaction, considering either CO_2, or H_2O or both CO_2 and H_2O as additional inhibiting terms besides the general CO inhibition term (Table 12.3, Table 12.4).

The following rate equation developed by Zimmerman & Bukur (1990) is given as an example for FT kinetics for Fe-based catalysts:

$$r_{FT} = k \frac{P_{CO} P_{H2}}{P_{CO} + bP_{H2O} + cP_{CO2}} \tag{2.10}$$

Similar to Co-based catalysts, higher P_{H2}/P_{CO} ratios promote the FT reaction rate. However, various authors report an inhibiting effect of H_2O partial pressure (e.g. Ledakowicz et al. 1985, Zimmerman & Bukur 1990, van der Laan & Beenackers 2000, Riedel 2003, Unruh 2006). This negative effect distinguishes the Fe-based from Co based catalyst. Figure 2.9 shows the response of the FT reaction rate to a change of H_2O partial pressure for different rate laws (Table 12.3), indicating the significant inhibiting effect of H_2O on the FT reaction rate. For highly shift active catalysts, Nettelhoff et al. (1985), Zimmerman & Bukur (1990) and van der Laan et al. (2000) derived a rate law with CO_2 inhibition and without H_2O inhibition. A series of rate equations incorporate a CO_2 inhibition term. Recent experimental studies conclude that CO_2 has a minor inhibiting effect (Riedel 2003). A comparison of inhibition constants reported by different authors (Zimmerman & Bukur 1990,

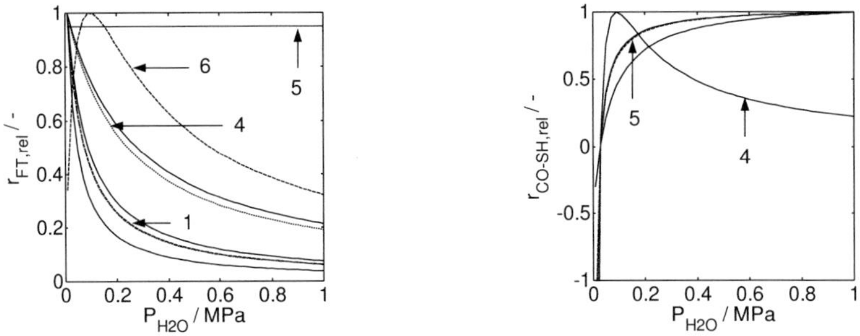

Figure 2.9. Relative reaction rates of the Fischer-Tropsch (left) and the CO/CO_2-shift reaction (right). Response of reaction rate to change of H_2O partial pressure at constant $H_2/CO/CO_2$ partial pressures (Rate laws acc. to Table 12.3-4).
Fischer-Tropsch reaction: $H_2/CO/CO_2/H_2O$ = 0.25/0.25/0.25/x MPa: 1: Huff et al. (1984), Zimmerman et al. (1990); 5: van der Laan (1999), gas-slurry; 4: Unruh (2003); 6: Riedel (2003)/ van Steen et al. (1999)
CO/CO_2 shift reaction: $H_2/CO/CO_2/H_2O$ = 0.5/0.1/0.5/x MPa, T = 250°C: 4: van der Laan (1999), gas-slurry; 5: Riedel (1999), Unruh (2003), Zimmerman et al. (1990)

Riedel 2003, Unruh 2006) for similar rate laws indicates that P_{H2O} has a much stronger effect on the reaction rate ($b_{H2O} > c_{CO2}$) than P_{CO2}.

The CO/CO_2 shift reaction is an equilibrium reaction and the forward CO/CO_2 shift reaction is mildly exothermic. The equilibrium constant $K_{P,CO\text{-}SH}$ can be calculated by the correlation given by Newsome (1980):

$$\log_{10} K_{P,CO-SH} = \log_{10}\left(\frac{P_{CO2}P_{H2}}{P_{CO}P_{H2O}}\right) = \frac{2073}{T} - 2.029 \tag{2.11}$$

The equilibrium composition is at the product side ($K_{P,CO\text{-}SH} > 1$), e.g. $K_{P,CO\text{-}SH}$ is 86 at 250°C and 20 at 350°C. Therefore, CO_2 will be formed with Fe-based catalysts under typical FT conditions. Table 12.4 summarizes representative macrokinetic models for the CO/CO_2 shift reaction on Fe-based FT catalysts, the rate equation of Zimmerman & Bukur (1990) should be given as an example:

$$r_{CO\cdot SH} = k\frac{P_{H2O}P_{CO} - P_{CO2}P_{H2}/K_{P,CO\cdot SH}}{P_{CO} + bP_{H2O} + cP_{CO2}} \tag{2.12}$$

The rate equations listed take the forward and reverse reaction into account. The ratio of the forward and reverse reaction rate is expressed by the equilibrium constant; the rate expressions are constructed such that if the gas phase reaches equilibrium compositions, the rate of the CO/CO_2 shift reaction approaches zero.

With regard to the conversion of CO_2 containing synthesis gas, an acceleration of the reverse CO_2/CO shift reaction is the key. The reverse reaction is favoured at higher temperatures or if CO and H_2O are removed from the equilibrium. The FT reaction removes CO, and therefore, the equilibrium composition cannot be reached (Riedel et al. 2001). A selective removal of H_2O during the FT reaction – e.g. via hydrophilic membranes – could lead to very low H_2O partial pressures, resulting in a negative rate of the forward CO/CO_2 shift reaction (Figure 2.9).

2.1.4 Catalyst stability

Catalyst deactivation problems in FTS arise from (a) poising of catalysts by sulphur and/or nitrogen compounds, (b) fouling of hard waxes and carbon, (c) formation of inactive catalytic phases as oxides, inactive carbides and metal-support compounds, (d) hydrothermal sintering, and (e) catalyst attrition including catalyst disintegrating by Boudouard carbon deposition (Bartholomew & Farrauto 2005). H_2O was identified to lead to deactivation of Fe- and Co-based catalysts.

Fe-based catalysts

Catalyst samples taken at different axial positions from a tubular fixed bed reactor (pilot plant) by Duvenhage & Coville (2006) confirm earlier findings reported in literature that the activity of the Fe-based catalyst gradually decreases with bed length. The following processes explain the loss of activity: The core of the catalyst particles is oxidized to inactive magnetite Fe_3O_4 (Dry 1990, Davis 2003) as the oxidation potential of reaction mixture increases with increasing penetration into the catalyst particle. The thermodynamic phase diagram in Figure 12.1 indicates that α-Fe can be oxidized to Fe_3O_4 at very low H_2O/H_2 ratios under FT conditions. The magnetite concentration increases along the bed – corresponding to increasing H_2O partial pressure in the gas phase – and with time on stream. Satterfield et al. (1986) found for a fused Fe catalyst that the addition of H_2O up to P_{H2O}/P_{H2} = 0.2 in the reactor inlet inhibited the FT activity reversibly. They observed only a slow formation of magnetite Fe_3O_4. However at P_{H2O}/P_{H2} = 0.3, the amount of Hägg carbide Fe_2C_5 significantly dropped and the fraction of magnetite Fe_3O_4 increased, followed by an irreversible decrease in FT activity.
Additionally, water vapour enhances the rate of sintering of Fe catalysts (Dry 1983, Dry 1990). Along the bed and with increasing time on stream, a decrease in surface area and an increase in the pore volume (hydrothermal sintering) and of the crystallite size are observed (Dry 1990, Duvenhage & Coville 2006).

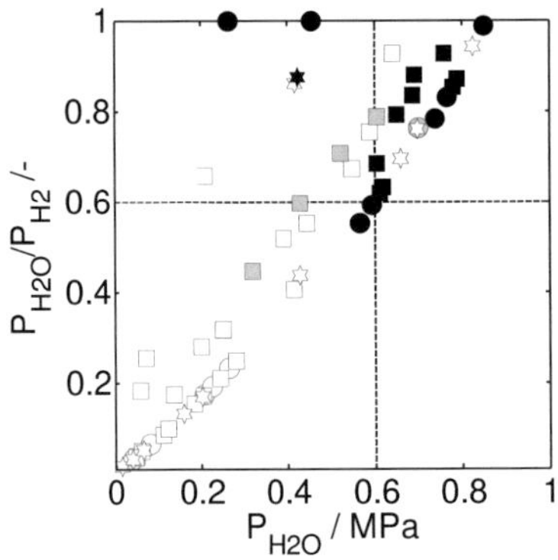

Figure 2.10. Effect of H_2O partial pressure and partial pressure ratio P_{H2O}/P_{H2} on supported Co-based catalysts stability.

Colours: white: no or reversible increase in activity, grey: reversible decrease in activity, black: irreversible decrease in activity

Symbols:
■ Co/Al_2O_3: Storsaeter et al. (2005), Borg et al. (2006), Saib (2006)
● Co/SiO_2: Chen et al. (2001), Krishnamoorthy et al. (2002), Kiss et al. (2003), Dalai et al. (2005), Das et al. (2005), Storsaeter et al. (2005)
★ Co/TiO_2: Bertole et al. (2002), Storsaeter et al. (2005)

Co-based catalysts

Various experimental studies indicate that H_2O plays a key role in multiple deactivation processes for Co-based catalysts. Storsaeter (2005) and Saib (2006) recently published extensive literature reviews regarding H_2O effects on Co catalysts.

Figure 2.10 summarizes experimental data of several experimental series published in literature with focus on the deactivation of supported Co catalyst. In these experiments, H_2O levels were deliberately manipulated by either H_2O co-feeding or by operating at high conversion levels (> 70%). The literature references were evaluated and grouped in three categories as follows: an increase in H_2O partial pressure has (1) no or a reversible, positive effect on catalyst activity, (2) a negative, but reversible effect on catalyst activity, and (3) a negative, irreversible loss of activity. The data is plotted versus the absolute H_2O partial pressure at reactor outlet and the partial pressure ratio P_{H2O}/P_{H2}. The partial pressure ratio determines the oxidation potential of the gas phase; metallic cobalt Co can be oxidized not only to inactive cobalt oxides and hydroxides, but also to cobalt-support compounds, so-called spinels such as cobalt aluminate, silicate or titanate ($CoTiO_3$, Co_2TiO_4) (Figure 12.1). Figure 2.10 correlates with the statement of Bartholomew & Farrauto (2005) to maintain the partial pressure ratio P_{H2O}/P_{H2} below 0.6 and the H_2O partial pressure P_{H2O} below 0.6 MPa for Co-based catalyst. Below these critical values, no enhanced deactivation was observed; if a Co-based catalyst is operated above these critical values there is increased likelihood that the catalyst activity will decay much faster. Various authors indicate that Co/Al_2O_3 catalysts irreversibly deactivate above P_{H2O}/P_{H2} ratios of 0.6 (Borg et al. 2006, Li et al. 2002). These partial pressure ratios and H_2O partial pressures can be easily exceeded under medium and high conversion levels. The rate and susceptibility for deactivation of Co catalysts depend strongly on the type of catalyst support and its pore structure. The deactivation tendency for different supported catalysts can be ranked as follows: Re/Al_2O_3> Al_2O_3> SiO_2>> TiO_2 (Storsaeter et al. 2005, 2005a), i.e. Co/TiO_2 is less susceptible to H_2O induced deactivation, however at high P_{H2O}/P_{H2} ratios, irreversible activity decay was found here as well (Li et al. 2002a, Bertole et al. 2002). The addition of reduction promoters can increase the deactivation tendencies as it was found for Re-promoted Co/Al_2O_3 catalysts (Schanke et al. 1995, 1996, Storsaeter 2005, 2005a). At least three mechanisms contribute to catalyst deactivation:

- *Surface oxidation and oxidation of highly dispersed cobalt phases* was identified by various authors as deactivation mechanism. The oxidation of bulk cobalt metal to cobalt oxides such as CoO and Co_3O_4 is thermodynamically not favoured under FT conditions ($P_{H2O}/P_{H2} < 2$, Figure 12.1) and can be ruled out as major deactivation mechanism for larger crystallites (Saib et al. 2006). However, Schanke et al. explained that highly dispersed crystallites deviate from bulk behaviour and that oxidation of crystallite surfaces and of highly dispersed cobalt is responsible for observed deactivation, which is supported by Mössbauer and XPS studies (van Berge et al. 2000, Schanke et al. 1995, 1996). Van Steen et al. (2005) showed in a thermodynamic analysis for spherical nanocrystals that metallic cobalt crystallites < 4.4 nm are expected to be unstable and be present as CoO. This was confirmed experimentally by Saib et al. (2006), who found that crystals > 6 nm are stable against oxidation under realistic FT conditions (Co/Pt/Al_2O_3 catalyst, pilot slurry reactor, P_{H2O}/P_{H2} = 1-1.5), while smaller crystallites were oxidized.

- *Formation of inactive metal-support compounds (spinels)* is another important deactivation mechanism. The formation of a spinel such as $CoAl_2O_3$ is thermodynamically spontaneous, but kinetically inhibited. However, at high H_2O concentrations, Co/SiO_2 catalyst deactivated due to formation of non-reducible Co silicate (Kiss et al. 2003). Qin & Ramkrishna (2004) tried to understand deactivation by considering size-dependent reoxidation of highly dispersed cobalt crystallites and size-dependent formation cobalt-support compounds. The negative impact of H_2O during catalyst reduction where it can induce the formation of these inactive catalyst-support compounds has been addressed (e.g. Jongsomjit, Panpranot & Goodwin 2001).

- *Hydrothermal sintering* of the support structure and of the cobalt crystallites induced by high H_2O partial pressure leads to activity decay as well. Dalai et al. (2005) explained deactivation of their Co/SiO_2 by loss of surface area and pore volume due to hydrothermal sintering and catalyst encapsulation by Co silicates. Jacobs et al. (2002) observed significant cobalt cluster growth in the initial phase of deactivation.

H_2O induced deactivation can be avoided or reduced by several methods on different scale levels such as catalyst, reactor and process design. Catalyst design – which involves different aspects such as the choice of the catalyst support, pore structure, catalyst dimensions, the addition of promoters and the degree of dispersion of the active phase – can lead to the development of catalysts which are less susceptible to H_2O induced deactivation. E.g. TiO_2 supported Co catalysts are more stable than Al_2O_3 and SiO_2 supported ones. The addition of transition metals or rare earth oxides such as La or Zr is beneficial as they can inhibit the growth of Co-support spinels (Bartholomew & Farrauto 2005). Font Freide & Hardy (2006) propose the incorporation of hydrophobic compounds into the catalyst surface to increase the catalyst lifetime. Highly dispersed Co-based catalysts are very active, but susceptible to oxidation and formation of metal-support compounds. Reoxidation and spinel formation can be prevented by properly selecting Co crystallite sizes (Saib 2006).

2.1.5 Fischer-Tropsch process

Process line-ups are discussed by Sie & Krishna (1999), Dry & Steynberg (2004) and Bartholomew & Farrauto (2005). The process line-up design depends on various factors such as feedstock, target products and FT catalysts. This chapter touches (a) the consideration of H_2O effects and (b) potential ways of CO_2 handling in FT process design.

Process design considering H_2O effects in the FT section

High H_2O partial pressures and high partial pressure ratios P_{H2O}/P_{H2} are detrimental to FT catalyst stability. The upper CO per pass conversion of a FT reactor is hence determined by catalyst constraints. Various patent applications discuss H_2O issues and process line-ups, as e.g. Espinoza (1999), Beer (1999), Clark & Walker (2000), Steynberg, Vosloo & van Berge (2003), Zhang et al. (2003), Zhang et al. (2006). For Co-based catalysts, the H_2O partial pressure and partial pressure ratio P_{H2O}/P_{H2} should be kept below 0.5-0.6 MPa and 0.6, respectively, to avoid hydrothermal sintering and the formation of oxides and spinels (Bartholomew & Farrauto 2005, Steynberg, Vosloo & van Berge 2003). Translated into CO conversion terms, this means to limit the per pass conversion to 50-60% (Bartholomew & Farrauto 2005, Steynberg et al. 2004, Beer 1999). For Fe-based catalysts, a limiting H_2O

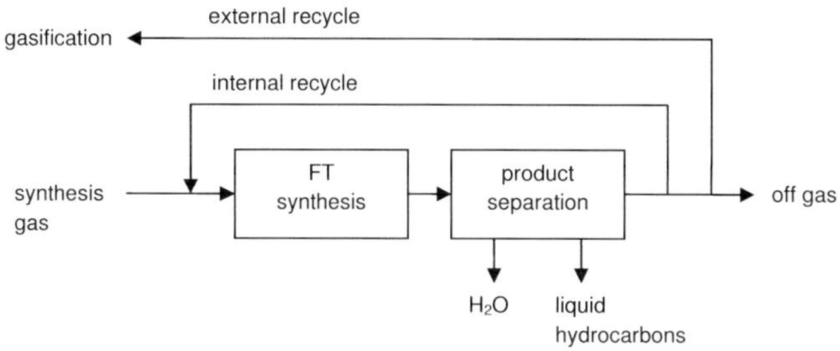

Figure 2.11. Recycle schemes around a Fischer-Tropsch reactor.

partial pressure of 0.3 MPa was reported to avoid rapid deactivation (Steynberg et al. 2004).

One option is to reduce the conversion per pass with recycle operation. Multi-tubular reactors – e.g. the ARGE fixed bed reactors – are usually operated with an internal gas loop (Figure 2.11) to obtain high linear gas velocities and to enhance the heat transfer (Sie & Krishna 1999, Steynberg et al. 2004, Dry & Steynberg 2004). Water and hydrocarbons are removed by cooling, condensation and phase separation before the recycle gas is sent back. Gas loops with water removal (dry recycle) are applied not only for multitubular, but also for slurry reactors to operate within the limits set by catalysts constraints (Steynberg, Vosloo & van Berge 2003, Steynberg et al. 2004, Bartholomew & Farrauto 2005). Steynberg et al. (2004) give a per pass conversion of 60% and ratio of the volumetric flow rates of feed and recycle gas of 1:1 as starting point for design, resulting in a overall stage conversion of about 90%.

Low conversion levels per reactor pass can also be realized by putting reactors in series (multi-stage concept). Water and hydrocarbons are removed between the stages by cooling, condensation and phase separation. Multistage concepts for increased catalyst lifetime were proposed e.g. by Beer (1999), Jess, Popp & Hedden (1999), Clark & Walker (2000), de Deugd, Kapteijn & Moulijn (2003). Shell's MDS process in Malaysia applies a two-stage concept; the first stage consists of three parallel multi-tubular reactors which are equipped with a gas recycle; the recycle tail gas is fed to a shared second stage (Steynberg et al. 2004). This concept allows high overall conversion levels (>90%) at limited per pass conversions. However, a low per-pass conversion is undesirable because it results in higher capital investment – increased number of reactors, installation of a recycle compressor – and operating costs. Here, *in-situ* H_2O removal could play a role as it may allow very high per pass conversions without running into the regime of H_2O induced deactivation.Various concepts for *in-situ* H_2O removal were considered such as (1) H_2O removal from the gas phase via hydrophilic membranes (Espinoza et al. 1999), (2) H_2O removal from a H_2O-rich liquid phase via an external water removal unit applying e.g. degassing, partial condensation, flash vaporization, azeotropic distillation, absorption, reboiled absorption, stripping, liquid-liquid extraction, centrifugation, decantation (Zhang & Espinoza 2003, Zhang et al. 2006 or Battista, Lo & Tatterson 1999). Besides these patent applications, a technical assessment was not published.

Process design considering CO_2 containing synthesis gases

CO_2 acts an inert in the FT section and therefore it is removed from the syngas stream beforehand in wash towers (Figure 2.1). Riedel (2003) conducted a case study for a Co-based slurry reactor configuration showing that 15 vol% CO_2 in the feed gas stream requires a reactor 12% larger to maintain the conversion level of the CO_2-free reference case.

Considering the potential costs for CO_2 abatement and the high investments required for synthesis gas production units, the utilization of CO_2 as feedstock for FT synthesis may become interesting. Low temperature gasification processes such as low temperature catalytic partial oxidation (LTCPO) of natural gas can be seen as low cost alternative, but these processes produce synthesis gases with a high CO_2 content (Rabe, Treong & Vogel 2007). Conversion of CO_2 containing synthesis gases to FT products requires CO_2/CO shift active FT catalysts and high temperatures or *in-situ* H_2O removal during the synthesis step. Both high temperatures and a low H_2O partial pressure in the reaction zone shift CO_2 to CO which is then converted to hydrocarbons. A process line-up for CO_2 utilization using *in-situ* H_2O removal was evaluated in the LTCPO-GTL project (LTCPO-GTL 2005).

2.2 Membrane reactors and membranes

The IUPAC definition (Koros, Ma & Shimidzu 1996) of a membrane reactor is 'a device for simultaneously carrying out a reaction and membrane-based separation in the same physical enclosure'. A membrane reactor is a *multifunctional* reactor and represents a sub-class of the area of *process intensification* (Stankiewicz & Moulijn 2000). The IUPAC definition refers to a classical membrane reactor concept where products or by-products are selectively removed from the reaction zone via a semi-permeable membrane to achieve higher conversion or selectivity levels. The most intensively studied concept in literature is the selective H_2 removal via dense metal membranes during dehydrogenation reactions of hydrocarbons. And indeed, it has been demonstrated in numerous publications that conversion levels above the thermodynamic equilibrium constraints were achieved in membrane reactors for reversible reactions (Dittmeyer 2001).

Various membrane reactor applications were proposed that go beyond the classical concept of an integrated separation function in the reaction zone. Membrane reactors are used to feed reactants into the reactions zone (e.g. for selectivity control) or non-permselective, but catalytically active membranes are used as defined reaction zone or as contactor for gas/liquid reactions. A membrane can retain a mobile catalyst e.g. enzymes, biological cells or fine catalyst powder in slurry phase, and the membrane can act as a filter, or it immobilizes the catalyst in its porous structure or it is catalytically active itself (Kemmere & Keurentjes 2006).

Although these type of reactors have gained a lot of attention, membrane reactors are only applied in low temperature applications such as biological production of fine, high-value chemicals in membrane bioreactors; esterification reactions in pervaporisation membrane reactors, which are commercially in operation, but the membrane-based separation step is in general not fully integrated into the reactor; or waste water treatment, where commercially available polymeric and inorganic meso- and microporous membranes can be used (Sanchez-Marcano & Tsotsis 2002). Membrane reactor applications at high temperatures (> 100℃ and significantly higher) are not yet commercial, but progress is made e.g. in membrane-enhanced methane reforming (MTR 2008). Availability of permselective, reliable and affordable membranes and the high initial investment remain the key factors for slowing down progress in large-scale petrochemical applications (Sanchez-Marcano & Tsotsis 2002, Baker 2002).

Detailed reviews on membranes and membrane reactor applications were published in the recent years, e.g. on membrane reactors in general (Dittmeyer 2001, Sanchez Marcano & Tsotsis 2002, Seidel-Morgenstern 2005, Nunes & Peinemann 2006), on inorganic membrane reactors (Saracco & Specchia 1998) on zeolite state-of-the-art zeolite membranes (Coronas & Santamaria 1999, Caro et al. 2000) and membrane reactors (Coronas & Santamaria 2004), on catalytic membrane reactors (Tsotsis et al. 1993), on applications for gas/ liquid reactions (Dittmeyer, Svajda & Reif 2004) and many more.

2.2.1 Membrane reactors

Membrane reactors can be categorized by their configuration or by their basic function. In the first case, one classifies according to the type of membrane – i.e. if the membrane is permselective or not – and the location of the catalyst – i.e. the catalyst is placed within the membrane or external to it. Tsotsis et al. (1993) derived on this basis four basic membrane reactor configurations: (1) catalytic membrane reactor with a permselective membrane, (2) catalytic non-

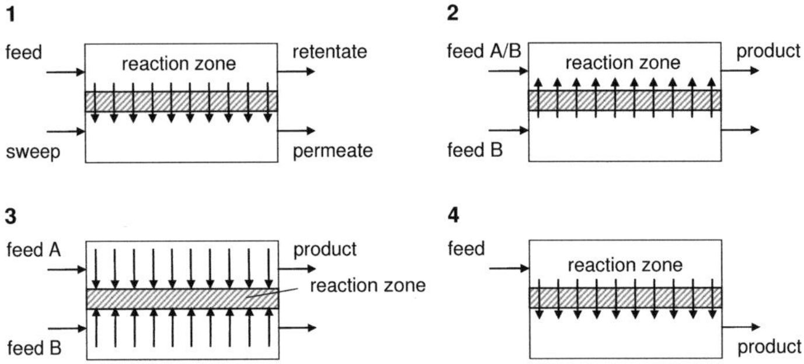

Figure 2.12. Classification of membrane reactors: 1: extractor-type membrane reactor, 2: distributor-type membrane reactor, 3: diffuser-/contactor-type catalytic membrane reactor, 4: forced-through flow type catalytic membrane reactor.

permselective membrane reactor, (3) packed bed catalytic membrane reactor and (4) packed bed membrane reactor. Besides that, the catalyst can be also located in a fluidized bed or in a slurry phase.

Membrane reactors can be classified according to their function. The extractor-type and distributor-type membrane reactors represent two standard concepts: in the extractor mode, products or by-products are selectively removed; in the distributor mode, reactants are co-fed controlled into the reaction zone. The four concepts shown in Figure 2.12 will be discussed briefly in the following paragraphs with a focus on the application of membrane reactors in Fischer-Tropsch synthesis as summarized in Table 2.1.

The extractor-type membrane reactor combines a reactor and a membrane separator in one single process unit (reactive separation). In this membrane reactor, a product or by-product is selectively removed from the reaction zone to achieve certain process goals such as (a) conversion enhancement by equilibrium displacement of equilibrium-limited reactions, (b) selectivity enhancement by removal of an intermediate in consecutive reactions, (c) conversion enhancement by removal of inhibitors or (d) catalyst protection by removal of catalyst poisons. Furthermore, this type of reactor can be applied as catalyst filter.

The selective removal of H_2 from the reaction zone via dense platinum or palladium membranes or porous inorganic membranes during equilibrium-limited dehydrogenation reactions of hydrocarbons such as ethane, propane, butane, cyclohexane or ethylbenzene were studied intensively in literature and significant increases in conversion levels were reported (Dittmeyer 2001, Sanchez-Marcano & Tsotsis 2002). However, the industrial break-through has not yet been achieved according to Dittmeyer (2001) – who gives a

Table 2.1. Concepts of membrane applications for FT synthesis, tested under reactive and non-reactive FT conditions. Controlled addition of [a] H_2 to CO feed gas and [b] CO to H_2 feed gas. X: conversion; S_{C2+}: hydrocarbons selectivity, Y_{C2+}: hydrocarbons yield.

	concept	membrane/ support	catalyst	X	S_{C2+}	Y_{C2+}
Unruh (2006) Rohde et al. (2005)	selective H_2O removal	amorphous silica/γ-Al_2O_3 on porous alumina	Fe/Al_2O_3 /5K/Cu	X_{CO2} ↑	→	↑
Espinoza et al. (1999a, 2000)	selective H_2O removal	MOR, ZSM-5, SIL-1 on porous alumina/ stainless steel tubes	n.a.			
Zhu et al. (2005)	selective H_2O removal	4A-zeolite/TiO_2 on porous stainless steel	n.a.			
Léonard et al. (2003) Vanhove et al. (2003) Guillou et al. (2004)	distributed feeding	γ-Al_2O_3/α-Al_2O_3, ZSM-5/α-Al_2O_3 on porous alumina	Co/Al_2O_3	X_{CO} ↓[a]/↑[b]	↑[a]/ ↓[b]	↑[a]/ ↓[b]
Khassin et al. (2005, 2005a)	forced-through flow catalytic membrane	porous catalyst & copper structure	Co/Al_2O_3	X_{CO} ↑	↑	↑
Bradford et al. (2005)	forced-through flow catalytic membrane	catalyst/γ-Al_2O_3 on porous alumina tube	P/Pt-Co/Al_2O_3	X_{CO} ↑	↓	↑
He et al. (2005)	control of residence time	ZSM-5 encapsulated catalyst pellet	Co/SiO_2	X_{CO} ↓	S_{C1-10} ↑	Y_{C1-10} ↑

comprehensive overview on the current status – due to the unavailability of membranes with sufficient high permeability and stability. Other interesting membrane application with high future potential are applications in membrane enhanced methane reforming (e.g. Nijmeyer 1999, Paturzo & Basile 2001, Patil, van Sint Annaland & Kuipers 2007) and membrane enhanced CO/CO_2 shift units (e.g. Bracht et al. 1995, Ma & Lund 2003) for the production of H_2-rich or ultra pure H_2 streams.

In-situ H_2O removal during Fischer-Tropsch synthesis was proposed by Espinoza et al. (1999a, 2000) as high H_2O partial pressures can inhibit the reaction rate or lead to accelerated catalyst deactivation. Espinoza et al. (1999a, 2000) carried out experiments under simulated non-reactive Fischer-Tropsch conditions, using microporous zeolite membranes. Unruh (Unruh 2006, Rohde et al. 2005) demonstrated the positive effect of *in-situ* H_2O removal in a packed-bed membrane reactor during the hydrogenation of CO_2 to Fischer-Tropsch products. H_2O removal via membranes was also proposed for other reactions such as CO_2 hydrogenation to methanol (Struis & Stucki 2001) or methanol to dimethyl ether (Lee, Youn & Sea 2006) or for reversible condensation reactions (Peters, Benes & Keurentjes 2005).

Distributor-type membrane reactor

Partial oxidations and hydrogenation reactions are driven to high selectivities by keeping oxygen or hydrogen concentrations rather low in the reacting mixture. Membrane reactors allow the controlled addition of one of the reactants into the reaction compartment to achieve (a) selectivity enhancement by distributed feeding and (b) temperature control by distributed feeding. Membranes applied are either non-permselective or permselective. In the first case, a pure reactant is co-fed across the membrane with the required rate as determined by the membrane permeability (profile) (Thomas 2003). In the second case, the membrane is permselective to the key component (e.g. H_2; O_2 in air), and the membrane reactor fulfils a separation function coupled with a chemical reaction. In particular the separation of O_2 from air via dense solid-oxide membrane coupled with e.g. a partial oxidation reaction is an emerging field.

With regard to the Fischer-Tropsch synthesis, this membrane distributor concept was picked up by Léonard, Miachon & Vanhove (2003), Vanhove & Léonard (2003) and Guillou et al. (2004) with the aim of controlling product selectivity and reaction exothermicity. As the hydrocarbons selectivity and the reaction rate depend on H_2/CO partial pressure ratio, Vanhove's group presented a conceptual study on distributed feeding of H_2 or CO. The controlled addition of H_2 via a non-permselective alumina membrane along the axis of a packed-bed reactor fed with CO led to a higher yield in long-chain hydrocarbons, but lower conversions as high CO partial pressures inhibit the reaction rate of Co-based catalysts. Overall, separate feeding of CO and H_2 into a membrane reactor does not seem to be a suitable concept as H_2 and CO need to be separated beforehand and therefore, feeding strategies of H_2 enriched and H_2 deficient streams should be discussed instead. And second, similar results can be obtained by co-feeding of H_2 along distinct points of a packed bed reactor (Sharifnia, Mortazavi & Khodadadi 2005).

Diffuser-/ contactor-type catalytic membrane reactor

Within these types of membrane reactors, the membrane does not fulfil a separation task, i.e. it is not permselective. In one application, a (symmetric) microporous catalytic membrane offers a defined reaction zone. The pure key reactants are fed from opposite sides to the membrane; in the case of a transport-limited reaction (fast kinetics), the reaction takes place only in a limited zone within the membrane (Saracco & Specchia 1998). A slip of reactants to the other side is therefore prevented. The conversion level is controlled by transport kinetics through the microporous membrane structure. This makes the concept attractive for combustion of hydrocarbons or hydrogen sulphide (Claus process, Sloot, Versteeg & van Swaaij 1990) and for reactions that require strict stoichiometric feeds (Harold et al. 1994).

In a second concept applicable for gas/liquid reactions, a thin catalytic layer is deposited on a porous membrane. Here, gaseous reactants are fed from the catalytic side, while the liquid reactants are fed from the support side. The catalytic active membrane provides high surface area for gas/liquid contacting; the operation is under isobaric conditions, i.e. the mass transport to the catalyst is by diffusion (Dittmeyer, Svajda & Reif 2004). Dittmeyer et al. provide various selective hydrogenation and oxidation reactions in their literature overview as examples. Harold et al. (1994) propose the diffuser type for rate enhancement of volatile-reactant-limited multiphase reactions.

Forced-through flow catalytic membrane reactor

In the forced-through flow concept, a pressure difference is applied across the catalytic membrane. This concept is applied e.g. in gas/liquid reactions to allow high liquid/solid and gas/solid mass and heat transfer rates with the aim of increasing reactor productivity. The catalyst is incorporated in the porous structure of a membrane, and the reactants are generally forced through the membrane with a short contact time. Intensive mixing occurs in the pore system. Two different forced-through catalytic membrane reactor (CMR) concepts were proposed recently by Khassin et al. (2005) and Bradford et al. (2005) for FT applications. These types of reactors may become interesting for the exploitation of small natural gas fields e.g. off-shore applications (flared gas on oil rigs) and the utilization of biomass-derived synthesis gas demand for small or medium-scale FT reactors that offer high specific production rates and safe and easy operability. Khassin et al. (2005, 2005a) developed a proprietary method (sintering in the presence of a pore-producing agent) to prepare catalytic membranes with a high load (up to 800-1000 kg/m^3) of Co/Al_2O_3 catalyst and a high thermal conductivity addressing the problem of efficient heat removal. The catalytic membranes have tubular geometry and they exhibit a distinct three-modal pore distribution. Large pores (3-7 µm) run through the membrane and enable high permeation rates. The transport pores are interconnected by narrower pores, which will be filled with liquid product due to capillary forces. The feed gas stream is forced through the membrane in radial direction either inwards or outwards. It has been suggested that the so-called plug-through contactor membrane (PCM) offers low pressure drop, flat temperature profiles, high reactor capacities, high gas-liquid mass transfer rates and low diffusive constraints. The specific C_{2+} production rate is given with 200 kg/m^3/h at 2 MPa and 210 °C.

Bradford, Te & Pollack (2005) proposed a monolithic loop catalytic membrane reactor (ML-CMR). Here, the mesoporous structure of a honeycomb monolith was coated with a microporous ceramic membrane (45 µm layer of Al_2O_3 with a 2 nm nominal pore size) and a 200 µm layer of $P/Pt\text{-}Co/\gamma\text{-}Al_2O_3$ catalysts. The idea is to pressurize the shell side of the honeycomb with synthesis gas. The synthesis gas permeates through the mesoporous structure and is pushed

through the thin, porous catalytic layer into the honeycomb channels. Products and heat are removed by recycling liquid product through the channels. The membrane acts in this case only as catalyst carrier. Experiments with a single channel showed that higher conversion levels can be obtained in the CMR than in a conventional packed bed reactor, but the C_{2+} selectivity was lower, presumably due to the presence of diffusion limitations in the catalytic layer. The potential production capacity of a ML-CMR was estimated with a simplified reactor model, based on the experimental results. The maximum C_{2+} production rate for a honeycomb membrane module (0.25m x 1.524m) with 11470 channels coated with a 300 µm catalyst layer is estimated with 270 $kg/m^3/h$ at 200°C and 2 MPa, but about 13 times the product mass flow rate has to be recycled to limit the temperature rise in the reactor to 25 K.
But the forced-through flow concept has significant drawbacks; large reactant streams have to be recycled (which is energy intensive); the volume specific catalyst amount is relatively low in these configurations; the catalyst cannot be exchanged as easily as in slurry or multi-tubular reactors.

Membrane-coated catalyst particles or membranes on particle level are new and interesting approach to control feed of reactants to or remove products from active sites. Membranes on particle level are attractive as reactor performance is not as sensitive to defects of single membranes. He et al. (2005) proposed such a membrane reactor concept with the objective to modify product selectivity by control of residence time (Coronas & Santamaria 2004). Here, Fischer-Tropsch catalyst pellets (Co/SiO_2, 0.38-1.7 mm) were encapsulated within ZSM-5 layer of about 10 µm thickness. The acidic zeolite layer represents a membrane on particle scale level: the reactants H_2 and CO permeate through the layer into the FT catalyst and react to hydrocarbons; the formed products are forced to diffuse through the zeolite layer in counter-current direction. The residence time in the pore system of the zeolite increases with the hydrocarbon chain length. Therefore, long-chain molecules undergo hydrocracking and isomerization reactions more frequently than short-chain hydrocarbons. By applying this concept, He et al. were able to produce hydrocarbon distributions with a sharp cut-off at C_9-C_{10}. Due to the forced secondary reactions in the zeolite layer, the selectivities to short-chain hydrocarbons, iso-paraffins and olefins are significantly higher compared to a physical mixture of FT catalyst and zeolite particles.

2.2.2 Membrane structure and materials

Membranes can be classified according to their morphology and to the membrane material. With regard to morphology, one distinguishes porous and dense membranes and symmetric and asymmetric membranes. Homogenous membranes consist of a single material. The membrane is symmetric if the membrane properties do not vary through the membrane layer. Composite membranes are built up of several different layers and therefore asymmetric. In general, these membranes have a very thin functional layer deposited on porous support structures, and combine the high transmembrane fluxes of a thin membrane layer with the mechanical stability of the support. Synthetic membranes are prepared either with polymers or inorganic materials. With regard to the membrane material, Dittmeyer (2001) differentiates between polymer, glass, ceramic, metal, carbon and liquid membranes. The most relevant membrane types for selective gas permeation – dense polymer membranes, amorphous microporous membranes and crystalline microporous membranes (zeolites) – will be discussed in the following paragraphs.

Polymer membranes

The majority of polymers applied in membrane preparation are thermoplastics. Thermoplastics are generally high molecular weight polymers, which chains are not cross-linked as it is the case for elastomers or thermosetting plastics, but associated by weak Van-der-Waals forces, dipole-dipole interactions or hydrogen bonds. The state of the polymer – which can be described by the glass transition temperature T_g and the degree of crystallinity – determines the mechanical, chemical, thermal and permeation properties and therefore the membrane performance. Thermoplastic materials are distinguished by the glass transition temperature T_g. For $T < T_g$, a thermoplastic exhibits a (partially) amorphous, glassy state. The polymer is in a non-equilibrium state within the experimental time scale. As the polymer backbone is stiff, the glassy polymer contains unrelaxed volume segments, called holes or microvoids of different sizes. These inhomogeneities on molecular scale are believed to cause different or multiple gas diffusion modes (Kimura & Hirose 1992, 1992a).
Diffusion takes place in the amorphous domains and microvoids of the polymer. For $T > T_g$, the secondary, non-covalent bonds between the polymer chains become weak, and the segments of the polymer chains have a high mobility. The polymer has a rubbery state i.e. the polymer can be deformed elastically or plastically without fracture. The permeability of the polymer in the rubbery state is much higher than for the amorphous, glassy state. The crystalline parts of the polymer are not altered on passing through T_g unless the melting point T_m is approached. Crystallites in the polymer increase mechanical stability due to physical crosslinking, but lower permeability.
The physical polymer properties including the glass transition temperature T_g and the degree of crystallinity are determined by the molecular weight and number of the monomer units, the sequence of the monomers, interaction and hydrophilicity of the functional side groups, the degree of branching and

crosslinking and the flexibility or stiffness of the resulting polymer backbone. Hence, one can tailor polymers with low glass transition temperatures such as isotactic polypropylene (-15°C) or polyamide (50°C) or polymers with high glass transition temperatures such as polyetherimide (210°C) or polyimide (300°C) (Dittmeyer 2001, Mulder 2000). Polymers successfully applied in commercial membrane applications are polysulfone, polyaramide, polyimide, polycarbonate, cellulose acetate and silicone rubber (Baker 2002).

Crystalline microporous or zeolite membranes

The most significant representatives of microporous ceramic membranes are zeolite membranes. Besides the standard textbooks, Caro et al. (2000) and Bowen, Noble & Falconer (2004) give a sound overview on the state of development and perspectives and fundamentals and application of zeolite membranes.

Zeolites are crystalline structures with uniform molecular-sized pores. These pores are made of rings in the framework and are designated by the number of oxygen atoms in the ring. Small pore zeolites are built up by eight-member oxygen rings, medium pore zeolites by 10-member rings and wide pore zeolites by 12-member rings. Chabazite (CHA) and Linde type-A (LTA) belong with 0.38 and 0.41 nm pore diameters to small pore zeolites. Examples for medium pore zeolites are ZSM-5 and silicalite-1 with MFI structure (0.51 x 0.55 nm, 0.53 x 0.56 nm). Mordenite (MOR) represents a large pore zeolite (0.65 x 0.70 nm, 0.26 x 0.57 nm) (Bowen, Noble & Falconer 2004). These molecular-sized pores offer a high selectivity by size exclusion or molecular sieving (shape selectivity). For comparison, the kinetic diameter of a H_2O molecule is 0.25 nm; organic molecules are larger than 0.4 nm. Additionally, the selectivity and the permeation rate are determined by the penetrant affinity to the membrane material (adsorption controlled permeation) and the mobility of the adsorbed species in the pore network (diffusion controlled permeation). With the focus on H_2O removal, the hydrophilicity of the zeolite is of major interest. Hydrophilicity increases with decreasing silicon/ aluminium (Si/Al) ratio of the zeolite framework; the Si/Al ratio can vary from > 1000 (silicalite-1) to 1 (type-A zeolite). However, zeolite membranes applied in low pH environments need to have relatively high Si/Al ratios to prevent destruction of the framework by acid leaching of aluminium.

Zeolite membranes are built up as asymmetric membranes, i.e. the zeolite layer synthesized on the surface of a porous substrate, consisting of support such as mesoporous alumina or porous stainless steel (e.g. Bernal et al. 2000, Espinoza et al. 2000) and various intermediate layers. Crystallite size and layer thickness vary from 100-500 nm up to 500 μm depending on the preparation method (Dittmeyer 2001). Preparation methods aim to reduce number and size of intracrystalline, non-zeolite pores, which have a detrimental effect on membrane permselectivity. Membranes with hydrophilic functional layers such as A-type, X-type (Si/Al = 1-1.5), Y-type (1.5-3), and mordenite (5) have successfully separated H_2O from organic compounds (Bowen, Noble & Falconer 2004). ZSM-5 and mordenite membranes were applied in *in-situ* H_2O removal under non-reactive Fischer-Tropsch conditions (Espinoza et al. 1999a, 2000).

Amorphous microporous membranes

Microporous membranes can be obtained by coating a porous support with a colloidal solution of dense oxide particles as Al_2O_3, SiO_2, TiO_2, ZrO_2 and subsequent drying and sintering (sol-gel process). Thin layers of only 30 nm thickness with a high permeability can be prepared (Vos & Verweij 1998). Another method is the deposition of microporous layers of SiO_x inside the pores of a support by chemical vapour infiltration. Tetra-ethyl orthosilicate (TEOS) is applied as a typical Si-precursor (Mooroka et al. 1995). Applying this preparation method, Kölsch et al. (1998, 2000) tailored amorphous membranes with pore diameters reduced below 1 nm and a SiO_x network structure with very hydrophilic surfaces rich in silanol groups. Unruh (2006) applied this type of membranes for in-situ H_2O removal during reactive Fischer-Tropsch synthesis.

2.2.3 Mass transport through dense membranes

The solution-diffusion model can describe the mass transport through dense, pore-free membranes such as polymer membranes or dense metal membranes (see e.g. Mulder 2000, Baker 2004, Nunes & Peinemann 2006). According to this standard model, the permeating species adsorb and dissolve at the surface of the membrane (sorption). The dissolved species diffuse through the membrane matrix and desorb at the opposite interface (desorption). The sorption/ desorption steps are assumed to be quick and in thermodynamic equilibrium, the diffusion step is slow and rate-determining. The driving force for the transport of the permeating species is the gradient in its chemical potential. The molar flux of component i yields the following:

$$j_i = -L_i \nabla \mu_i \tag{2.13}$$

with L_i as proportionality factor linking the chemical driving force to the flux. Restricting to applications where the driving forces are generated by concentration or pressure gradients, the chemical potential is written as:

$$d\mu_i = RT\, d\ln a_i + \upsilon_i dP \tag{2.14}$$

The solution-diffusion model assumes that the pressure within the membrane is constant (at the highest pressure) and that the chemical potential gradient across the membrane is expressed only as a concentration gradient (Baker 2004). With these assumptions, equation (2.13) can be rewritten by combining equations (2.13) and (2.14). Considering a constant activity coefficient and introducing a diffusion coefficient D_i of component i in the membrane matrix, one yields Fick's law:

$$j_i = -D_i \nabla c_i^* \tag{2.15}$$

The concentrations of the species i in the membrane matrix c_i^* at the gas/ membrane interfaces can be correlated to the partial pressure in the gas phase by the following expression:

$$c_i^* = S_i P_i \tag{2.16}$$

where S_i represents the solubility coefficient (or gas phase sorption coefficient) of component i in the membrane material. Equation (2.16) can be derived by equating the chemical potentials of component i in the compressible gas and the incompressible membrane phase (neglecting the Poynting correction) assuming thermodynamic equilibrium at the membrane interface. Integration of Fick's law (2.15) across the thickness of the membrane Δs then gives together with equation (2.16) the gas separation transport equation, which is widely used to rationalize the properties of gas permeation membranes (Baker 2004):

$$j_i = \frac{S_i D_i}{\Delta s}\left(P_{i,FS} - P_{i,SS}\right) \tag{2.17}$$

where $P_{i,FS}$ and $P_{i,SS}$ is the partial pressure of component i on the feed and sweep side, respectively. The expression S_iD_i is referred to as permeability; the expression $S_iD_i/\Delta s$ is called permeance Q_i:

$$Q_i^{SD} = \frac{S_i D_i}{\Delta s} \tag{2.18}$$

Gas separation transport equation (2.17) dictates the molar flux is proportional to the partial pressure difference (driving force) and inversely proportional to thickness of the dense membrane layer. For high-pressure applications, the partial pressure difference may need to be replaced by the fugacity difference to accommodate non-ideal gas behaviour. One can recognize that the molar flux directly depends on the solubility of the permeating species in the membrane matrix, which is a thermodynamic parameter, and on diffusivity or mobility of the penetrant in the membrane matrix, which is a kinetic parameter. Therefore, the selective mass transport is based on differing solubilities and

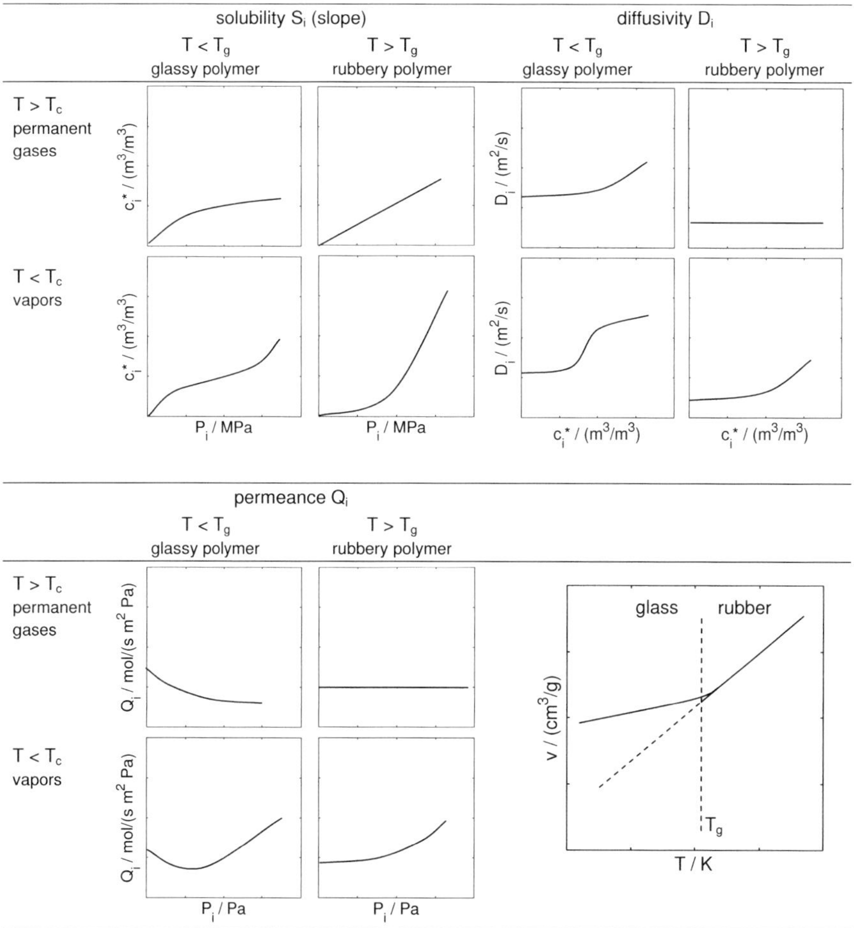

Figure 2.13. Schematic representation of solubility, diffusivity and permeance characteristics in polymer membranes categorized according to type of polymer membrane (glassy or rubbery) and penetrant (permanent gas or vapour) (based on Kimura & Hirose 1992, 1992a). c_i^*: concentration of species i in polymer, P_i: partial pressure, T_g: glass transition temperature, T_c: critical temperature, v: specific polymer volume.

diffusivities of the permeating species in the membrane, expressed by the permselectivity, the ratio of the respective permeances:

$$S_{ij} = \frac{Q_i}{Q_j} = \frac{S_i}{S_j}\frac{D_i}{D_j} \tag{2.19}$$

The ratio of the solubility coefficients S_i/S_j can be seen as solubility selectivity; the ratio of diffusion coefficients D_i/D_j can be seen as mobility selectivity.

Depending on nature of the penetrant and the nature and state of the polymer, the solubility coefficients and the diffusion coefficients can be complex functions of penetrant concentration, pressure and temperature. Kimura & Hirose (1992, 1992a) use the state of the polymer (amorphous, glassy state or the rubbery state) and the nature of the permeating species (gas or vapour) to classify the different permeation characteristics. Hereby, the contribution of solubility and diffusion (mobility) are considered separately (Figure 2.13).

In an amorphous, glassy polymer, the small gas molecules do not only dissolve in the polymer matrix (according to Henry's law), but also adsorb onto the surface of the microvoids (which can be described by a Langmuir isotherm), resulting in a concave sorption isotherm (dual-sorption model, Stern 1972, Koros & Chern 1987). In a rubbery polymer, the polymer chains are distributed homogeneously due to the high mobility of the polymer segments and penetrant concentration increases linearly with pressure (Henry's law). If vapors are involved ($T < 0.7\ T_c$), the sorption isotherms become more complex, and concave-convex isotherms for glassy polymers and convex isotherms for rubbery polymers are found. Depending on the nature of polymer, the convex part of an isotherm can be explained by swelling and plasticization – i.e. by an increase of free volume between the polymer chains resulting in a significantly decreased glass transition temperature – or by clustering – i.e. by non-random aggregation of penetrants (Detallante et al. 2001). These types of sorption isotherms have been observed and mathematically described for water vapour which tends to cluster in polymers (Barrie 1968, Stern & Saxena 1980, Mauze & Stern 1982).

The diffusion coefficient of small gas molecules or permanent gases ($T > T_c$) in a rubbery polymer is independent of the penetrant concentration. Due to the high segmental motion of the polymer, the mobility of penetrants is generally low. In the case of a glassy polymer, the diffusion coefficient is strongly dependent on penetrant concentration. The penetrants dissolved in the polymer matrix (according to Henry's law) are fully mobile, while the penetrants adsorbed in the microvoids are only partially mobile (dual-mobility model, Paul & Koros 1976, Koros & Chern 1987). In the case of vapours ($T < T_c$), the diffusion coefficients in rubbery and glassy polymers become strongly (linearly or exponentially) dependent on penetrant concentration due to effect of plasticization.

The resulting dependence of the permeance on partial pressures for glassy and rubbery polymers is shown in Figure 2.13. In an amorphous, glassy polymer, diffusion is most often the predominant mechanism and separation is based on mobility selectivity. The smaller the molecules, the faster they permeate. However, solubility decreases with increasing mobility, yielding that

the molecules with the lowest molecular weight do not need to have the highest permeability (Mulder 2000, Baker 2004). Also species adsorbed at the membrane surface and within the free volume (free-volume filling) can alter the permeation significantly, so that otherwise highly permeable species are blocked (Stookey 2006). Plasticization and membrane swelling can increase the solubility of a less soluble component significantly.
In a rubbery polymer, the contribution of mobility selectivity is low due to the high segmental motion of the polymer chains. Solubility is the predominant mechanism and separation is based on solubility selectivity. The solubility increases with increasing critical volume (or condensability) of the penetrant (Baker 2004). Therefore, rubbery membranes are applied in separation of organic vapors from gas streams (Ohlrogge & Stürken 2006), while glassy membranes are applied for H_2 separation (Stookey 2006).

For small temperature ranges, the temperature dependence of the solubility coefficient and of the diffusion coefficient can be described by a van't Hoff-type relation (2.20) and an Arrhenius-type relation (2.21), respectively.

$$S_i = S_{i,0} e^{(-\Delta H_S / RT)} \tag{2.20}$$

$$D_i = D_{i,0} e^{(-E_{A,D} / RT)} \tag{2.21}$$

The activation energy of the diffusion step $E_{A,D}$ is positive. The solution enthalpies ΔH_S for vapors and larger molecules are negative, while the enthalpies are positive for smaller gas molecules as H_2 and He. The temperature behaviour of the permeance depends therefore on the contributions of the solution and diffusion steps: permeances of larger molecules and vapors decrease with increasing temperature as $(-\Delta H_S - E_{A,D} > 0)$ and the permeances of small molecules generally increase with increasing temperature as $(-\Delta H_S - E_{A,D} < 0)$.

2.2.4 Mass transport in porous membranes

Transport through porous media is discussed in various monographs (Burggraaf & Cot 1996, Saracco & Specchia 1998, Mulder 2000). The selective mass transport through a porous membrane layer can be governed by several different transport mechanisms (Figure 2.14). In macro- and mesopores of the support (d_{pore} = 1-15 μm) and intermediate layers (100-1500 nm), non-selective viscous flow plays a dominant role. In mesoporous separation layers (3-100 nm), Knudsen diffusion plays a role. In the microporous separation layer (0.5-2 nm), configurational diffusion and, if gas molecules strongly adsorb, surface diffusion are prevailing (Burggraaf & Cot 1996). The Knudsen number Kn helps to distinguish between various transport

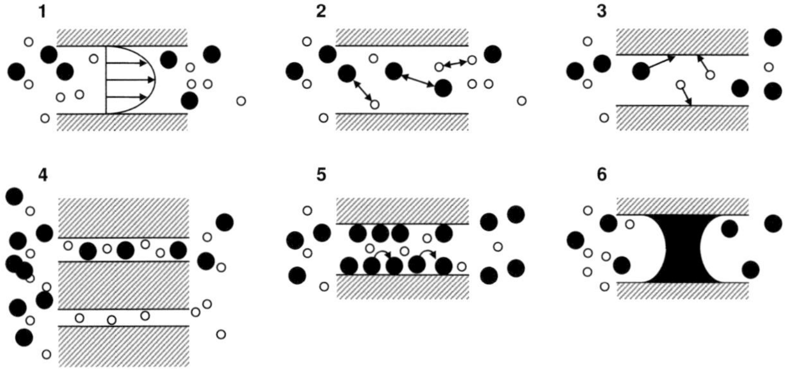

Figure 2.14. Transport mechanisms through porous membranes 1: viscous flow, 2: molecular diffusion, 3: Knudsen diffusion, 4: configurational diffusion, molecular sieving, 5: surface diffusion, 6: capillary condensation.

numbers. It relates the mean free path Λ of gas molecule to the pore diameter d_{pore}.

Viscous flow (Kn << 1: $\Lambda << d_{pore}$)

In the presence of a total pressure gradient across the membrane, a non-selective laminar convective flow is established if the pore diameters are significant larger than the mean free path of the gas molecules e.g. in meso- and macroporous support layers. Under isobaric conditions, viscous flow does not occur. The molar flux of component i can be described by a Hagen-Poiseuille type law (Burggraaf & Cot 1996, Tuchlenski 1998):

$$j_i^V = -\frac{1}{RT}\frac{B_0}{\eta}P_i\nabla P \quad \text{with} \quad B_0 = \frac{\varepsilon}{\tau}\frac{d_{pore}^2}{32} \tag{2.22}$$

$$Q^V = \frac{1}{\Delta s}\frac{\varepsilon}{\tau}\frac{d_{pore}^2}{32}\frac{P_{mean}}{RT} \tag{2.23}$$

The permeability constant B_0 is only a function of the porous material. E.g. for a porous Cerasiv support tube, Tuchlenski (1998) found a permeability constant B_0 of $1.37 \cdot 10^{-15}$ m^2 and porosity/ tortuosity ε/τ ratio of 0.1. The permeance Q^V for viscous flow is proportional to the mean pressure and to the pore diameter squared. If cracks and holes in the functional layer of a porous membrane are present or in the case of failed membrane sealing, the non-selective viscous flow will be detrimental to selective gas separation.

Molecular diffusion (Kn < 1: $\Lambda < d_{pore}$)

In the presence of concentration gradient across the membrane, molecular diffusion will take place if the pore diameters are larger than the mean free path of the gas molecules. Molecule-molecule interactions dominate. In the case of binary or pseudo-binary mixtures, the first Fick's law describes the molar flux of component i along the concentration gradient:

$$j_i^D = -\frac{1}{RT}\frac{\varepsilon}{\tau}\delta\nabla P_i \tag{2.24}$$

$$Q_i^D = \frac{1}{\Delta s}\frac{\varepsilon}{\tau}\frac{\delta}{RT} \tag{2.25}$$

with Q_i^D as permeance for molecular diffusion and δ as binary diffusion coefficient. The binary diffusion coefficient for H_2/H_2O at 250°C and 1 MPa is about $2.5 \cdot 10^{-5}$ m^2/s (according to the equation of Fuller et al. in Reid, Prausnitz & Poling 1986). Molecular diffusion in multi-component mixtures is generally described by the classical Maxwell-Stefan equations for diffusion (Taylor & Krishna 1993, Tuchlenski 1998), with $\mathbf{D^F}$ as matrix of concentration dependent Fickian diffusion coefficients:

$$\mathbf{j^D} = -\frac{1}{RT}\mathbf{B}^{-1}\mathbf{\Gamma}\nabla P = -\frac{1}{RT}\mathbf{D}^F\nabla P \tag{2.26}$$

Knudsen diffusion (Kn > 1: $\Lambda > d_{pore}$)

Knudsen diffusion occurs when the mean free path of gas molecules is larger than the pore diameter. Molecule-wall interactions dominate over molecule-molecule interactions. This mechanism becomes important at small pore dimensions at low pressures and high temperatures. The molar flux of component i can be described by first Fick's law:

$$j_i^K = -\frac{D_i^K}{RT}\frac{\varepsilon}{\tau}\nabla P_i \quad \text{with} \quad D_i^K = \frac{1}{3}d_{pore}\sqrt{\frac{8RT}{\pi \tilde{M}_i}} \tag{2.27}$$

$$Q_i^K = \frac{1}{\Delta s}\frac{\varepsilon}{\tau}\frac{2}{3}d_{pore}\sqrt{\frac{2}{\pi \tilde{M}_i RT}} \tag{2.28}$$

The permeating species diffuse almost independently of one another through the pores, and the transmembrane flux is proportional to the reciprocal square root of the molar weight. The ratio of transmembrane fluxes of two components is inversely proportional to the square root of the ratio of the molecular weights, which is the highest attainable selectivity. The permeance is inversely proportional to the square root of the temperature and independent of pressure. This means that H_2 molecules will permeate 3 times faster through a Knudsen membrane than H_2O molecules.

Surface diffusion

High permselectivities can be obtained by surface diffusion, when one of the permeating species physisorbs preferentially on the pore walls. Though surface diffusion coefficients are lower than molecular diffusion coefficients, significant transport rates can be obtained due to high surface concentrations. This can be the case for membranes with sufficiently small pores (high specific surface area) and at high partial pressures (multilayer adsorption). In pores of a few nanometers, adsorbed molecules can reduce the pore diameter and diminish the mobility of the molecules in the gas phase (Sarraco & Specchia 1998). The surface flux for a single gas can be described by the two-dimensional Fick law (Burggraaf & Cot 1996):

$$j_i^S = -\frac{\varepsilon}{\tau} D_S \nabla c_s \tag{2.29}$$

where c_s is the surface concentration (mol/m^2). With increasing temperature, the surface diffusion coefficient and hence the mobility of the adsorbed phase increases, however the surface concentration declines, as the adsorptive bonds become weaker. Therefore, at high temperatures (T > 300℃), the contribution of surface diffusion will become negligible while molecular diffusion will prevail.

Configurational diffusion

In microporous materials, the kinetic diameters of the molecules are in the range of pore or channel dimensions. Configurational or translational diffusion becomes the dominating mechanism (Burggraaf & Cot 1996, van de Graaf, Kapteijn & Moulijn 1998). The molecules do not adsorb at the pore walls and keep their gaseous character, but due to the strong energetic interactions with atoms of the solid, configurational diffusion is an activated transport process. The diffusion coefficients of configurational diffusion are in a range of 10^{-16} -10^{-8} m²/s (van de Graaf, Kapteijn & Moulijn 1998) and depend strongly on the ratio of kinetic diameter of the molecule to the channel diameter.

For microporous materials, larger molecules may have difficulties entering the small pores directly from the gas phase compared to smaller molecules. The larger molecules will adsorb first at the external surface and move via surface diffusion to the pore entrance (Barrer 1990). Competitive adsorption at the external surface or blockage of the pore entrances by non-penetrating species can severely alter flux and permselectivity (Burggraaf & Cot 1996).

If the pores are small enough to hinder large molecules from permeating the membrane by size exclusion, one speaks of molecular sieving. This highly selective mechanism plays an important role for zeolite membranes as the channel apertures are in the range of kinetic diameters of the molecules. The adsorption cut off diameter is the maximum kinetic diameter of a molecule that can adsorb within the zeolite. The adsorption cut-off diameter for zeolite X/Y is 0.95 nm, for ZSM 0.65 nm and for zeolite 4A 0.4 nm (van de Graaf, Kapteijn &

Moulijn 1998). E.g. permanent gases and water vapour could permeate through all zeolites, but zeolite 4A membranes would retain hydrocarbons higher than propane.

Capillary condensation

Vapors are able to condensate in the pores of mesoporous membranes or in the intercrystalline microdefects of zeolite membranes (Coronas & Santamaria 1999). The condensate fills the pores and blocks the transport of small molecules. High transport rates and very high selectivities with respect to the condensing species can be obtained. However, small pores with a very homogenous pore radius distribution and a minimum vapour pressure at low temperatures are pre-requisites to promote this mechanism (Saraccho & Specchia 1998), and therefore, capillary condensation as selective transport mechanism is not relevant in inorganic membrane reactor applications.

The transport through porous media was successfully quantified applying the so-called dusty gas model (DGM). In the DGM, the pore walls contribute to the momentum transfer as they are considered as a very heavy species (dust molecules) (Burggraaf & Cot 1996). The flux contributions of Knudsen diffusion and molecular diffusion are coupled in series, while the contributions of viscous flow and surface diffusion are in parallel. For further details, refer to Burggraaf & Cot 1996 and Tuchlenski (1998). For a single gas, the DGM simplifies to a combination of Knudsen diffusion and viscous flow:

$$j = -\frac{1}{RT}\left(\frac{\varepsilon}{\tau}D_i^K + \frac{B_0}{\eta}P\right)\nabla P \tag{2.30}$$

2.2.5 Permeation characteristics of hydrophilic membranes

In-situ H_2O removal by means of membranes has already been applied on pilot or technical scale; examples are the dehydration of natural gas (Hammer 2006, Löwe 2001), air (Stookey 2006, Hammer 2006) and organic process streams by pervaporisation (e.g. solvent recovery, ethanol dehydration, coupling of distillation and pervaporisation). Membranes are either on polymer or zeolite basis and the operating temperature range of these membrane applications is rather low (<150°C). Membrane reactor applications combined with fuel-related synthesis reactions require much higher temperatures, but the number of publications with regard to H_2O removal at temperatures above 200°C is limited. These key publications listed in Table 12.5 were evaluated in this thesis with regard to permeances and permselectivities to obtain an overview on the state-of-the-art hydrophilic membranes.

H_2O permeance

Figure 2.15 (top left) summarizes measured H_2O permeances Q_{H2O} for various types of membranes. Zeolite membranes offer the highest H_2O permeance at elevated temperatures; for various types of zeolite membranes, H_2O permeances are reported in a range between 10^{-6} and 10^{-7} mol/(s m² Pa). Zeolites such as mordenite or zeolite A with high aluminium content are promising candidates for selective H_2O removal from gas mixtures due to their high hydrophilicity. The current upper boundary of the H_2O permeance of hydrophilic state-of-the-art membranes is about 10^{-6} mol/(s m² Pa), represented by very hydrophilic, but pH sensitive zeolite 4A membranes (Zhu et al. 2001). With a permeance of 10^{-6} mol/(s m² Pa), a mass flux of 4.3 kg H_2O/(h m²) could be obtained at 1 bar water partial pressure difference across the membrane. Due to activated adsorption and diffusion steps of competing species, the response of the permeance to changes in temperature can be complex. Various authors observed that the H_2O permeance increases with temperature and levels out at higher temperatures (Piera et al. 1998, Bernal et al. 2000, Lindmark & Rezai 2007). Overall, the temperature dependence appears to be weak for H_2O (e.g. $E_{A,P,H2O}$ = 3.2 kJ/mol for zeolite 4A, Zhu et al. 2001).

The H_2O permeances of amorphous microporous membranes (Rohde, Unruh & Schaub 2005, Unruh 2006, Lee, Youn & Sea 2006) are with $3 \cdot 10^{-8}$ and $1 \cdot 10^{-7}$ mol/(s m² Pa) lower than for zeolite membranes. Though the functional layer of amorphous membranes can be much thinner than for zeolite membranes, the deposition of amorphous SiO_x and Al_2O_3 layers in the pore structure of the upper support layer reduces the permeability significantly. At temperatures above 150°C, the fragile hydrophilic network structures are irreversibly destroyed (Kölsch et al. 1998, Unruh 2006).

At low temperatures (< 50°C), H_2O permeances of polymer membranes vary over a wide range from $3 \cdot 10^{-10}$ to $3 \cdot 10^{-6}$ mol/(s m² Pa), strongly depending on the type of the polymer (Metz et al. 2005). Hammer et al. (2006) report a H_2O permeance of $4.4 \cdot 10^{-5}$ mol/(s m² Pa) for a commercial membrane unit for natural gas dehydration. With focus on fuel cell applications, polymer membranes for high-temperature (>130°) are developed (Nunes & Peinemann 2006, Vielstich 2003) such as polybezimidazole (PBI) and Nafion® membranes. In the case of the Nafion® membranes, the initially high H_2O permeance drops with increasing temperature due to the exothermic nature of H_2O vapour solution in the polymer layer (Struis & Stucki 2001, Löwe 2001, Metz et al. 2002). Polymer membranes are limited in operating temperature to avoid changes or decay of the polymer structure. In the case of the Nafion®

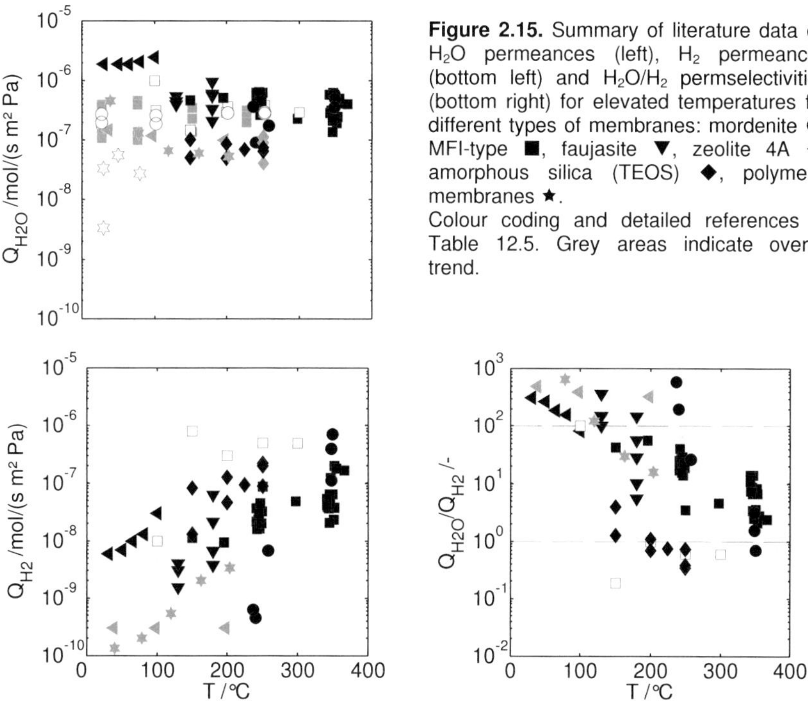

Figure 2.15. Summary of literature data on H_2O permeances (left), H_2 permeances (bottom left) and H_2O/H_2 permselectivities (bottom right) for elevated temperatures for different types of membranes: mordenite ●, MFI-type ■, faujasite ▼, zeolite 4A ◄, amorphous silica (TEOS) ◆, polymeric membranes ★.
Colour coding and detailed references in Table 12.5. Grey areas indicate overall trend.

membrane, the upper temperature limit is given with 205°C (Struis & Stucki 2001).

H_2O/H_2 permselectivity

At low temperatures (<150°C), the H_2 permeance is significantly lower than the H_2O permeance, indicating that a selective H_2O transport is possible through all membranes (Figure 2.15, bottom left). Certain membranes achieve high permselectivities between 10 and 500 (Figure 2.15, bottom right). A general increase of the H_2 permeance Q_{H2} with temperature is observed for almost all types of membranes with the zeolite 4A membrane from Aoki et al. and the ZSM5 membrane from Lindmark at al. as exceptions. Due to the strong dependence of the H_2 permeance on temperature, the H_2O/H_2 permselectivity Q_{H2O}/Q_{H2} drops with increasing temperature, resulting in a loss of permselectivity.

The ZSM-5 membranes by Espinoza et al. offer still fairly high permselectivities under simulated FT conditions between 200-350°C. On the other hand, microporous membranes (Unruh 2006, Lee, Youn & Sea 2006) reveal H_2O/H_2 permselectivities below 1 as high temperatures lead to a degradation of the hydrophilic network structure in the membrane pores. The measured H_2O/H_2 permselectivities come very close to Knudsen permselectivity of 0.33. Therefore, these membranes are not suitable for H_2O removal under high temperatures, but rather for H_2 recovery (Kölsch et al. 1998, Caro et al. 2000).

H_2O/CO and H_2O/CO_2 permselectivity

The H_2O/CO and H_2O/CO_2 permselectivities drop as well with increasing temperature (Figure 2.16). For zeolite membranes, the respective permselectivities are not significantly higher than for H_2O/H_2, though one could expect that larger kinetic diameters of CO (0.380 nm) and CO_2 (0.330 nm) will restrict the transport. The Nafion® membrane withholds CO much better than H_2, expressed in very high permselectivities which decline with temperature.

H_2O/alkanol permselectivity

Considering methanol, DME and DEE synthesis as potential reactions for a membrane reactor, the permeability of alcohols and ethers are of interest. Various authors investigated alkanol permeation through membranes (Piera et al. 1998, Sato et al. 2007, Lee, Park & Jo 2006). Very high H_2O/alkanol permselectivities (> 100 for methanol and > 1000 for ethanol at 105°C) are predicted on basis of single gas experiments for ZSM5 membranes (Noack et al. 2000). Figure 2.17 summarizes literature data from permeation experiments with binary and ternary mixtures. Piera et al. (mordenite, ZSM5) and Sato et al. (faujasite) found H_2O/methanol permselectivities between 2 and 5 (100-250°C). Similar values are reported by Stucki & Struis (2001) for the polymer Nafion® membrane. Lee et al. measured higher permselectivities (about 10) for a microporous hydrophilic silica membrane at 250°C; dimethyl ether permeated around 60 times slower through the membrane than methanol. The alcohol permeance drops with decreasing polarity and increasing size of the molecule (Piera et al. 1998).

Hydrocarbons

Noack et al. (2000) found in single gas experiments at 105°C with ZSM-5 membranes, that methane and butane fluxes were more than five orders of magnitude smaller than for H_2O. Kölsch et al. (1998) observed also that the fluxes of methane and butane were severely restricted in a microporous amorphous membrane (about 800 smaller than H_2O flux). However, various multi-component permeation experiments with membranes applying wide-pore zeolites as functional layers such as ZSM-5 or mordenite yielded much lower permselectivities. Bernal et al. (2000) obtained for ZSM-5 only H_2O permselectivities of 1-3 for methane, propane and butane (30-230°C),

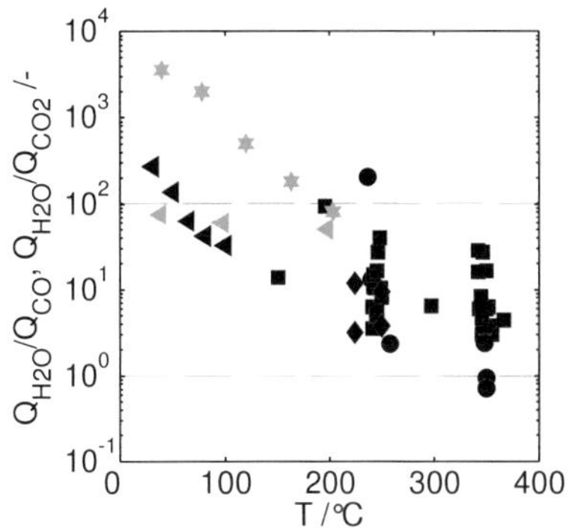

Figure 2.16. Summary of literature data on H_2O/CO and H_2O/CO_2 permselectivities for elevated temperatures for different types of membranes: mordenite ●, MFI-type ■, zeolite 4A ◄, amorphous silica (TEOS) ◆, polymeric membranes ★.

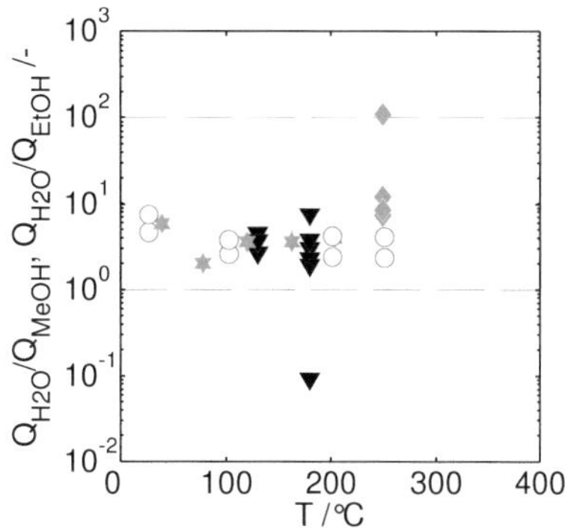

Figure 2.17. Summary of literature data on H_2O/methanol and H_2O/ethanol permselectivities for elevated temperatures for different types of membranes: mordenite ●, faujasite ▼, amorphous silica (TEOS) ◆, polymeric membranes ★.

Colour coding and detailed references in Table 12.5. Grey area indicates overall trend.

Espinoza et al. (1999, 2000, 2002) measured mean H_2O/CH_4 and H_2O/i-octane permselectivities of about 35 for ZSM-5 layers deposited on porous stainless steel tubes at 250°C. Tests with this type of membrane immersed in liquid hydrocarbons revealed significant fluxes of hydrocarbons across the membrane, about 2-8 kg/(h m²) compared to 0.5-6 kg H_2O/(h m²). Various authors observed that permeability correlates with the pore size of the chosen zeolite type, i.e. the measured hydrocarbons flux was higher for large-pore mordenite compared to medium-pore ZSM-5 or silicalite-1. The most promising results were achieved with small-pore zeolite 4A membranes. Aoki, Kusakabe & Morooka (2000) and Zhu et al. (2005) measured H_2O/CH_4 permselectivities significantly above 100 (at 100°C), Aoki et al. found a clear retention of molecules larger than ethane.

In summary, H_2O can be separated selectively from H_2 via membranes. To put the H_2O permeances into relation, H_2 permeances of palladium membranes dedicated for selective H_2 transport at high temperatures (250-800°C) should be given here. For different types of Pd-composite membranes, Dittmeyer (2001) reports H_2 permeances in the range from $1 \cdot 10^{-7}$ to $5 \cdot 10^{-6}$ mol/(s m² Pa) and H_2/inert gas permselectivities of about 500-1000. Therefore, the attainable H_2O permeances are in the range of H_2 permeances for Pd-composite membranes. The ability of selective separation declines with increasing temperature. In the temperature range interesting for FT synthesis (200-

350 °C) none of the membranes achieves H_2O/H_2, H_2O/CO or H_2O/CO_2 permselectivities above 100. Realistic permselectivities are in the range of 5 to 20. Zeolite membranes look the most promising; microporous membranes are not suitable; the data set for polymer membranes is still very limited.

3 Mass transfer in hydrophilic membranes

High H_2O permeances and high permselectivities are essential for membrane applications for in-situ H_2O removal. A new ceramic supported polymer (CSP) membrane was tested in non-reactive experiments with regard to its separation performance. The data set for hydrophilic polymer membranes for temperature above 200 °C is very limited. This chapter describes the experimental approach to determine these transport parameters. The results allow a ranking of the CSP membrane in comparison to other state-of-art membranes such as zeolite and microporous membranes.

The permeances are determined from single-gas permeances as well as from experiments with gas mixtures with and without vapors such as H_2O, methanol and ethanol. Representative sets of permeances are derived which then can be used in membrane reactor models for in-situ H_2O removal during CO_2 hydrogenation and methanol and ethanol dehydration.

3.1 Experimental

3.1.1 Applied membranes

Within the joint European project (LTCPO-GTL 2005), the Energy research Centre of The Netherlands (ECN) developed several new hydrophilic membranes, ranging from silica- over zirconium- to polymer-based membranes. On basis of benchmark tests, ECN discarded the modified amorphous silica- and zirconium-based membranes due to their weak hydrothermal stability at elevated temperatures. Three polymers suitable for applications at higher temperatures - polyimide Matrimid®, polyimide P84® and polyimide-amide Torlon® - were tested (Kreiter et al. 2008) and polyimide P84® membrane revealed the best permeation characteristics and a high stability in the benchmark tests. Therefore, the membrane of choice was the polyimide P84® ceramic supported polymer membrane.

In the remainder of this thesis, this membrane will be referred to as CSP2 membrane. ECN delivered in total 14 of these membranes. These newly developed asymmetric membranes base on a well-described support system (Bonekamp 1996, Bonekamp et al. 2006). This four-layer system consists of a commercially available macroporous extruded tube, which is coated with two thick (~ 40 µm) macroporous α-Al_2O_3 layers and a thin (~ 2 µm) mesoporous γ-Al_2O_3 layer. The mean intrinsic pore size of the γ-Al_2O_3 layer is ~ 4 nm. The selective functional layer is deposited on the exterior of the γ-Al_2O_3 layer. In Figure 3.1, the asymmetric structure of the membrane – coarse support tube, two α-Al_2O_3 layers, thin γ-Al_2O_3, and polymer layer – is clearly visible. The structural and geometrical specifications of the support are given in Table 3.1.

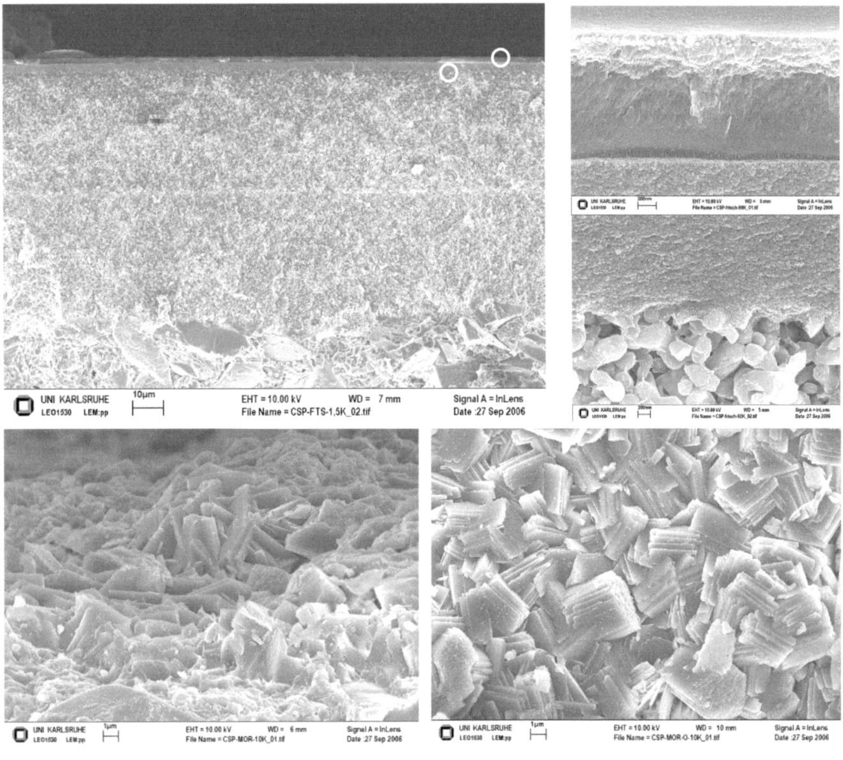

Figure 3.1. Top row: Electron microscopic (SEM) pictures of an asymmetric ceramic supported polymer (CSP) membrane, developed and provided by ECN (NL), consisting of a coarse support tube, two α-Al_2O_3 layers, one thin γ-Al_2O_3 layer, and polymer layer; (upper right corner, detail) polymer coating on γ-Al_2O_3 layer; (bottom right corner, detail) interface of γ-Al_2O_3 / α-Al_2O_3 layers.
Bottom row: Electron microscopic pictures of an asymmetric crystalline zeolite/ mordenite (MOR) membrane, prepared by the University of Zaragoza (Spain)/ ECN (NL), cross-section, zeolite coating on support tube (left) and top view (right).

The functional layer, which is about 1 µm thick, consists of the hydrophilic high performance polymer polyimide P84®, which exhibits a high chemical and thermal stability with an acceptable solubility (Kreiter et al. 2008). The glass transition temperature T_g is above 300°C. Membrane performance depends mainly on the thickness of the layer which can be steered through varying the preparation parameters as polymer concentration and curing temperature.

The evaluation of literature data (Chapter 2.2.5) showed that zeolite membranes are the most promising membrane type for H_2O removal. Therefore, it was agreed with Prof. M. Menendez from the University of Zaragoza, Spain, and ECN to prepare a mordenite (MOR) membrane using

Table 3.1. Structural and geometrical specifications of the applied asymmetric ceramic supported polymer (CSP) membranes developed and provided by ECN (NL). CSP membrane based on ultrafiltration support (Bonekamp et al. 2006). d_{50}: pore diameter at 50% of the pore volume of the specific layer. $L = 150$ mm, $r_o = 6.75$ mm, $r_i = 3$ mm, $A_m = 6.91 \cdot 10^{-3}$ m^2.

layer	coating type	name	compound	thickness	porosity	pore d_{50}
				µm	-	nm
support	-	E/ extruded tube	α-Al_2O_3	3000	0.35	4000
1	suspension	A1	α-Al_2O_3	30 – 50	0.22	180
2	suspension	A6	α-Al_2O_3	30 – 40	0.34	170
3	sol-gel	γ	γ-Al_2O_3	1.5 – 2.0	0.5	3 – 5
4		polymer	P84®	≈ 1	dense	-
4		zeolite	mordenite	>1		

the commercial tubular ceramic tube provided by ECN as support. The mordenite top layer was deposited in an autoclave according procedure described by Salomón et al. (1998). As expected the appearance of the CSP and the mordenite membrane are distinct. While the functional layer of the CSP2 membrane appears dense and smooth, the functional layer of the mordenite membrane consists of single crystals, which are grown together (Figure 3.1, bottom).

3.1.2 Experimental set-up

The integration of a membrane into a reactor system is a challenging task, in particular the design of a reliable gas-tight sealing between the inorganic (ceramic) membrane tube and the reactor shell. ECN developed and patented a special sealing and interconnection system to address this issue. All delivered membranes were equipped with a closed end cap and a specially developed stainless steel flange with a ¼" fitting for sweep inlet and permeate outlet (Figure 3.2). Compressed carbon gaskets fix the stainless steel parts to the membrane tube. This sealing system is proofed to be applicable at high

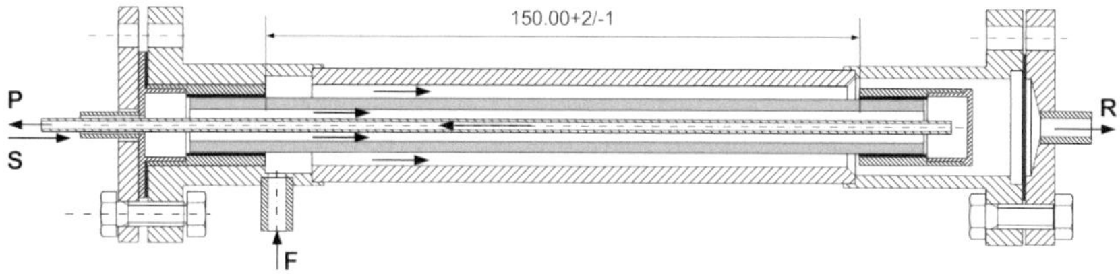

Figure 3.2. Permeation cell and ceramic supported polymer (CSP) membrane as delivered by ECN, with stainless steel fittings (flange and end cap). Permeation cell is operated in vertical position, integrated in set-up see Figure 3.3. **F**: feed, **R**: retentate, **S**: sweep, **P**: permeate.

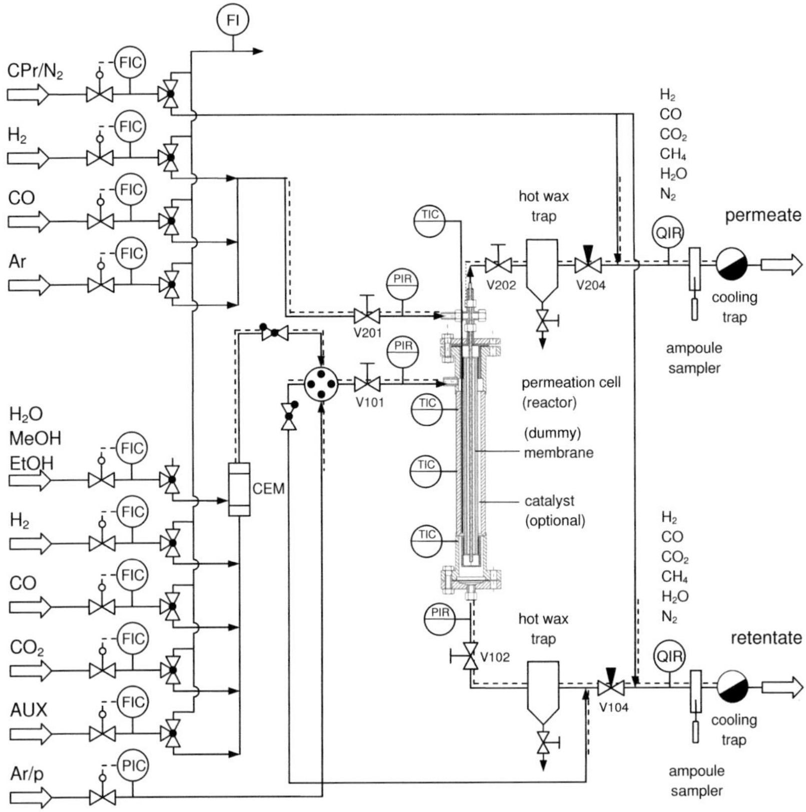

Figure 3.3. Flow scheme of the experimental set-up that is used as (a) permeation test cell, (b) reactor without membrane (with non-permeable insert/ dummy membrane), (c) reactor with integrated membrane.

temperatures and under reducing atmosphere.

The membrane is inserted from the top into the permeation cell (Figure 3.2). Top and the bottom flanges are sealed with thin copper gaskets. The feed gas enters the permeation cell from the side, flows downwards through the annular gap between membrane and shell and leaves the bottom of the cell as retentate. The sweep gas enters the tubular membrane from the top. It flows downwards in a co-current manner and it leaves the permeation cell at the top as permeate through an inserted tube (tube-in-tube).

Figure 3.3 shows the flow scheme of the experimental set-up of the permeation cell. Separate mass flow controllers (Bronkhorst) meter the feed gases H_2, CO, CO_2 and Ar (BASI, Air Liquide) and the sweep gases Ar or H_2 (BASI, Air Liquide). The feed gas mixture flows through an evaporator (Bronkhorst), where optionally a pressurized liquid stream of H_2O, methanol or ethanol is metered by a Liquiflow (Bronkhorst) and evaporated. The feed gas is fed over a 4-way valve either to the reactor or directly the product analysis (reactor bypass). Any liquid products in the retentate and permeate stream are collected in high-pressure traps (1 MPa, 180 °C) and are removed periodically. A fine-metering valve (Swagelok) and an auxiliary Ar flow which pressure is adjusted by a two-stage pressure reducer (Druwa) control the feed-side pressure. The sweep-side pressure on the sweep side is adjusted by a fine-metering valve (Swagelok).

Pressure gauges (WIKA, relative to atmosphere) measure the feed, retentate and permeate pressure. Three separate heating zones – three aluminium blocks with four heating cartridges each (120 W, Helios) – controlled by Eurotherm temperature controllers keep the reactor on temperature. Axial temperature profile in annular gap between shell and tubular membrane can be determined by a movable thermocouple installed in a thermowell. Temperature and pressure data is registered and displayed on a personal computer (Beckhoff interface, LABVIEW software).

The retentate and permeate streams are expanded to atmospheric pressure after the high-pressure traps across the fine-metering valves. An accurately metered flow (Bronkhorst mass flow controller) of reference gas (0.5 vol% cyclopropane in N_2, BASI) is added and mixed either to the retentate or permeate stream. Then, a small side stream of the retentate or permeate is sent to the heated injection valves of the on-line gas chromatograph (6890N Agilent/ JAS) and analyzed. Before the off-gas is vented, any condensables are collected either in a cooling trap cooled by a cryostat (-15 °C) or adsorbed on silica gel (Merck).

The line-up of the columns of the on- and off-line GCs, the analysis of GC results and the methods to measure the H_2O content in the permeate and retentate streams are discussed in detail in Chapter 12.3 in the appendix.

3.1.3 Data analysis and definitions

In various experiments, the transport of the reactants and products across the membrane is measured under different non-reactive conditions to determine the transport characteristics of the membrane. The experiments and experimental conditions are summarized in the appendix in Table 12.6 (experimental plan). The permeation cell was operated in the different permeation modes, which are shown in Figure 3.4.

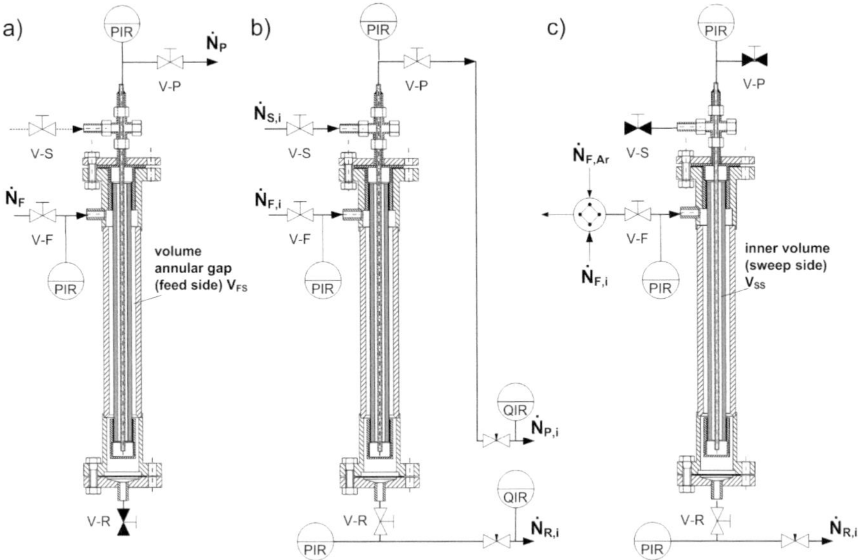

Figure 3.4. Schematic illustration of the experimental set-ups for permeation experiments: a) single gas permeation experiments (steady-state and transient), b) steady-state permeation experiment with gas mixtures and c) transient permeation experiment. Valve open (▷◁), valve closed (▶◀). PIR: Pressure gauge, QIR: quality indicator (GC, silica gel trap, cooling trap).

Single gas permeation experiments in dead-end configuration

In this configuration (Figure 3.4, a), the feed inlet is open and the retentate exit is closed. The sweep side of the tubular membrane may be swept with a sweep gas at ambient pressure, but a least the permeate exit is fully open. A single gas with a constant flow rate is forced through the membrane. The pressure on the feed side (shell side) will rise until the flow rate across the membranes equals the inflowing feed flow (transient pressure-rise experiment). If the feed inlet is closed at t = 0, the pressure on the feed side (shell side) will drop (transient pressure-drop experiment). From these experiments, it is possible to determine the permeances of single gases.

The single gas permeance Q can be calculated either from the steady-state pressure difference (3.1), which is obtained, when a fixed molar flow rate of a specific component is forced across the membrane, or it can be derived from the rate of pressure change during the transient pressure-drop experiments (3.2). If the permeance is not a (strong) function of pressure, the permeance can be calculated from the half-life time $t_{1/2}$, i.e. the time period, in which the initial pressure at t_0 dropped by the half (3.3).

$$J_{tmb} = A_m Q(P_F - P_P) \quad \text{with} \quad J_{tmb} = \dot{N}_F = \dot{N}_P \tag{3.1}$$

$$Q = \frac{J_{tmb}(t)}{(P_F(t) - P_P)} = -\frac{1}{RT}\frac{V_{FS}}{A_m}\frac{dP_F/dt}{(P_F - P_P)} \tag{3.2}$$

$$Q_{t1/2} = -\frac{1}{RT}\frac{V_{FS}}{A_m}\ln(0.5)\frac{1}{t_{1/2}} \tag{3.3}$$

Steady-state permeation experiments in gas separation configuration

In the gas separation configuration (Figure 3.4, b), gas mixtures without and with vapours (e.g. H_2O, methanol, and ethanol) are fed to the permeation cell. The aim is to understand the permeation behaviour of multi-component mixtures, which in general cannot be predicted from single-gas experiments.

The tubular membrane is swept in co-current mode with Ar or H_2. The feed-side and the sweep-side pressure are adjusted via the fine-metering valves in the retentate and permeate lines, respectively. The composition and the flow rates of the inlet streams (feed, sweep) and outlet streams (retentate, permeate) are measured by GC; the amount of H_2O in the respective streams is determined either by GC or gravimetrically by adsorption on silica gel or by freezing out in cooling traps (cryostat, -15°C). On basis of the molar flow rates of the in- and out-going streams, the individual permeances Q_i are determined by least-square fitting. This approach assumes constant permeances along the axial coordinate, but considers the development of non-linear partial pressure profiles along the axis of the permeation cell. The set of equations given in (3.4) represents the differential material balance of the permeation cell for each species expressed in terms of molar flow rates.

$$\begin{aligned} \frac{d\dot{N}_{i,FS}}{dz^+} &= A_m Q_i (P_{i,FS} - P_{i,SS}) \\ \frac{d\dot{N}_{i,SS}}{dz^+} &= -A_m Q_i (P_{i,FS} - P_{i,SS}) \end{aligned} \quad \text{with } Q_i \text{ such that} \quad \sum_{i=1}^{N} (\dot{N}_{i,R,exp} - \dot{N}_{i,R,model})^2 + (\dot{N}_{i,P,exp} - \dot{N}_{i,P,model})^2 = \min \tag{3.4}$$

$$\begin{aligned} \dot{N}_{i,FS}(z^+ = 0) &= \dot{N}_{i,F} & \dot{N}_{i,R} &= \dot{N}_{i,FS}(z^+ = 1) & i &= 1....N \\ \dot{N}_{i,SS}(z^+ = 0) &= \dot{N}_{i,S} & \dot{N}_{i,P} &= \dot{N}_{i,SS}(z^+ = 1) & i &= 1....N \end{aligned} \tag{3.5}$$

Besides sweep gas composition, the sweep ratio Ψ and the pressure ratio Φ are the key experimental variables as they determine the driving force across the membrane:

$$\psi = \frac{\dot{N}_S}{\dot{N}_F} \tag{3.6}$$

$$\Phi = \frac{P_{SS}}{P_{FS}} \tag{3.7}$$

Dynamic permeation experiments

This type of experiments is well described by Tuchlenski (1998). Tuchlenski applied this experiment to analyse surface diffusion effects for porous membranes. In this thesis, transient experiments are only used to test the quality of a simplified mass transfer model. In this configuration (Figure 3.4, c), the sweep inlet and permeate exit are closed i.e. the tube side volume is isolated and mass can only be exchanged across the membrane. A single gas or a gas mixture is fed to the permeation cell and leaves through the retentate exit. Under steady-state conditions, the pressure and gas composition in the sweep side volume equals the pressure and gas composition on the feed side. At t = 0, a step-change in feed composition is induced by switching the upstream 4-way valve. This step-change in concentration initiates non-equimolar counter diffusion between feed and sweep side, resulting in a pressure change in the enclosed sweep side volume. The rate of pressure change and the obtained pressures depend on the membrane transport properties. The results of these experiments are discussed in Chapter 12.4.3 in the appendix.

The overall performance of the gas permeation cell (and of the membrane reactors) is assessed by:

- degree of the removal of component i (3.8) from the feed side or (reaction zone), where i is the dedicated species (e.g. H_2O) or a (by-)product of a reaction.
- undesirable loss of component i (3.9) to the permeate side, where i is a valuable component e.g. a reactant.
- flux of key component i across the membrane (3.10) to rank the performance for a dedicated application (e.g. dehydration applications).

$$R_i = \frac{\dot{N}_{i,P}}{\dot{N}_{i,P} + \dot{N}_{i,R}} \quad (3.8) \qquad L_i = \frac{\dot{N}_{i,P} - \dot{N}_{i,S}}{\dot{N}_{i,F}} \quad (3.9)$$

$$\dot{j}_{i,tmb} = \frac{\dot{N}_{i,P} - \dot{N}_{i,S}}{A_m} \quad (3.10)$$

3.2 Results and discussion

First, the results for the mordenite membranes should be discussed briefly as they highlight a typical problem and challenge for membrane reactors.

Two improved mordenite membranes were tested in single gas and steady-state permeation experiments with gas mixtures with and without H_2O co-fed. The single gas permeation experiments revealed a strong hysteresis with temperature and a high permeability. Permeances > $8 \cdot 10^{-7}$ mol/(s m^2 Pa) were

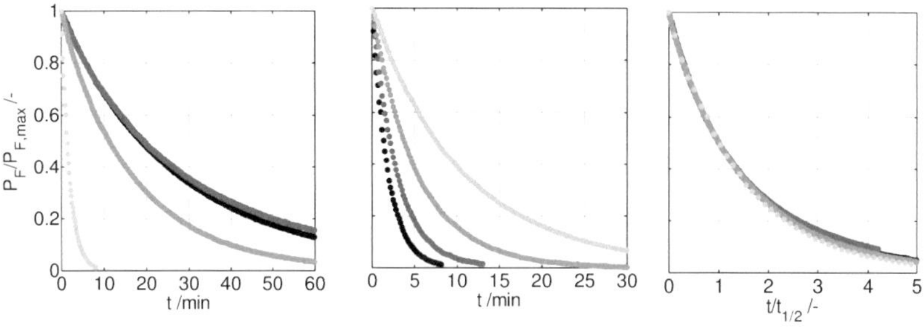

Figure 3.5. Transient pressure drop experiments with single gases for CSP2 membrane, (left) with different gases at 200°C: ( ) H_2, (●) CO_2, (●) CO, (●) Ar; (centre) with H_2 at different temperatures: ( ) 100, (●) 150, (●) 200, (●) 250°C; (right) dimensionless presentation of Figure 3.5 (left) using characteristic time $t_{1/2}$.

found for H_2 at 270°C, which are higher than for microporous silica membranes, but still lower than for ultra-filtration supports. In permeation experiments with gas mixtures ($H_2/CO/CO_2/H_2O$), one membrane achieved only Knudsen permselectivities; the second membrane was not permselective at all. Post-mortem analysis revealed that the permeance determined in a reference experiment with N_2 was one order of magnitude higher than before the carbon gaskets were installed. Membrane tests under water indicated a high gas flow near the sealing. The most probable cause of the high leak rate can be found in the high surface roughness of the mordenite membrane layer. The relative stiffness of the graphite gasket does not allow the sealing of sharp edges as can be found between zeolite crystals. This highlights that (a) the installation of a gas-tight sealing between membrane and the equipment shell is still a challenging task and that (b) small failures in the functional layer can jeopardize the membrane reactor concept. As a consequence, further experiments with mordenite membranes were discarded. The ceramic supported polymer had a much smoother surface and similar problems did not occur.

3.2.1 Permeation experiments with single gases

Steady-state and transient single gas permeation experiments were carried out with the permanent gases H_2, CO, CO_2 and Ar, covering a temperature range from 100 to 250°C. Figure 3.5 (left) indicates that H_2 is the fastest permeating species, followed by CO_2, Ar and CO. The permeability of the CSP2 membrane for permanent gases increases with temperature (Figure 3.5, centre). The feed-side pressure covered in these experiments ranges from 0.1 to 2 MPa, depending on the tested gas. When the results are plotted versus the dimensionless time $t/t_{1/2}$, all the data falls onto a single curve described by

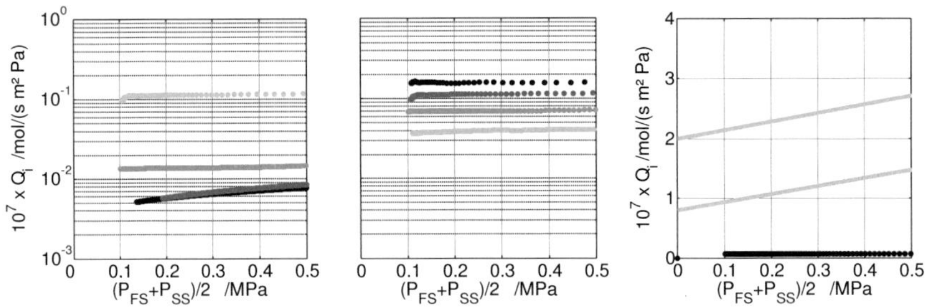

Figure 3.6. Permeances Q_i derived by numerical differentiation of transient pressure drop data as function of the mean pressure P_{mean}, (left) with different gases at 200 °C: (●) H_2, (●) CO_2, (●) CO, (●) Ar; (centre) with H_2 at different temperatures: (●) 100, (●) 150, (●) 200, (●) 250 °C. (Right) Dependence of the H_2 permeance on the mean pressure P_{mean}, measured at 150 °C in single gas experiments; CSP2 membrane (●) in comparison with microporous silica membranes M5 and M4 (—, experimental data fit, Unruh 2006).

exponential decline (Figure 3.5, right). Although Ar and CO deviate slightly from ideal exponential curve at lower pressures, this plot already indicates that the permeances are not a strong function of pressure.

The single gas permeances Q_i are either determined by the characteristic half-life time (3.3) or on basis of the rate of pressure drop (3.2). Hereunto, the data is smoothened and numerically differentiated by means of dedicated MATLAB® functions. Figure 3.6 (left) indicates that H_2 and CO_2 permeances are not a function of the mean pressure, i.e. solubility and diffusion coefficients are independent of pressure (ideal solution-diffusion model) and what is even more important, that viscous flow does not occur. The thin polymer layer is free of defects and the transport takes place via diffusion. CO and Ar show a slight dependence on mean pressure, resulting from a weak pressure dependence of the solubility and diffusion coefficients or from an experimental artefact.

The H_2 permeance is with $1 \cdot 10^{-8}$ mol/(s m^2 Pa) about one order of magnitude higher than for CO_2, CO and Ar. However, it is much smaller than the H_2 permeance reported by Unruh (2006) for hydrophilic microporous silica membranes. Figure 3.6 (right) compares both types of membranes and one recognizes that the transport through the microporous silica membrane is determined by selective Knudsen diffusion and by unselective viscous flow, which contribution increases linearly with mean pressure (2.30). For the CSP2 membrane, viscous flow is virtually absent and the limiting transport step is clearly located in the polymer layer.

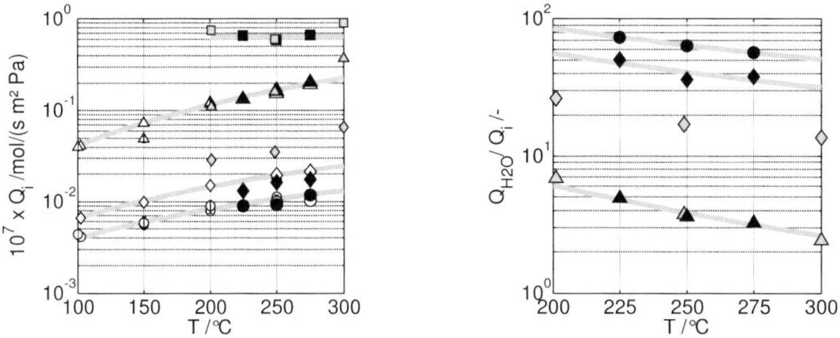

Figure 3.7. Comparison of permeances (left) and permselectivities (right) of the CSP2 membrane determined in (△◇○) transient/ steady-state single gas experiments and permeation experiments with dry gas mixture, (■▲◆●) permeation experiments with gas mixture with 25 vol% H_2O co-fed, and (■▲◆●) ECN permeation experiments with gas mixture with 30 vol% H_2O co-fed (Vente 2006). Curves: calculated on basis of Arrhenius law.

Q_i: (■■) H_2O, (▲▲△) H_2, (◆◆◇) CO_2, (●●○) CO; Q_{H2O}/Q_i: (▲▲) H_2, (◆◆) CO_2, (●●) CO

* Dry gas mixture: $H_2/CO/CO_2$ (50/30/20 vol%), 25 vol% H_2O, P_F = 1 MPa, Ψ = 2, Φ = 0.15, $(Ar)_S$, * ECN: CSP2/dry gas mixture: $H_2/CO/CO_2/CH_4$ (67/0/29/4 vol%), 30 vol% H_2O, ΔP_{tmb} = 0.9 MPa

3.2.2 Permeation experiments with gas mixtures

In the gas separation configuration (Figure 3.4, b), a dry gas mixture consisting of $H_2/CO/CO_2$ (50/30/20 vol%) was fed to the permeation cell. The membrane was swept with Ar or H_2 in co-current mode; the sweep ratio Ψ and the pressure ratio Φ were varied within the relevant operating range of the membrane reactor experiments.H_2O, methanol, and ethanol were co-fed to determine the permeability of the CSP2 membrane for polar vapors and to investigate the effect of vapors on the overall permeation characteristics. As ECN carried out similar permeation experiments with H_2O vapour during the benchmark tests (Vente 2006); their results are used as reference.

Figure 3.7 combines the permeances measured in (a) transient and steady-state single gas experiments, (b) gas permeation experiments with a multi-component gas mixture without H_2O addition and (c) with H_2O addition. The following conclusions can be drawn:

- H_2O is the fastest permeating component with a permeance of $7 \cdot 10^{-8}$ mol/(s m^2 Pa). The hydrophilic polymer layer favours the selective transport of the polar H_2O molecules over H_2, CO_2 and CO molecules.
- H_2O addition does not alter the transport of the other species significantly. Experiments with single gases, with gas mixtures and gas mixtures with H_2O addition yield similar permeances. Individual components of a gas

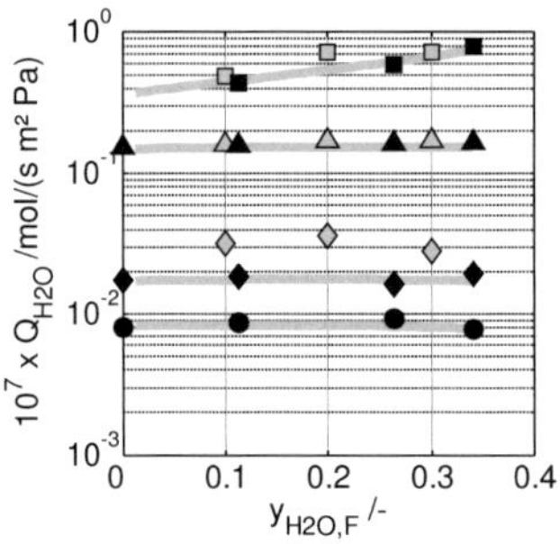

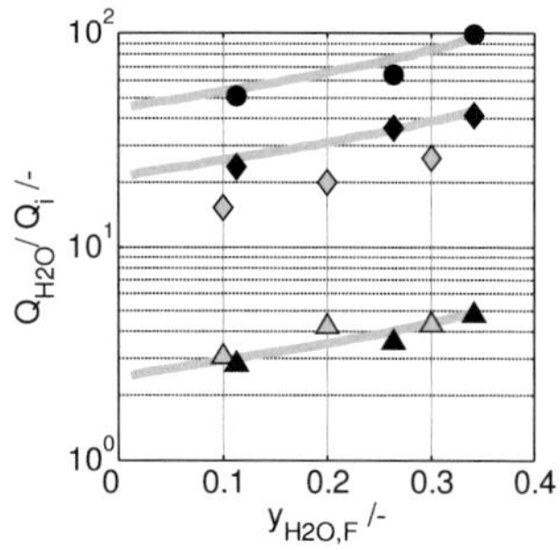

Figure 3.8. Effect of H_2O co-feeding on permeances (left) and permselectivities (right) of the CSP2 membrane. Grey symbols: ECN experiments. Conditions see Figure 3.7. Curves: trends.
Q_i: (■ ■) H_2O, (▲ ▲) H_2, (◆ ◆) CO_2, (● ●) CO; Q_{H2O}/Q_i: (▲ ▲) H_2, (◆ ◆) CO_2, (●) CO

mixture dissolve and diffuse through the polymer independently of each other. In this case, single gas experiments could predict the permeances for gas mixtures. Generally, the permeances of porous membranes for gas mixtures cannot be derived from single gas experiments, in particular if vapors or strongly adsorbing species are involved.

- The permeances of H_2, CO_2, CH_4, CO and Ar increase with temperature. The overall activation energy for permeation $E_{A,P}$ found is about 10-20 kJ/mol, which should represent the activation energy of the diffusion step $E_{A,D}$. H_2 has the highest activation energy with 15-20 kJ/mol.
- The permeance of H_2O remains constant with increasing temperature. The solution enthalpy ΔH_S of H_2O vapour in the polymer matrix is negative. Therefore, the negative solution enthalpy ΔH_S and the positive activation energy of the diffusion step $E_{A,D}$ cancel each other out, resulting in an overall activation energy for permeation $E_{A,P}$ around zero.
- As consequence, the H_2O permselectivities with regard to other species drop with increasing temperature due to the different activation energies (Figure 3.7, right). The permselectivities in the temperature range between 225 and 275°C are found as (5...2) for H_2O/H_2, (50...40) for H_2O/CO_2, (60...40) for H_2O/CH_4, (85...60) for H_2O/Ar and (90...60) for H_2O/CO.
- The permeances and trends found were in agreement with the data generated by ECN in a different experimental set-up with different gas mixtures. ECN found systematically higher permeances for CO_2.

Figure 3.8 confirms that an increasing H_2O fraction in the gas mixture does not alter the transmembrane transport kinetics of the other species. A preferential

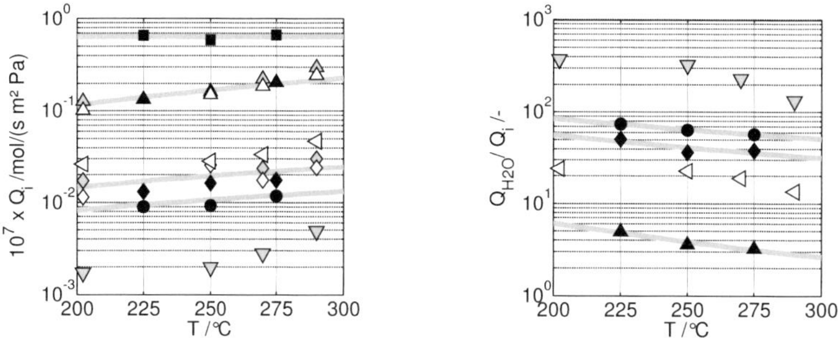

Figure 3.9. Effect of H_2O (black, 25 vol%), methanol (white, 25 vol%) and ethanol (grey, 26 vol%) co-feeding on CSP2 permeances Q_i (left) and permselectivities Q_{H2O}/Q_i (right). Curves: calculated on basis of Arrhenius law.
Q_i: (■) H_2O, (◁) methanol, (▽) ethanol, (▲ ▲ △) H_2, (◆ ◆ ◇) CO_2, (● ● ○) CO
Q_{H2O}/Q_i: (▲) H_2, (◁) methanol, (▽) ethanol, (◆) CO_2, (●) CO
* CSP2/dry gas mixture: $H_2/CO/CO_2$ (50/30/20 vol%), P_F = 1 MPa, Ψ = 2, Φ = 0.15, $(Ar)_S$

transport of H_2O accompanied with an inhibition of the transport of competing species does not take place. However, it is found that the H_2O permeance itself increases with increasing molar H_2O fraction in the feed, indicating a concentration dependence of the solubility or diffusion coefficient. Higher H_2O concentrations result in a faster H_2O permeation and therefore in higher permselectivities. This means with regard to dehydration applications, the more H_2O is removed; the less permselective will be the membrane transport.

Methanol and ethanol co-feeding experiments show as well that the transport of the permanent gases is not affected by the presence of vapours (Figure 3.9). Permeances of the polar components rank as follows: H_2O > methanol > ethanol. The methanol permeance is about two times larger than the one of CO_2; the ethanol permeance is significant lower than the one of CO or Ar. It can be expected that H_2O permeates 20 times faster than methanol and 200 times faster than ethanol across the membrane. In contrast to H_2O, the permeances of the alcohol vapours increase with temperature, first slowly, and above 250 °C steeper.

3.3 Membrane assessment

The experimental results – which are summarized in the appendix in Table 12.19 - Table 12.21 – suggest that the permeances of the CSP2 membrane (based on the polyimide P84®) for the permanent gases H_2, CO_2, CO, Ar, CH_4 exhibit no or only a very weak partial pressure dependence. On the other hand, the permeance for H_2O vapour increases with increasing H_2O concentration in the feed gas. Furthermore, transport kinetics of a single component is not affected by the presence of other permanent gases or even vapors. With regard to the categorization of Kimura & Hirose (1992) in Figure 2.13, the CSP2 membrane shows rather the characteristics of a rubbery than a glassy membrane, though the glass transition temperature of the polymer is expected to be above 300°C (Kreiter et al. 2008). The transport-limiting step is located in the functional polymer layer; significant transport limitations in the porous support structure or in the external boundary layers do not occur and are therefore neglected (see appendix). The simple relationship between partial pressures and molar fluxes justifies the application of the ideal solution-diffusion model to describe the transport kinetics across the CSP2 membrane:

$$j_{i,tmb} = Q_i \left(P_{i,FS} - P_{i,SS} \right) \tag{3.11}$$

where $P_{i,FS}$ is the feed-side partial pressure and $P_{i,SS}$ the sweep-side partial pressure. The permeance Q_i is assumed to be only a function of temperature. The set of respective permeances Q_i at a given temperature can be taken from Table 12.19 - Table 12.21 (appendix) which summarize the collected permeation data for the CSP2 membrane.

A study on the dehydration performance of the gas permeation cell (membrane unit) equipped with a CSP2 membrane (appendix Chapter 12.4) shows that the membrane model with a set of constant permeances yields satisfying agreement between model calculations and experimental data, which is a sound basis for the assessment of membrane reactor applications.

The previous experiments in the gas permeation cell illustrated that more than 90% of the H_2O in the feed gas stream can be removed across the membrane. If a high driving force is applied, also up to 60% of H_2 in the feed gas stream is lost to the permeate side. This result demands clearly higher permselectivities. Figure 3.10 shows how CSP2 membrane ranks compared to other hydrophilic membranes as zeolite and micro-porous alumina and silica membranes:

- The H_2O permeance of the CSP2 membrane does not exceed $1 \cdot 10^{-7}$ mol/(s m^2 Pa) and is in the range of the polymer Nafion® membrane tested by Struis & Stucki (2001) or of the microporous silica membranes tested by Unruh (2006); higher H_2O permeances would require an even thinner polymer layer, which is already only 1 μm thick. Attempts to make thinner membranes resulted in the presence of defects. Polymer or microporous membranes cannot compete with crystalline zeolite membranes which exhibit with $1 \cdot 10^{-7}$-$1 \cdot 10^{-6}$ mol/(s m^2 Pa) up to 10 times higher H_2O permeances. In that respect, it was a disappointment that the planned experiments with mordenite membranes failed due to damage of the brittle functional layer by the sealing system.

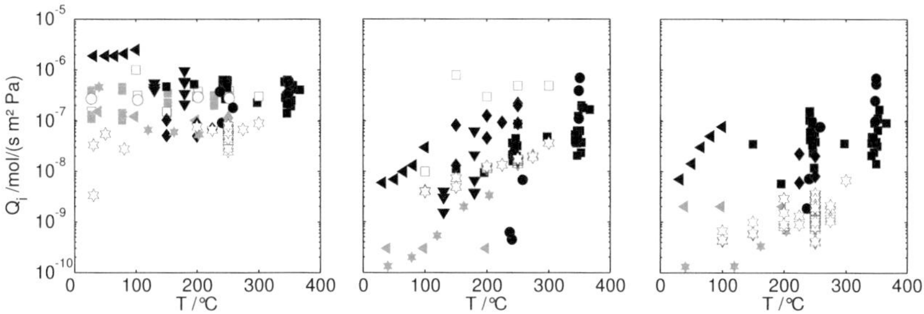

Figure 3.10. Comparison of the CSP2 membrane performance (☆) with data from literature on H_2O permeances (left), H_2 permeances (centre) and CO_2 and CO permeances (right) for different types of membranes: mordenite ●, MFI-type ■, faujasite ▼, zeolite 4A ◀, amorphous silica (TEOS) ◆, polymeric membranes ★. Detailed literature references in Table 12.5.

- The CSP2 membrane shows a clear permselectivity of H_2O towards H_2, which declines with increasing temperature. The measured H_2O/H_2 permselectivities are higher than for microporous membranes, but lower than for zeolite membranes, in particular, zeolite membranes of the MFI/ZSM-5 type. However, the CSP2 permselectivities for H_2O towards CO_2 and CO (and CH_4 and Ar) are outstanding, also in comparison to zeolite membranes.

The CSP2 membrane offers defined permselective properties, but the permeability of H_2O is limited. The performance is similar to that of the Nafion® membrane tested by Struis and Stucki (2001), but the operating window can be stretched to higher temperatures. The CPS2 membrane still has to prove its applicability and stability under reactive conditions, i.e. in the presence of long-chain hydrocarbons (FT synthesis) and higher partial pressures of alcohol and ethers (DME/DEE synthesis).

4 Kinetics and selectivity of the Fischer-Tropsch reaction

The assessment of membrane reactor applications requires not only the knowledge of membrane permeance and permselectivity data, but also a fair understanding of the kinetics of the involved chemical reactions. This chapter describes the experimental approach to determine the rate parameters of Co- and Fe-based Fischer-Tropsch (FT) catalysts. Various catalysts were tested and benchmarked with regard to their applicability for in-situ H_2O removal experiments and CO_2 hydrogenation reactions. Appropriate catalyst systems should fulfil certain prerequisites such as high FT activity, high CO_2/CO shift activity and reasonable stability.

With regard to additional in-situ H_2O removal applications, rate parameters were also determined for the dehydration of methanol and ethanol to the respective ethers and for CO/CO_2 shift reaction for a low-temperature shift catalyst. The interested reader is referred to the appendix.

4.1 Experimental

4.1.1 Applied catalysts

Three Fe-based and two Co-based FT catalysts were tested for their applicability for *in-situ* H_2O removal experiments (Table 4.1):

- The Fe-5K catalyst is a K-promoted Fe-based catalyst prepared by co-precipitation and impregnation according to the method described by Riedel (2003).
- The Fe-GTL3 was provided within the European project LTCPO-GTL (2005) as reference catalyst and was prepared by Euro Support BV, Amersfoort, The Netherlands.
- The Fe-GTLX is a Fe-based FT catalyst provided from a German catalyst supplier. The catalyst composition is unknown. This catalyst Fe-GTLX was tested by Mena (2009) in parallel.
- The Co-GTL1 was provided within the European project LTCPO-GTL (2005) and was prepared by Euro Support BV, Amersfoort, The Netherlands. This cobalt/manganese catalyst was tested as it should exhibit a certain CO/CO_2 shift activity. Riedel (2003) describes the preparation method in detail.
- The Co-GTL4 was provided within the European project LTCPO-GTL (2005) as reference catalyst and was prepared by Euro Support BV, Amersfoort, The Netherlands. Co-GTL4 is a simple unpromoted alumina supported catalyst, prepared by aqueous incipient wetness impregnation on a commercial alumina support (Oukaci, Singeleton & Goodwin 1999).

Table 4.1. Applied catalyst systems and their characteristic properties.

	composition		surface area	pore volume	average mesopore diameter	crystallite size
		/wt% wrt active metal	/ m^2/g	/ cm^3/g	/ nm	/ nm
Fe-5K	$Fe/Cu/K_2O/Al_2O_3$	100/8.9/5.0/14.1	180	-	~5	-
Fe-GTL3	$Fe/Cu/K_2O/SiO_2$	100/4.2/2.3/23.7	222	0.43	8.3 (7.0)	-
Fe-GTLX	$Fe/Cu/K_2O/SiO_2$	not known	93.7	0.75	-	-
Co-GTL1	$Co/Mn/SiO_2/Pt$	100/49/145/0.31	321	0.81	14.7	-
Co-GTL4	Co/Al_2O_3	100/515	131	0.31	9.7	0.49

All catalysts are crushed and sieved. The $100 < d_P < 160$ µm fraction is mixed with inert material within a similar particle diameter range. The small particle diameters ensure that internal mass transfer limitations and channelling effects do not become prevailing in the catalyst bed. The Fe- and Co-based catalysts were reduced in the lab-scale fixed-bed reactor according to the reduction procedure given in Table 4.2. The H_2 consumption during temperature programmed reduction (TPR) of the Fe- and Co-based FT catalysts is plotted in Figure 12.17.

After catalyst reduction, the reactor is cooled down to the activation temperature under a steady Ar flow and pressurized to 1 MPa. At t=0, the feed stream is switched from Ar to H_2/CO (2/1) synthesis gas by turning a 4-way valve. From this point of time, the formation of the FT regime starts; this is a process that runs through several transient kinetic regimes (episodes) until the state of the highest FT activity is reached. It can take up to 100 hours until steady-state conditions are achieved. Figure 12.18 compares the activation curves of the four Fe- and Co-based catalysts.

4.1.2 Experimental set-up

Two fixed-bed lab-scale reactors were used for kinetics measurements. The first reactor – the so-called fixed-bed glass reactor - is a dedicated reactor concept that has been used at the EBI for several years for kinetic experiments (van Steen 1993, Claeys 1997, Riedel 2003, Unruh 2006). The fixed-bed glass reactor (Figure 4.1) is built up as follows: The catalyst bed is

Table 4.2. Conditions for catalyst reduction and activation of Fe- and Co-based FT catalysts.

	gas	P	$\tau_{mod,n}$	T_{start}	heating rate	T_{end}	holding time
		/ MPa	/ kg s/m^3	/ °C	/ K/min	/ °C	/ h
reduction	H_2/Ar (1/3)	0.1	800	20	2	400	16
activation	H_2/CO (2/1)	1	2000-4000	230 (Co) 250-270 (Fe)	-	-	100

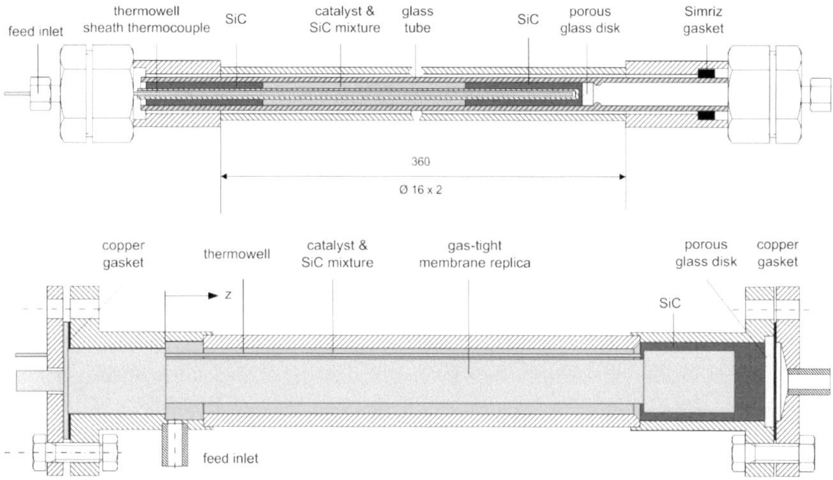

Figure 4.1. Lab-scale fixed-bed reactors for kinetic experiments, top: dedicated kinetic reactor, bottom: (membrane) reactor equipped with gas-tight membrane replica for kinetic measurements (H_2O co-feeding experiments, DME/DEE synthesis); reactors are operated in vertical position.

located in a glass tube which is inserted into a tubular stainless steel reactor shell (diameter: 16 mm, wall thickness: 2 mm). A high temperature gasket between glass tube and the reactor shell at the lower end prevents gas by-passing (sealing ring: Simriz® 75 FFPM 495, Freudenberg, inner diameter: 9.5 mm, width: 3 mm). Two reducing Swagelok® tube fittings are mounted on the stainless steel reactor shell, providing ¼" fittings for feed inlet and product outlet.

A thin glass tube that serves as thermocouple well is inserted into the larger glass tube (glass reactor: outer diameter: 11 mm, wall thickness: 1.5 mm, length: 400 mm; thermocouple well: outer diameter: 4 mm, wall thickness: 0.9 mm, length: 300 mm). A porous glass disk supports the thermocouple well and the catalyst bed. The temperature profile along the catalyst bed is measured by moving the thermocouple along the reactor axis. The catalyst bed in the annular gap between the thermowell and the glass tube consists of three layers, first, an inert layer of silica carbide β-SiC (d_P > 250 μm), followed by a mixture of 2-4 g of catalyst (100 < d_P < 160 μm) and 6 cm^3 SiC (200 < d_P < 250 μm) and finally another inert layer of silica carbide SiC (d_P > 250 μm), which allows preheating and the formation of an even flow pattern.

The second reactor uses the pressure shell of the permeation cell (Figure 4.1). Here, a gas-tight stainless steel replica of the membrane replaces the tubular membrane. In this configuration, the kinetic experiments are carried out under

similar geometrical conditions and flow patterns as encountered in the membrane reactor.
The gas-tight membrane replica is inserted from the top into the stainless steel reactor shell. The top flange is sealed with a copper gasket. Ground and sieved catalyst ($100 < d_P < 150$ µm) is mixed with β-SiC ($200 < d_P < 250$ µm) and filled into the annular gap between reactor shell and inner steel tube. The amount of SiC is chosen such that the catalyst/SiC mixture (27 cm^3) fills up the isothermal middle section of reactor (Figure 4.1). The rest of the free volume of the reactor is filled up with coarse SiC ($d_P > 250$ µm). A porous glass disk supports the catalyst /SiC mixture and the inert bed. The reactor is closed with a metal flange sealed with a copper gasket.
A detailed assessment of the lab-scale reactors is compiled in chapter 12.7. In summary, these lab-scale fixed-bed reactors are appropriate for kinetic measurements as:

- internal and external mass transfer limitations do not occur. For catalyst particles with $d_P < 250$ µm, the catalyst is fully utilized (catalyst efficiency η = 1, Claeys 1997).
- the diluted catalyst bed is isothermal. Axial temperatures vary under reactive conditions maximum ± 2 K along the reactor axis.
- the pressure drop across the catalyst is with less than 2% of the total pressure negligible.
- a significant effect of axial dispersion do not occur as indicated by residence-time distribution (RTD) measurements and calculations.

Kinetic measurements are carried out in the experimental set-ups described in Figure 3.3 and Figure 4.2. The latter flow scheme is dedicated to the fixed-bed glass reactor. Here, separate mass flow controllers (Brooks) meter the feed gases H_2, CO, CO_2 and optional Ar (BASI, Air Liquide). A defined H_2O stream - metered by a high pressure liquid pump - can be vaporized and mixed into the feed stream. The feed gas is fed over a 4-way valve either to the reactor or directly the product analysis (reactor bypass).
Any liquid products in the reactor effluent are collected in a high-pressure trap (1 MPa, 180℃) and are removed periodically. The reactor pressure is controlled by a fine-metering valve (Swagelok) and an auxiliary Ar flow pressure is adjusted by a two-stage pressure reducer (Druwa). Pressure gauges (WIKA, relative to atmosphere) measure the pressure up- and downstream of the catalyst bed. Three separate heating zones – three aluminium blocks with four heating cartridges each (120 W, Helios) – controlled by Eurotherm temperature controller keep the reactor on

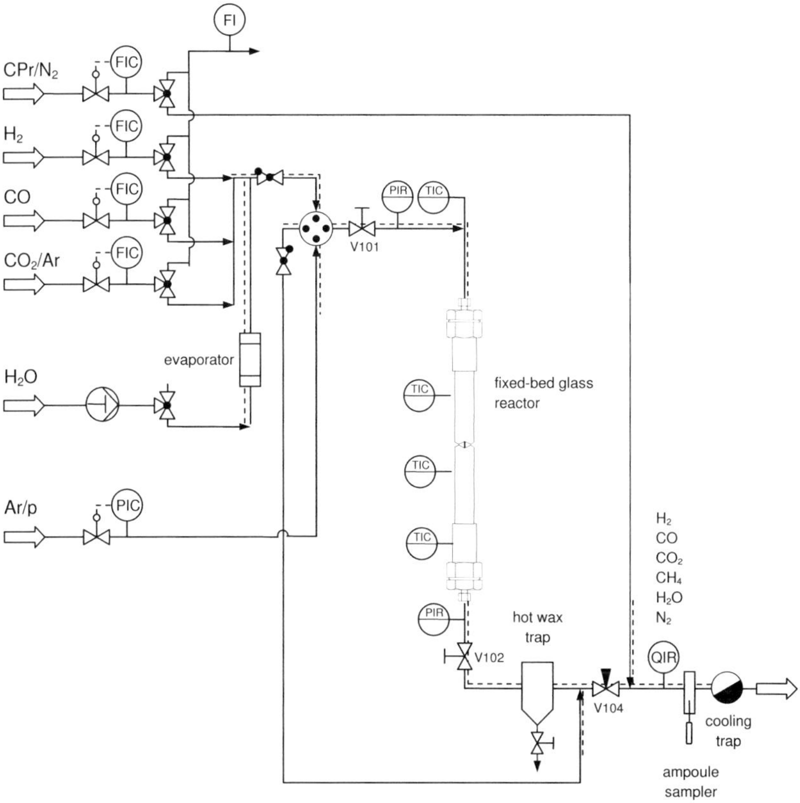

Figure 4.2. Flow scheme of the experimental set-up for kinetic experiments using the fixed-bed glass reactor.

temperature. The axial temperature profile is determined by a movable thermocouple installed in a thermowell.

The reactor effluent stream is expanded to atmospheric pressure after the wax trap via a fine-metering valve. An accurately metered flow (Bronkhorst mass flow controller) of reference gas (0.5 vol% cyclopropane in N_2, BASI) is mixed into the effluent stream. A small side stream is sent to the heated injection valves of an on-line gas chromatograph (6890N Agilent/ JAS) and analyzed. Additional samples can be taken by the ampoule method and are analyzed off-line in a dedicated gas chromatograph.

Before the off-gas is vented, any condensable products are collected in a cooling trap cooled by a cryostat (-15 °C). The line-up of the columns of the on-

and off-line GCs and the evaluation of the GC results discussed in detail in the appendix (Chapter 12.3).

4.1.3 Data analysis and definitions

The modified residence time, feed gas composition and temperature were varied in the kinetic experiments. The experiments and experimental conditions are summarized in the appendix in Table 12.7- Table 12.10 (experimental plan). In fixed-bed reactor experiments, the modified residence time is the key parameter to generate conversion and yield data. A systematic variation of the residence time allows the construction of conversion, yield and partial pressure profiles along the reactor axial coordinate. As the catalyst bed was diluted with inert material, a modified residence time was employed which relates catalyst mass (instead of reactor volume) to the volumetric feed flow rate under norm conditions:

$$\tau_{mod,n} = \frac{m_{cat}}{\dot{V}_{F,n}} \tag{4.1}$$

The feed gas composition for FT experiments is described by two parameters; the first one defines the CO_2 content in the synthesis gas (4.2). $z_{CO2,C,F}$ varies between 0 and 1 for H_2/CO and H_2/CO_2 synthesis gases, respectively.

$$z_{CO2,C,F} = \frac{\dot{N}_{CO2,F}}{\dot{N}_{CO2,F} + \dot{N}_{CO,F}} \tag{4.2}$$

The second parameter defines the H_2 availability (4.3) for full conversion of available carbon to hydrocarbons. A synthesis gas with $z_{H2,F} < 1$ is H_2 deficient, synthesis gas with $z_{H2,F} = 1$ is H_2 balanced according to stoichiometry.

$$z_{H2,F} = \frac{\dot{N}_{H2,F}}{3 \cdot \dot{N}_{CO2,F} + 2 \cdot \dot{N}_{CO,F}} \tag{4.3}$$

The feed and the effluent streams are analyzed by on- and off-line gas chromatography. The methods applied and the procedures to calculate molar flow rates of individual species on basis of GC data are explained in detail in the appendix in Chapter 12.3.

Kinetic rate parameters are determined on basis of conversion (4.4) and yield (4.5) data. The conversion of educt i and the yield of the product j are defined as follows:

$$X_i = 1 - \frac{\dot{N}_i}{\dot{N}_{i,F}} \tag{4.4}$$

$$Y_{j,i} = \left|\frac{\nu_i}{\nu_j}\right| \frac{\dot{N}_j - \dot{N}_{j,F}}{\dot{N}_{i,F}} \tag{4.5}$$

with ν_i and ν_j as the coefficients in the respective stoichiometric equation. The molar flow rates of the feed have to be determined prior in bypass runs. For the FT reaction with H_2/CO and H_2/CO_2 syngases, the CO, CO_2 and in particular the total carbon conversion to hydrocarbons are important. Note that the total carbon conversion (4.6) equals the overall hydrocarbon yield (4.7).

$$X_{CO} = 1 - \frac{\dot{N}_{CO}}{\dot{N}_{CO,F}} \qquad X_{CO2} = 1 - \frac{\dot{N}_{CO2}}{\dot{N}_{CO2,F}} \qquad X_C = 1 - \frac{\dot{N}_{CO} + \dot{N}_{CO2}}{\dot{N}_{CO,F} + \dot{N}_{CO2,F}} \tag{4.6}$$

$$Y_{HC,C} = X_C \tag{4.7}$$

Catalyst performance can be assessed by the weight time yield (WTY), which relates the mass of product j produced per time unit to the mass of catalyst in the reactor:

$$WTY = \frac{\dot{m}_j}{m_{cat}} = \frac{\dot{N}_{i,F} \cdot Y_{j,i} \cdot \tilde{M}_j}{m_{cat}} \tag{4.8}$$

The selectivity to the product j is generally defined as in equation (4.9) and can be calculated on basis of conversion and yield data.

$$S_{j,i} = \frac{Y_{j,i}}{X_i} = \left|\frac{\nu_i}{\nu_j}\right| \frac{\dot{N}_j - \dot{N}_{j,F}}{\dot{N}_{i,F} - \dot{N}_i} \tag{4.9}$$

Within this thesis, two definitions for product selectivity are applied for the FT synthesis. The first definition uses all carbon atoms converted to hydrocarbons and inorganic compounds as CO and CO_2 as basis for the calculation (4.10):

$$\begin{aligned} S_{CO2,CO} &= \frac{Y_{CO2,CO}}{X_{CO}} = \frac{\dot{N}_{CO2} - \dot{N}_{CO2,F}}{\dot{N}_{CO,F} - \dot{N}_{CO}} \qquad & S_{CO,CO2} &= \frac{Y_{CO,CO2}}{X_{CO2}} \\ S_{HC,CO} &= 1 - \frac{Y_{CO2,CO}}{X_{CO}} \qquad & S_{HC,CO2} &= 1 - \frac{Y_{CO,CO2}}{X_{CO2}} \end{aligned} \tag{4.10}$$

The second definition considers only carbon which is converted to organic compounds. Note that the selectivity is now calculated on basis of the total carbon conversion (4.11) and not on basis of CO or CO_2 conversion (4.10). This selectivity definition is labelled with the index HC for hydrocarbons as it focuses solely on the product distribution within the hydrocarbon spectrum. The methane (4.12) and the C_{5+} selectivity (4.13) are key parameters to describe a FT product distribution.

$$S_{j,HC} = \frac{Y_{j,C}}{X_C} = \frac{\dot{N}_j N_{C,j}}{\dot{N}_{CO,F} + \dot{N}_{CO2,F} - \dot{N}_{CO} - \dot{N}_{CO2}} \tag{4.11}$$

$$S_{CH4,HC} = \frac{Y_{CH4,C}}{X_C} = \frac{\dot{N}_{CH4}}{\dot{N}_{CO,F} + \dot{N}_{CO2,F} - \dot{N}_{CO} - \dot{N}_{CO2}} \tag{4.12}$$

$$S_{C5+,HC} = 1 - \sum_{j=1}^{4} \frac{Y_{j,C}}{X_C} \tag{4.13}$$

4.1.4 Fixed-bed reactor balance and kinetic analysis

In integral reactor systems, the reaction rates cannot be derived directly from the experimental data. Therefore, fully back-mixed or differentially operated lab-scale reactor systems are recommended and preferred to derive kinetic rate equations and their rate parameters. However, rate parameters can be extracted from experimental data from integral (fixed-bed) reactor experiments by means of differential reactor models and software packages with ODE (ordinary differential equation) solvers and tools for nonlinear least square analysis (regression).

The integral lab-scale fixed-bed reactor is represented by a set of differential material balances expressed in terms of molar flow rates for each species. The reactor model can be reduced to the ideal design equation of a plug-flow reactor (4.14):

$$\frac{d\dot{N}_i}{dz^+} = m_{cat} \sum_j \nu_{ij} r_j \qquad \text{with } \dot{N}_i(z^+ = 0) = \dot{N}_{i,F} \tag{4.14}$$

$$r_j = r_j(T, P_i) \qquad \text{with } P_i = P \cdot \dot{N}_i / \sum_{i=1}^{N} \dot{N}_i \tag{4.15}$$

with z^+ as dimensionless reactor length, r_j as catalyst mass specific reaction rate of reaction j, ν_{ij} as stoichiometric coefficient and P_i as partial pressure of species i. The underlying assumptions are discussed in detail in chapter 12.7 the appendix.

The stoichiometric coefficients of the FT reaction ν_{i1} and CO_2/CO-shift reaction ν_{i2} are taken from the stoichiometry of the simple reaction network, which considers only the FT and the CO_2/CO shift reaction (Table 4.3). Additional reactions such as methanation, carbon formation or direct CO_2 hydrogenation are not taken into account for Fe-based catalysts. As the carbon-based average chain length found in the experiments is about three (α = 0.67), the complex product distribution of FTS with hydrocarbons ranging from methane

Table 4.3. Stoichiometry of the FT and CO_2/CO shift reaction with propene (C_3H_6) as pseudo-component for the FT product distribution.

	i = 1: CO	2: CO_2	3: H_2	4: H_2O	5: C_3H_6	6: inert
j = 1: FTS	- 1		- 2	+ 1	+ 1/3	
j = 2: CO_2/CO shift	+1	-1	- 1	+ 1		

to waxy products is represented in the stoichiometric equation by the pseudo-component propene (C_3H_6). The usage ratio of the FT reaction, i.e. the ratio of the H_2 and CO consumption, is set to two.
The reactor model together with the rate equations of the FT and CO_2/CO shift reaction and the initial conditions form a set of N ordinary differential equations (ODE), where N is the number of the species. This set of ODEs is solved numerically with the MATLAB® ODE-solver `ode15s`. The kinetic rate parameter values (k, a, b, c) of the rate equations are determined by nonlinear regression (nonlinear least square analysis) of the experimental data, i.e. the sum of the squared differences between the measured (experimental) values and the values calculated by the model for a set of data points is minimized (4.16). Nonlinear regression is carried out in MATLAB® using the function `nlinfit` (Gauss-Newton method).

$$\sigma^2 = \frac{s^2}{N_{exp} - N_{para}} = \sum_{k=1}^{N_{exp}} \frac{(X_{k,exp} - X_{k,model})^2 + (Y_{k,exp} - Y_{k,model})^2}{N_{exp} - N_{para}} \tag{4.16}$$

with σ^2 = variance
with s^2 = sum of squared differences (sum of squares)

4.2 Benchmarking of Fe- and Co-based catalysts

Medium or high FT activity and CO_2/CO shift activity are prerequisites for conversion of CO_2 containing synthesis gases to hydrocarbons as a high CO_2/CO shift activity enables the reverse reaction of CO_2 to CO and a high FT activity a swift CO consumption to hydrocarbons. With respect to the complex set-up of membrane reactor experiments and to the potential decay of membrane properties over time it is crucial that catalysts exhibit only a slow rate of deactivation.
The catalysts were tested according to their performance and stability under similar reaction conditions for FT synthesis (feed gas composition: H_2/CO = 2) and CO_2 hydrogenation (feed gas composition: H_2/CO_2 = 3). The decisive parameters to determine the CO/CO_2 shift activity are the CO_2 selectivity for H_2/CO syngases and the hydrocarbon yield for H_2/CO_2 syngases. The weight-time-yield (WTY) ranks the overall catalyst productivity (Figure 4.3 and Table 12.22). The Fe-5K catalyst characterized by Claeys (1997) and recently investigated by Unruh (2006) was used as a reference for Fe-based catalysts. Experimental data on the Fe-GTLX performance collected by Mena (2009) was consulted to support own findings.

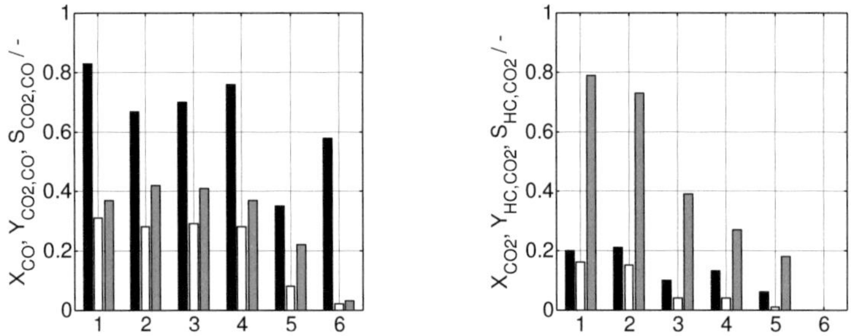

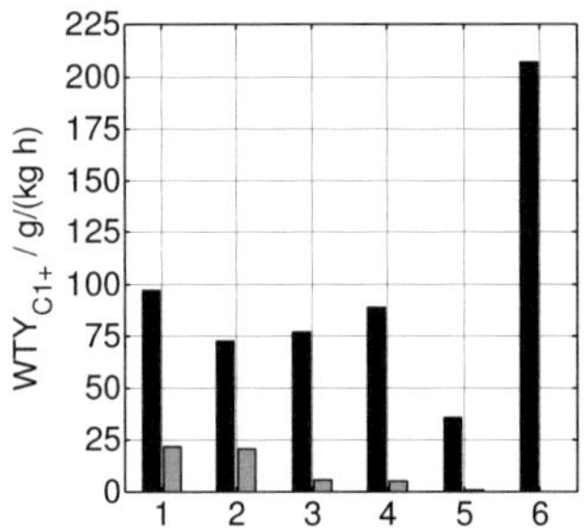

Figure 4.3. Benchmarking of Fe- and Co-based catalysts listed in Table 4.1, test conditions specified in Table 12.22.
Top left: FT synthesis, $H_2/CO = 2$:
■ X_{CO} □ $Y_{CO2,CO}$ ■ $S_{CO2,CO}$
Top right: CO_2 hydrogenation, $H_2/CO_2 = 3$:
■ X_{CO2} □ $Y_{HC,CO2}$ ■ $S_{HC,CO2}$
Left: catalyst productivity:
■ $H_2/CO = 2$ ■ $H_2/CO_2 = 3$

1: Fe-5K, **2**: Fe-5K (Unruh 2006)
3: Fe-GTLX, **4**: Fe-GTLX (Mena 2009)
5: Fe-GTL3
6: Co-GTL4 (#2)

The formation of CO_2 under FT conditions starting with a dry H_2/CO synthesis gas is a coupled parallel reaction. Carbon monoxide reacts with hydrogen in the FT reaction to hydrocarbons and H_2O and CO reacts then with the formed H_2O to CO_2 and H_2 (Figure 4.4). The CO_2 selectivity $S_{CO2,CO}$ cannot exceed 50% for a dry synthesis gas due to this direct coupling of the CO/CO_2 shift reaction to the H_2O formation of the FT reaction. The closer the measured CO_2 selectivity comes to 50%, the higher is the CO/CO_2 shift activity.

The Fe-5K and Fe-GTLX catalysts were tested under comparable conditions (250 °C, 1 MPa, 4000 kg s/m^3) and both catalysts achieve CO conversion levels between 65-80% and CO_2 selectivities around 40%. They represent active catalyst systems with a high CO/CO_2 shift activity compared to the Fe-GTL3 catalyst with a CO_2 selectivity of just 22% (Figure 4.3, top left). Both Co-based catalysts did not show any significant CO_2 formation under the given conditions.

Figure 4.4. Conversion of H_2/CO synthesis gases (left) and H_2/CO_2 synthesis gases (right) to hydrocarbons on shift active FT catalysts.

Starting with H_2 and CO_2, the formation of hydrocarbons runs through the reverse shift of CO_2 to CO. CO reacts in a subsequent step through the FT reaction to hydrocarbons; the reaction network represents a consecutive reaction with CO as intermediate (Figure 4.4, right). Therefore, a high CO_2/CO shift activity and a FT catalyst able to run at low CO partial pressures are essential for CO_2 hydrogenation (which is not the case for Co-based catalysts).

Under reverse shift conditions, the Fe-5K achieves 20% CO_2 conversion with a hydrocarbon selectivity of 70-80% at moderate residence times and at 250°C and 1 MPa. For the Fe-GTLX system, the CO_2 conversion and the hydrocarbon selectivity are with 10% and 39%, respectively, significantly lower (Figure 4.3, top right). This is an interesting finding as both catalysts showed similar results under typical FT conditions. Compared to the Fe-5K, the Fe-GTLX is not able to convert the intermediate CO fast enough, resulting in low hydrocarbon or high CO selectivities.

The weight time yield (WTY) is a decisive factor for the industrial applicability of a catalyst as it determines how much catalyst mass is required to achieve a certain production level (Figure 4.3, left). The Co-based catalyst systems outnumber the Fe-based catalysts Fe-5K and Fe-GTLX at lower residences times and temperatures, achieving WTY-values above 200 g C_{1+} per kg catalyst and hour. Under conditions of CO_2 hydrogenation, WTY-levels are much lower with about 22 for the Fe-5K and about 5 for the Fe-GTLX catalyst, indicating that 5-20 times more catalyst is required to produce the same amount of hydrocarbons when CO_2 is used as carbon source instead of CO. *In-situ* H_2O removal could help increasing reactor productivity by accelerating the reverse shift and FT reaction.

Before we move on, the findings of the benchmarking study on FT and shift activity should be generalized. Figure 4.5 introduces two typical conversion-selectivity diagrams, which are often used in chemical reaction engineering to understand reaction networks. In the final discussion of the results of the *in-situ* H_2O removal experiments by membrane and by chemical reaction, we will come back to these diagrams.

The left-hand diagram is dedicated to FT synthesis starting with a dry feed gas with H_2/CO = 2 and plots the reactor CO_2 selectivity $S_{CO2,CO}$ versus the CO conversion X_{CO}. As the CO/CO_2 shift reaction is coupled to the H_2O formation of the FT reaction, the maximum attainable CO_2 selectivity is 50% unless full conversion is approached. This stoichiometric boundary is more or less invariant towards changes in H_2/CO feed ratio and temperature. However, if H_2O is added to the fresh synthesis gas, the boundary shifts to much higher values.

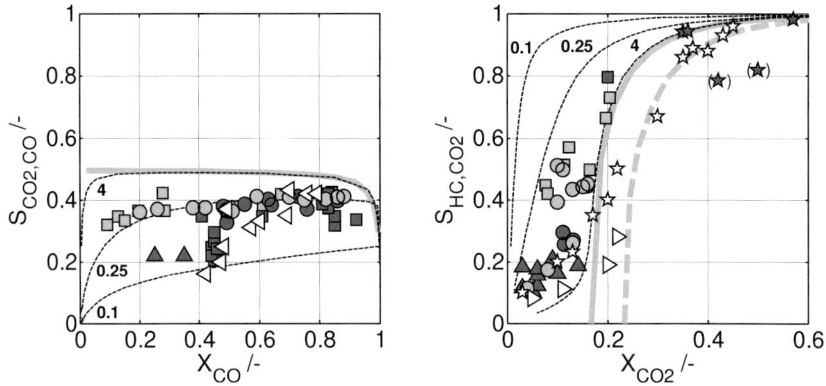

Figure 4.5. $S_{CO2,CO}$–X_{CO} plot for H_2/CO synthesis gases (left) and X_{CO2}–$S_{HC,CO2}$ plot for H_2/CO_2 synthesis gases (right). Stoichiometric and thermodynamic boundaries calculated at 250 °C (—), at 300 °C (- - -). Selectivities (---) calculated on basis of Fe-5K kinetics (Unruh 2006) at 250 °C with varied CO/CO_2 shift activity: CO/CO_2 shift rate constant multiplied with 0.1 (low), 0.25 (intermediate) and 4 (high CO/CO_2 shift activity). Data from own experiments and from literature as indicated:

■ Fe-5K (Unruh 2006)	● Fe-GTLX (Mena 2009)	▷ Fe (Lee et al. 1999, 300 °C)
■ Fe-5K	● Fe-GTLX	☆ Fe (Riedel 2003, 300 °C)
▲ Fe-GTL3	◁ Fe (van der Laan 1999)	★ Fe (Choi et al. 1996, 300 °C)

The formation of CO_2 is strongly favoured under FT conditions. The proximity of the data points is therefore determined by reaction kinetics or the CO/CO_2 shift activity of the catalyst. None of the Fe-based catalysts – either tested in own experiments or data chosen from literature – is able to reach the theoretical CO_2 selectivity of 50%. The catalysts with the highest shift activity obtain selectivities of about 40% at medium-high conversion levels. Within this group, these catalysts exhibit a 'high' shift activity. The Fe-GTL3 catalyst and the deactivated Fe-5K catalyst (due to coke formation) represent catalysts with 'low' CO/CO_2 shift activity.

The second conversion-selectivity diagram (Figure 4.5, right) illustrates the thermodynamic boundary set to CO_2 hydrogenation. Here, the reactor hydrocarbon selectivity $S_{HC,CO2}$ is plotted versus the CO_2 conversion X_{CO2}. The indicated thermodynamic boundary is determined by the CO_2/CO shift equilibrium and shifts with increasing temperature to the right and with decreasing H_2/CO_2 feed ratio to the left.

The boundary separates the diagram into two areas; the reaction pathways of CO_2 hydrogenation – examples are indicated here as broken lines – run only to the left of the boundary which they finally approach at higher conversions. This fact is confirmed by data from own experiments and from selected publications on CO_2 hydrogenation. Choi et al. (1996), Lee et al. (1999) and

Riedel (2003) present examples for FT catalysts with high to very high potassium promotion, resulting in a high CO_2/CO shift activity and high FT activity.

Three different Fe-based catalysts were tested. In summary, the catalyst can be ranked as follows under FT conditions (H_2/CO = 2): Fe-5K ≈ Fe-GTLX >> Fe-GTL3; and under CO2 hydrogenation conditions (H_2/CO_2 = 3): Fe-5K > Fe-GTLX >> Fe-GTL3. The benchmarking indicates that the Fe-5K catalyst exhibits the best performance under FT and CO_2 hydrogenation conditions. However, one aspect has not been discussed so far, and that is catalyst stability. Figure 12.18 (appendix) plots the conversion and yield data of four tested catalyst over time on stream. After >2000 minutes on stream, the Fe-5K in-house catalysts showed a decay in activity, ranging from slow deactivation to steep drop in conversion levels and coke formation depending on the catalyst batch. Unruh (2006) assumes that the instability derives from a misdistribution of the potassium (K) promoter on the catalyst surface.
Due to this lack of stability, the Fe-5K catalyst was discarded with regard to further experiments. Though the Fe-GTLX cannot reach the performance of the Fe-5K reference catalyst, it was the catalyst of choice for membrane experiments, as it operates stably and shows a satisfactorily activity under H_2/CO_2 conditions compared to the other tested catalysts.
The Mn-promoted Co-GTL1 was discarded right away as it did not show any shift activity and due to its difficult physical-mechanical properties in fixed-bed reactor experiments (high pressure drop). The Co-GTL4 catalyst represents a simple and stable FT catalyst and was used as reference Co catalyst in *in-situ* H_2O removal experiments by chemical reaction.

4.3 Kinetics and reference experiments

The following experiments were carried out to derive a set of kinetic rate parameters for the Fe-based catalyst Fe-GTLX and Co-based catalyst Co-GTL4. The knowledge of the kinetic rate equations will allow describing the *in-situ* H_2O removal experiments with a mathematical reactor model. Furthermore, these experiments serve also as reference cases, which enable a direct comparison between experiments with and without *in-situ* H_2O removal.

4.3.1 Fe-based catalyst Fe-GTLX

Figure 4.6 illustrates the experimental results of residence time variation under FT synthesis (H_2/CO = 2) and CO_2 hydrogenation (H_2/CO_2 = 3) conditions. For the H_2/CO feed gas, the CO conversion and CO_2 yield increase with increasing residence time as it is expected for a parallel reaction.

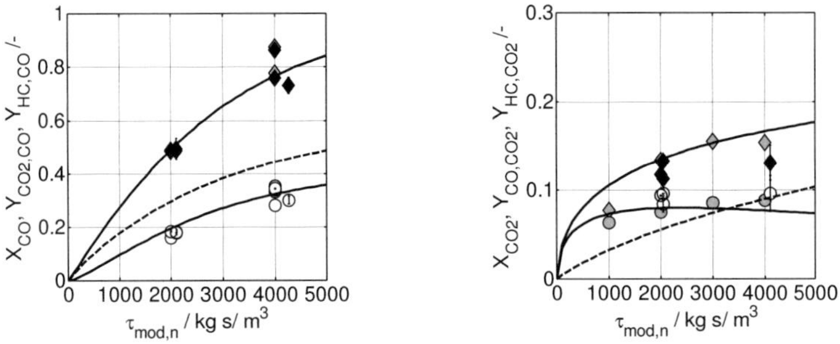

Figure 4.6. Measured conversions (◆◆◇) and yields (○●●) for H_2/CO = 2/1 (left) and H_2/CO_2 = 3/1 (right) synthesis gas as function of the modified residence time. Curves: calculated with kinetic model, broken line: hydrocarbon yield. ◆●: data taken from Mena (2009).
* Fe-GTLX, T = 270℃, P = 1 MPa

The hydrogenation of CO_2 to hydrocarbons runs via CO as intermediate. The yield of the intermediate CO should pass through a maximum, and decline with increasing residence time while the CO_2 conversion and the hydrocarbon yield steadily increase. For the Fe-GTLX catalyst, the maximum is flat and stretched over wide residence time range. The shape of the $Y_{CO,CO2}$ maximum depends on the relative reaction rates of the FT and CO_2/CO shift reaction, and it is more distinct at elevated temperatures (Riedel 2003, 300-360℃) or for catalysts with high FT activity under CO_2 hydrogenation conditions (Unruh 2006). However, the shape of the curves is generic as shown in a dimensionless plot in Figure 12.19.

From thermodynamic equilibrium calculations (H_2/CO_2 = 3 at 270℃), a maximum $Y_{CO,CO2}$ of 19% (H_2/CO_2 = 3 at 270℃) could be expected. However, this value is not reached for the Fe-GTLX catalyst, and neither Unruh (2006) nor Riedel (2003) found values far above 10%. This indicates that the CO_2/CO shift reaction is limited in its rate and that equilibrium composition cannot be reached as the FT reaction steadily consumes CO.**Error! Reference source not found.** plots CO, CO_2 and total carbon conversion during a gradual transition from FT synthesis (H_2/CO = 2, $z_{CO2,C,F}$ = 0) to CO_2 hydrogenation (H_2/CO_2 = 3, $z_{CO2,C,F}$ = 1) conditions, while keeping the synthesis gas H_2-balanced. The total carbon conversion and hence the hydrocarbon yield drops quickly with increasing $z_{CO2,C,F}$. This fact was also reflected in the weight-time yield data from the benchmarking experiments (Figure 4.3). Under the given conditions, CO_2 can only be utilized as carbon source when $z_{CO2,C,F} > 0.6$ ($X_{CO2} > 0$). Negative CO_2 conversions indicate that additional CO_2 is formed. Therefore, CO_2 hydrogenation takes place only if the synthesis gas is very rich

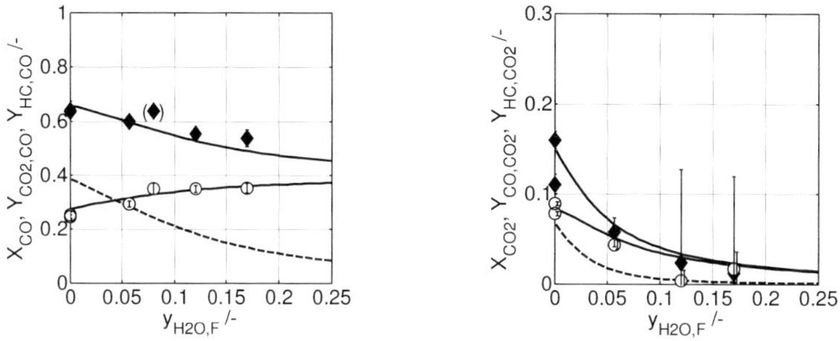

Figure 4.7. Measured conversions (◆) and yields (○) for H_2/CO = 2/1 (left) and H_2/CO_2 = 3/1 (right) synthesis gas as function of the H_2O fraction co-fed in the fresh feed. Pressure is increased to keep $P_{H2}/P_{CO}/P_{CO2}$ in the feed constant. Curves: calculated with kinetic model, broken line: hydrocarbon yield.
* Fe-GTLX, T = 270 °C, $P_{H2,F}+P_{CO,F}+P_{CO2,F}$ = 1 MPa, $\tau_{mod,n}$ = 2000 kg s/m^3 (x 1.53) †

in CO_2. Higher temperatures and higher residence times are beneficial for CO_2 utilization.

For H_2 balanced synthesis gases rich in CO ($z_{CO2,C,F}$ < 0.6), it is therefore not advisable to apply shift active catalysts as carbon will be lost in form of CO_2; instead, non-shift active Co-based catalyst should be used. However, for H_2 deficient gases, shift active catalysts may be an interesting option as additional H_2 can be produced *in-situ* during the reaction.

To understand the role of H_2O in the kinetic rate equation, different amounts of H_2O were co-fed with the feed gas. The total pressure was increased so that the inlet partial pressures of H_2, CO and CO_2 were kept constant. Figure 4.8 shows the response of conversion and yield to an increasing H_2O fraction in the feed. Under FT synthesis conditions (H_2/CO = 2), CO conversion drops with increasing H_2O fraction in the feed, though the CO_2 yield increases. Higher H_2O partial pressures favour the formation of CO_2 under equilibrium considerations. And H_2O must inhibit the FT reaction rate; otherwise, the decreasing CO conversion cannot be explained. This observation is in agreement with various rate laws published in literature.

Under FT synthesis conditions (H_2/CO = 3), H_2O co-feeding disrupts CO_2 hydrogenation and the formation of hydrocarbons more or less completely. Higher H_2O partial pressures prevent the formation of the intermediate CO

† The conversion and yield data in Figure 4.8 suggests that the Fe-GTLX catalyst batch used in the H_2O co-feeding experiments was about 1.5 times more active than the other batch tested in Figure 4.6. The experiments were carried out in two different experimental set-ups. Possible differences in activation procedures cannot be tracked back, but may have resulted in different activity levels. Temperature misreadings and non-inert diluents can be excluded as error sources.

Table 4.4. Tested kinetic rate equations for FT and CO_2/CO-shift reaction. FTS1 and CO2-SH1: Zimmerman & Bukur (1990), FTS2 and CO2-SH2: slightly modified version.

	FTS		CO_2/CO-shift
FTS1	$r_{FT} = k\frac{P_{CO}P_{H2}}{aP_{CO} + bP_{H2O} + cP_{CO2}}$	CO2-SH1	$r_{CO2-SH} = k\frac{P_{CO2}P_{H2} - P_{H2O}P_{CO}K_{P,CO\cdot SH}}{aP_{CO} + bP_{H2O} + cP_{CO2}}$
FTS2	$r_{FT} = k\frac{P_{CO}P_{H2}}{1 + aP_{CO} + bP_{H2O} + cP_{CO2}}$	CO2-SH2	$r_{CO2-SH} = k\frac{P_{CO2}P_{H2} - P_{H2O}P_{CO}K_{P,CO\cdot SH}}{1 + aP_{CO} + bP_{H2O} + cP_{CO2}}$

through the reverse shift reaction; additionally, the FT reaction rate is slowed down due to potential H_2O inhibition. This emphasizes that *in-situ* H_2O removal during CO_2 hydrogenation can reverse the detrimental effects and accelerate CO_2 conversion.

The observed loss of activity during the H_2O co-feeding experiments was reversible. After the co-feeding was interrupted, the catalyst needed a regeneration period of several days, but it recovered its initial activity. However, above $y_{H2O,F}$ > 20 mol%, the catalyst deactivated irreversibly. Nevertheless, such high H_2O fractions should not be encountered during operation of shift active Fe-based catalysts, even at high conversion levels.
The set of experimental data was used to determine the kinetic rate parameters for the FT and CO_2/CO shift reaction. For the Fe-GTLX catalyst, the rate equations of Zimmerman & Bukur (1990) and a slightly modified version (Table 4.4) were tested in the reactor model (4.14). Riedel (2003), Unruh (2006) and Mena (2009) applied these rate equations proposed by Zimmerman & Bukur for K-promoted Fe catalysts. Rohde et. al (2004) and Unruh (2006) showed that this set of rate equations is able to describe experimental data covering the entire range from H_2/CO to H_2/CO_2 syngases. The kinetic rate parameters (k, a, b, c) of the rate equations were determined by nonlinear regression (nonlinear least square analysis) of the experimental data (4.16). The results of the regression analysis in Table 4.5 show that the experimental data can be best described with a significant H_2O inhibition and a negligible CO_2 inhibition term for both FT and CO_2/CO shift reaction. This is in agreement with the rate laws published by Unruh (2006) or Riedel (2003) for different Fe-based catalysts. The experimental data set could not be fitted satisfactorily with other combinations of governing inhibition terms (FT/CO2-SH: H_2O/CO_2, CO_2/H_2O, CO_2/CO_2), confirming that H_2O inhibition plays a significant role. The negligible CO_2 inhibition can be removed from the rate equations if the modified version is applied.

Table 4.5. Kinetic parameter values of the Fischer-Tropsch and CO_2/CO-shift reaction over the K-promoted Fe-GTLX catalyst, determined for two different sets of kinetic rate equations. Parity plot is given in the appendix, Figure 12.21.
P = 1 MPa, T = 270 °C, $\tau_{mod,n}$ = 0 - 4000 kg s/m³, $z_{H2,F}$ = 1, 0 < $z_{CO2,C,F}$ < 1, 0 < $y_{H2O,F}$ < 0.2

		Fischer-Tropsch FTS1	CO_2/CO-shift CO2-SH1
$k_{i,\,270°C}$	/mol / (s·kg·Pa)	$7.04 \cdot 10^{-9}$	$5.12 \cdot 10^{-9}$
a_{CO}	/-	1	1
b_{H2O}	/-	10.5	43.9
c_{CO2}	/-	$1 \cdot 10^{-6}$	$1 \cdot 10^{-6}$
		Fischer-Tropsch FTS2	CO_2/CO-shift CO2-SH2
$k_{i,\,270°C}$	/mol / (s·kg·Pa²)	$7.75 \cdot 10^{-9}$	$4.44 \cdot 10^{-9}$
a_{CO}	/ Pa^{-1}	1	1
b_{H2O}	/ Pa^{-1}	11.6	38
c_{CO2}	/ Pa^{-1}	-	-

4.3.2 Co-based catalyst Co-GTL4

Figure 4.8 and Figure 4.9 illustrate the experimental results of residence time and feed gas variations for the Co-GTL4 (#1) catalyst. Conversion increases with increasing residence time and $z_{CO2,C,F}$ or H_2/CO feed ratio. Higher H_2/CO feed ratios accelerate the reaction rate, but are detrimental to product selectivity. The product spectrum shifts to methane and short-chain hydrocarbons and the FT regime cannot be maintained at very high feed ratios.

Under FT conditions with H_2/CO = 2, the CO_2 make of the catalyst was minor, confirming the negligible shift activity. CO_2 in the feed acts as inert; only small amounts are converted, presumably directly to methane. These results clearly

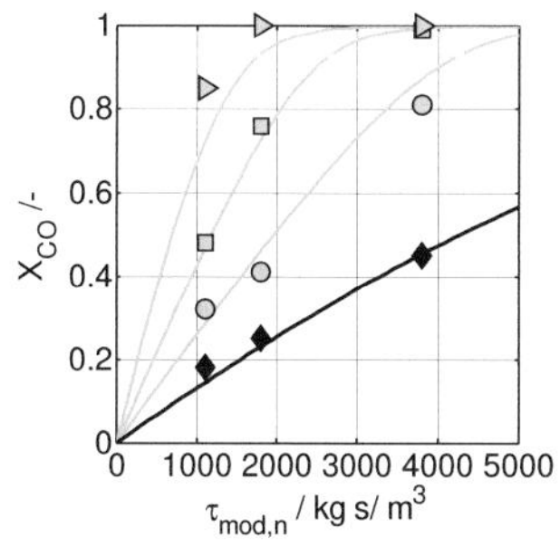

Figure 4.8. Co-GTL4 (#1): Measured conversions (◆●■▶) for various H_2/CO/CO_2 syngases as function of the modified residence time. Curves: calculated with kinetic model, $z_{CO2,C,F}$ ◆: 0, ●: 0.3, ■: 0.5, ▶: 0.7.

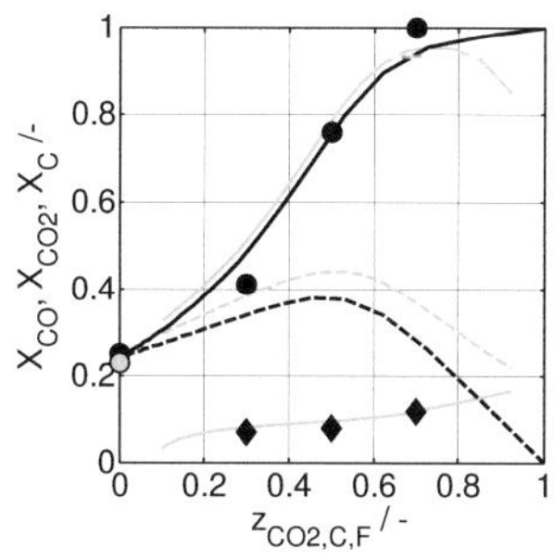

Figure 4.9. CO (●), CO_2 (◆) and total carbon conversion (---) as function of the stoichiometric syngas composition; lines: (black) calculated with kinetic FT model, (grey) calculated with FT kinetics and simple CO/CO_2 shift and CO_2 methanation kinetics. (●) repro.

* Co-GTL4 (#1), T = 230 °C, P = 1 MPa, $\tau_{mod,n}$ = 1900 kg s/m³

Table 4.6. Kinetic parameter values of the Co-GTL4 (#1) catalyst. Kinetic rate parameters of FT reaction according to the rate equation of Yates & Satterfield (1991).
* $P = 1$ MPa, $T = 220\text{-}240\,°C$, $\tau_{mod,n} = 0 - 4000$ kg s/m^3, $z_{H2,F} = 1$, $0 < z_{CO2,C,F} < 1$
* Co-GTL4 (#1): $E_A = 219$ kJ/mol, $k_0 = 7.4 \cdot 10^9$ mol/(s·kg·Pa^2)

		Fischer-Tropsch FTS
$k_{220°C}$	/mol / (s·kg·Pa^2)	$5.1 \cdot 10^{-14}$
$k_{230°C}$	/mol / (s·kg·Pa^2)	$1.5 \cdot 10^{-13}$
$k_{240°C}$	/mol / (s·kg·Pa^2)	$4.1 \cdot 10^{-13}$
a_{CO}	/ Pa^{-1}	$9.2 \cdot 10^{-6}$

point out that Co-based catalysts are not applicable for synthesis gases with H_2/CO feed ratios significant higher than 2 and not suitable for CO_2 hydrogenation applications.

The kinetic analysis considered only the CO conversion to organic products, and the rate equation developed by Yates and Satterfield (1991) was tested in the reactor model (4.14):

$$r_{FT} = k \frac{P_{CO} P_{H2}}{(1 + aP_{CO})^2} \quad (2.9)$$

and the kinetic rate parameters k and a of the rate equation were determined by nonlinear regression (nonlinear least square analysis) of the experimental data (4.16). Data was employed from experiments varying residence time (1000-4000 kg s/ m^3), feed gas composition ($0 < z_{CO2,C,F} < 0.7$) and temperature (220-240 °C). The kinetic rate parameters are summarized in Table 4.6.

H_2O co-feeding experiments were carried out with Co-GTL4 catalyst. The experiments with the Co-GTL4 (#2) lasted about 140 h (6d) and the amount H_2O co-fed was increased stepwise until the critical H_2O/H_2 ratio (~0.7) and H_2O partial pressure (~0.6 MPa) were exceeded (Figure 12.25). At the end, the CO conversion dropped to 53% compared to the reference conversion prior co-feeding of 58%. Further experiments are needed to investigate H_2O driven deactivation of Co-based catalysts, in particular at lower temperatures.

4.4 Product selectivity

It was not the primary task of this thesis to investigate FT product distributions in detail as *in-situ* H_2O removal will affect mainly reaction kinetics rather than product selectivity. However, the principle differences between the product-spectra of Fe- and Co-based catalyst should be illustrated here by means of experimental data. Figure 4.10 shows the C_1-C_{15} product distributions (expressed as carbon selectivities) of the Fe-GTLX catalyst obtained under FT ($H_2/CO = 2$) and CO_2 hydrogenation conditions ($H_2/CO_2 = 3$), which can be compared directly to the Co-GTL4 (#2b) product distribution obtained under FT

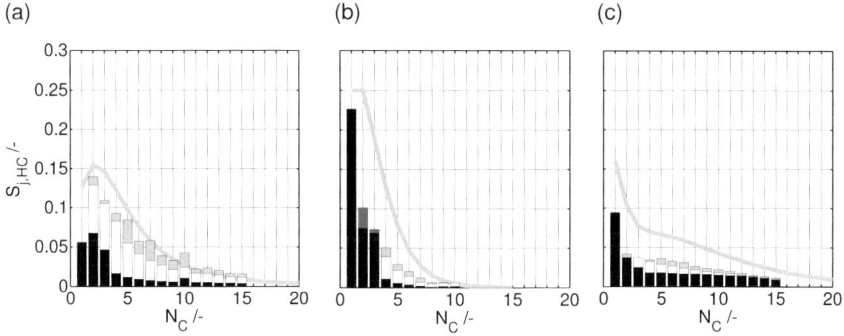

Figure 4.10. Comparison of the distribution of volatile hydrocarbons (carbon selectivity) for (a) Fe-GTLX (H_2/CO =2), (b) Fe-GTLX (H_2/CO_2 =3), (c) Co-GTL4 (#2b) (H_2/CO =2). ■: n-alkanes, □: n-alkenes, ▒: branched hydrocarbons and oxygenates. Note: (a) and (c): samples analyzed by ampoule method/ offline GC, (b): online-GC (not able to separate $C_2/C_{2=}$ and $C_3/C_{3=}$).
Envelope (grey line) calculated on basis of ASF distributions, see equation (2.8): (a) H_2/CO = 2: $\alpha_1 = 0.6$, $\alpha_{2,C9+} = 0.8$; (b) $H_2/CO_2 = 3$: $\alpha = 0.5$; (c) $H_2/CO = 2$: $\alpha_1 = 0.2$, $\alpha_{2,C2+} = 0.8$
* Fe-GTLX, T = 270°C, P = 1 MPa, $\tau_{mod,n}$ = 2000 kg s/m^3
* Co-GTL4, T = 230°C, P = 1 MPa, $\tau_{mod,n}$ = 2000 kg s/m^3

conditions (H_2/CO = 2). The organic products are broken down to n-alkanes, n-alkenes and branched hydrocarbons including oxygenates.

A comparison of the product distributions of Co-based and Fe-based catalysts for H_2/CO =2 feed gas yields that the unpromoted Co-GTL4 catalyst generates the heavier product though it exhibits a relatively high methane selectivity. Overall, the catalysts tested showed a relatively low C_{5+} selectivity $S_{C5+,HC}$ of about 80% for Co-GTL4 and 65% for the Fe-GTLX catalyst. The product of the Co-based catalyst consists mainly of linear alkanes and the fraction of alkenes drops fast with increasing carbon number. In comparison, the Fe-based product contains a much higher fraction of alkenes and branched hydrocarbons and oxygenates.

The Fe-based product shifts to lighter hydrocarbons under CO_2 hydrogenation conditions. While a Co-based catalyst is not able to produce short-chain hydrocarbons under these conditions, Fe-based catalyst still produce hydrocarbons and the product distribution is similar to that one under FT conditions. E.g. it could be shown that the relative olefin/ paraffin distribution remained the same. Though the chain growth probability declined with increasing $z_{CO2,C,F}$, it should be mentioned that waxy hydrocarbons were still removed from the hot wax trap during the experiments. The appendix contains more information on the product distributions of the various tested catalysts.

5 *In-situ* H_2O removal during Fischer-Tropsch and other fuel-related synthesis reactions

Three experiments with regard to in-situ H_2O removal are presented in this chapter. The focus is on the application of a membrane reactor for CO_2 hydrogenation using the hydrophilic CSP2 membrane (Chapter 3) and the Fe-based FT catalyst Fe-GTLX (Chapter 4). The results are compared to a reference case from literature (Unruh 2006) for a less selective membrane. The etherification of methanol and ethanol is chosen as second example reaction to test the membrane and the membrane reactor concept. As membrane stability was identified as a critical issue during the experiments, the observations will be discussed in more detail. Finally, the combination of two catalytic functions – i.e. the combination of non-shift active FT catalyst with low temperature shift catalyst – is presented as different approach to in-situ H_2O removal.

The experimental results are described and discussed by means of a reactor model with and without membrane. The required transport parameters of the applied CSP2 membrane and the kinetics of the involved reactions have been determined in separate experiments beforehand.

5.1 Experimental

The objectives of *in-situ* H_2O removal are manifold: selective removal can boost reaction rates, can shift compositions beyond equilibrium limitations and it can reduce the rate of catalyst deactivation. In this chapter, the potential and the limits of *in-situ* H_2O removal should be demonstrated experimentally by means of three example reactions and two different approaches (Table 5.1).

Table 5.1 Overview of experiments with respect to *in-situ* H_2O removal during FTS and other fuel-related reactions. Experiments planned and carried out (●). Data from Unruh (2003) serves as reference.

	by membrane			**by reaction**
reaction/ catalyst	CSP2	MOR [a]	$SiOH_x$	$CuZnO/Al_2O_3$
CO_2 hydrogenation/ Fe-GTLX	●●	membrane defect		
CO_2 hydrogenation/ Fe-GTL3	●			
CO_2 hydrogenation/ Fe-5K	catalyst not stable		reference exp. Unruh (2006)	
FT synthesis/ Co-GTL4	membrane not stable	membrane defect		●
DME/DME/ α-Al_2O_3	●●	membrane defect		

[a] any experiments with hydrophilic mordenite (MOR) membrane provided by the University of Zaragoza/ ECN were stopped after severe problems with the membrane sealing had been discovered

The first approach is the application of a hydrophilic CSP2 membrane in a membrane reactor configuration. Two different reactions, namely the conversion of CO_2 containing synthesis gases to long-chain hydrocarbons (CO_2 hydrogenation) and the etherification of methanol and ethanol to DME and DEE, are tested in the membrane reactor configuration.
The second approach is the addition of a CO/CO_2 shift function to a non-shift active Co-based FT catalyst. This concept of *in-situ* H_2O removal by chemical reaction can be applied to all reactions based on synthesis gas chemistry.

5.1.1 Introduction and execution of *in-situ* H_2O removal experiments

The three example reactions chosen differ in various aspects such as H_2O partial pressures encountered, gas phase composition or H_2O related limitations. The following paragraphs give a brief introduction to the specifics of each example reaction including the description of the experimental set-up:

***In-situ* H_2O removal by a hydrophilic membrane during CO_2 hydrogenation**

Membrane-enhanced CO_2 hydrogenation was seen as an option for improving the conversion of CO_2 containing synthesis gases (LTCPO-GTL 2005). The objectives to test this type of reaction in the membrane reactor were the following:

- H_2O removal from the reaction zone should improve conversion and yield as the CO_2/CO shift reaction is equilibrium-limited and the reaction rates of both the FT and the shift reaction are H_2O inhibited.
- The CSP2 membrane should demonstrate its applicability to remove H_2O with regard to permselectivity, permeability and stability. Relative high temperatures, low H_2O partial pressures, H_2 rich gas phase and presence of hydrocarbons create challenging conditions for a polymer-based membrane.
- Findings from these experiments should be transferred to Co-based FT applications.

The membrane reactor experiments were carried out in the experimental set-up described in Chapter 0. The permeation cell was equipped with a CSP2 membrane (Figure 5.1), and the annular between the tubular membrane and the reactor shell was filled with a mixture (27 cm^3) of pre-activated Fe-based FT catalyst ($100 < d_P < 150$ µm) and β-SiC ($200 < d_P < 250$ µm). The remaining free volume of the reactor is filled up with coarse SiC ($d_P > 250$ µm). A porous glass disk supports the catalyst/ SiC mixture and the inert bed.

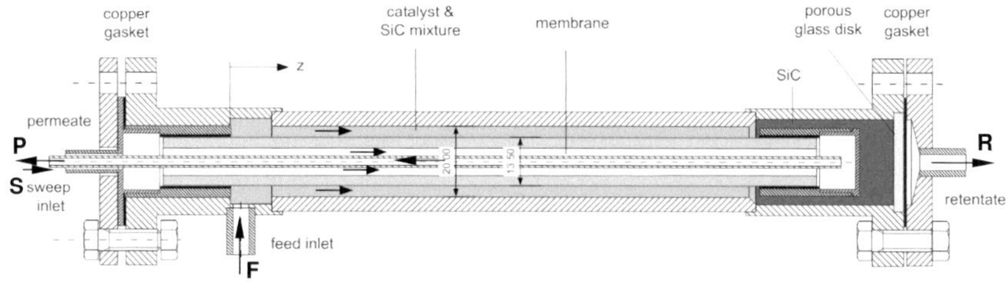

Figure 5.1. Lab-scale fixed-bed membrane reactor for experiments with *in-situ* H_2O removal by hydrophilic membranes; reactor is operated in vertical position.

Due to the limited thermal stability of the CSP2 membrane, the FT catalyst is pre-activated before it is filled into the membrane reactor. The reduction and activation of the FT catalyst is carried out according to the standard procedures beforehand either in the fixed-bed glass reactor or in the membrane reactor equipped with a gas-tight membrane replica (Figure 4.1). After sufficient time on stream, the catalyst has produced enough wax to fill up its own pores and to cover its surface. These pre-activation experiments generate also the reference points for the fixed-bed reactor without membrane. As the wax coating inhibits a fast oxidation of the activated catalyst in air; the FT catalyst can be removed from the reactor.

The membrane reactor was operated in co-current sweep mode. H_2 and Ar were applied as sweep gases. The feed side conditions were kept constant in the most cases at 1 MPa and 270°C. Sweep ratio Ψ and the pressure ratio Φ were systematically varied (Table 12.13 - Table 12.15).

$$\psi = \frac{\dot{N}_S}{\dot{N}_F} \qquad (5.1)$$

$$\Phi = \frac{P_{SS}}{P_{FS}} \qquad (5.2)$$

In-situ H_2O removal by a hydrophilic membrane during DME/DEE synthesis

Dimethyl ether (DME) and diethyl ether are attractive fuel additives which received increasing interest in the recent years (see introduction, chapter 1). The etherification of alcohols such as methanol or ethanol takes place on an acidic catalyst and represents a weakly exothermic equilibrium reaction. The equilibrium conversion of methanol to DME is about 85% at 300°C.

$$2\ \text{R-OH} \rightleftharpoons \text{R-O-R} + H_2O \qquad \Delta H_{R\,300^\circ C}^{0} = -23\ \text{kJ/mol} \qquad (5.3)$$

The reaction rate is strongly inhibited by H_2O according to Bercic & Levic (1992). Furthermore, Kim et al. (2006) report that acidic catalysts deactivated in presence of high H_2O partial pressures. This led in the case of the DME to the development of the *direct* synthesis route from synthesis gas (Ogawa 2003, de Mestier du Bourg 2003). The general objective to test this type of reaction in the membrane reactor was three-fold:

- Due to the strong rate inhibition by H_2O, the degree of *in-situ* H_2O removal should be directly linked to an increase in conversion level and therefore easily measurable.
- Secondly, preceding permeation experiments with a CSP2 membrane under non-reactive conditions suggested that H_2O permeates about 10 times faster than methanol and about 100 times faster than ethanol (Figure 3.9); therefore higher H_2O permselectivities should be expected than in the CO_2 hydrogenation experiments, which makes the membrane more suitable for this application.
- And thirdly, higher H_2O partial pressures are encountered and larger H_2O amounts need to be removed. Higher fluxes and the presence of oxygenates expose the membrane to a different operating regime.

Commercial available γ-Al_2O_3 (Merck KG, TA1329695) was used as acidic catalyst; methanol (Merck KG, CAS 64175, purity > 99.5%) and ethanol (Merck KG, CA 67561, purity > 99.5%) were used as educts. 8 g crushed γ-Al_2O_3 was mixed with β-SiC ($200 < d_P < 250$ µm) as inert diluent and filled either in the kinetic (Figure 4.1) or in the membrane reactor (Figure 5.1). The experimental set-up shown in Figure 3.3 and described in detail in Chapter 0 was used. Methanol or ethanol were metered by a Liquiflow and then evaporated into a carrier gas stream of argon.
In a separate series of experiments a set of rate parameters for the DME and DEE synthesis on γ-Al_2O_3 as catalyst were determined. The rate equation of Bercic & Levec (1992) was used as reference, and the rate parameters were adjusted accordingly. The kinetic experiments and kinetics are summarized in Table 12.12 and Chapter 12.5.3 in the appendix. These experiments serve also as reference experiments without membrane.
The membrane reactor was operated in co-current sweep mode. Ar was applied as sweep gas. Residence time and sweep ratio Ψ were kept constant and temperature and pressure ratio Φ were varied (Table 12.16 - Table 12.17). Due to concerns about the stability of the CSP2 membrane, the upper temperature was limited to 290 °C.

In-situ H_2O removal by chemical reaction during FT synthesis

A potentially more robust alternative to the application of hydrophilic membranes is *in-situ* H_2O removal by chemical reaction. Considering the production of fuel-components or chemicals from synthesis gas, the CO/CO_2

shift reaction is a promising candidate for this task. A proven example is the direct DME synthesis from H_2 deficient synthesis gas, where methanol synthesis, dehydration and CO/CO_2-shift functions are combined in a multifunctional catalyst system (Ogawa et al. 2003, de Mestier du Bourg 2006, Renk et al. 2006). H_2O partial pressure and hence H_2O induced catalyst deactivation can be reduced significantly, and DME conversion is shifted beyond the equilibrium conversion (Ogawa et al. 2003). Coal-based FT synthesis using Fe-based catalyst should also utilize the internal shift function to produce H_2 and to remove H_2O during the reaction. On lab-scale basis, non-shift active Co-based FT catalysts were combined with a low temperature shift catalyst (Post & Sie 1985, Chanenchuk, Yates & Satterfield 1991), whereas the focus was rather on *in-situ* H_2 production (internal shift) rather than on *in-situ* H_2O removal.
Low temperature (LT) shift catalysts operate in a temperature range from 210-250℃ (Twigg 1989) or 210-270℃ (Higman & Supp 2006), which overlaps with the operating window of Co-based FT catalysts. The minimum temperature is determined by catalyst activity and by the dew point of steam to prevent condensation. Under FT conditions, issues on H_2O condensation will not occur unlike in commercial shift units which run on H_2O saturated synthesis gas. The upper temperature limit is set by the onset of accelerated sintering and recrystallization of the active material (Twigg 1989). LT shift catalysts consist of copper crystallites finely dispersed on ZnO/Al_2O_3 support. These catalysts are very sensitive to sulphur and chloride poisoning.

The general objective to test this type of reaction was the following:

- Co-based FT catalysts are susceptible to H_2O induced deactivation phenomena. Though the impact on catalyst stability can only be measured in long-term experiments, these experiments should demonstrate the applicability of LT shift catalyst under FT conditions and the potential for *in-situ* H_2O removal.
- Furthermore, the combination of a Co-based FT catalyst with a LT shift catalyst is interesting as one yields a multifunctional catalyst system with the CO/CO_2 shift activity of a Fe-based FT catalyst and the high FT activity and the high selectivity towards long-chain paraffinic hydrocarbons of a Co-based FT catalyst.

The experiments on *in-situ* H_2O removal by chemical reaction were based on a physical mixture of the Co-based Co-GTL4 (#1) catalyst with a commercial LT shift catalyst. This catalyst was provided by BASF, Ludwigshafen, and had the following composition: $CuO/ZnO/Al_2O_3$ 48.1/30.5/13.2 wt%. The kinetics of the Co-GTL4 (#1) and the LT shift catalyst were determined in separate

Table 5.2. Conditions for catalyst reduction of the low temperature shift catalyst ($CuZnO/Al_2O_3$).

	gas	P	$\tau_{mod,n}$	T_{start}	heating rate	T_{end}	holding time
		/MPa	/ kg s/m^3	/ °C	/ K/min	/ °C	/ h
reduction	H_2/N_2 (1/9)	0.1	460	20	1.5	210	16

experiments described in Chapter 4.3.2 and in Chapter 12.5, respectively. The experiments with the physical mixture of the FT and LT shift catalyst were carried out in the fixed-bed glass reactor (Figure 4.1 and Figure 4.2).
Due to the limited thermal stability of the $CuZnO/Al_2O_3$ shift catalyst, the Co-GTL4 (#1) catalyst needed to be pre-activated. The FT catalyst was reduced and activated according to the standard procedures in the glass reactor beforehand. The reaction was stopped after sufficient wax had been produced to coat the catalyst, and the pre-activated catalyst was removed from the reactor. Then a mixture of 2 g of pre-activated Co-GTL4 ($100 < d_P < 160$ µm) catalyst, 2 g of $CuZnO/Al_2O_3$ shift catalyst ($100 < d_P < 160$ µm) and 8 g of calcined sand ($100 < d_P < 160$ µm) was prepared and filled into the glass reactor. The mixture was exposed to a mild activation procedure to reduce the copper crystallites of the LT shift catalyst (Table 5.2). After the reduction step, FT synthesis conditions were applied and synthesis gas with $H_2/CO = 2$ was fed to the bifunctional catalyst system. The experimental plan is given in Table 12.11. The performance was monitored with regard to CO conversion, CO_2 yield and selectivity and H_2O partial pressure.

5.1.2 Data analysis and definitions

The flow rates and molar composition of the ingoing (feed and sweep) and the outgoing (retentate and permeate) streams are determined by on-line GC analysis as described in Chapter 12.3.1. The molar flow rates can be measured directly for all key components except for H_2O and the inert Ar.
First, the performance of the reactor configurations with *in-situ* H_2O removal is assessed on basis of conversion, yield and selectivity level. An alteration with respect to reference cases without membrane implies an effect of H_2O removal.
In the case of a membrane reactor, conversion and yield can be affected by reactant loss, co-feeding or dilution. The CSP2 membrane is much more permselective than e.g. amorphous silica membranes that had been applied under similar conditions (Unruh 2006), but still reactants, products and inert components can permeate from the reaction zone to the sweep side or vice versa. Therefore, the envelope of the material balance to calculate conversion and yield has to be drawn around the entire membrane reactor taking into

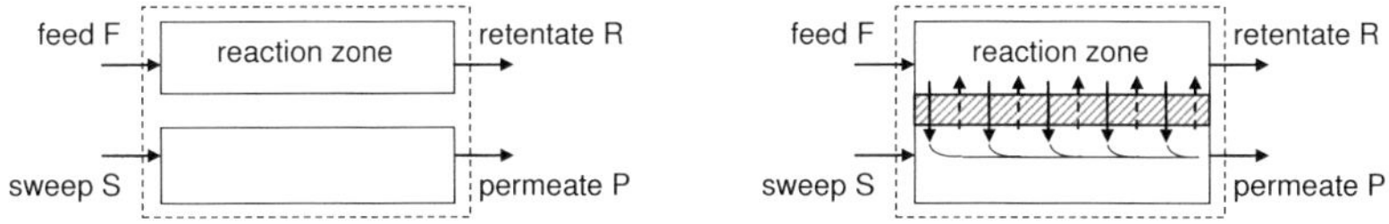

Figure 5.2. Schematic of the material balance for conversion and yield calculations for a fixed-bed reactor without (left) and with membrane (right).

account feed and sweep as ingoing and retentate and permeate as outgoing streams (Figure 5.2).

The conversion definition (5.4) ensures that loss of the key reactant – e.g. CO_2 in the CO_2 hydrogenation experiments and methanol and ethanol in the etherification experiments – will not lead to an apparent increase in conversion. In the yield calculation (5.5), the loss of intermediates or products to the sweep is taken into account; this is worth discussing as these species will generally not recovered from the sweep stream and are lost. In this case, the yield calculation should be corrected by (1-R_j), equation (5.10).

$$X_i = \frac{\dot{N}_{i,F} - \dot{N}_{i,R} - \dot{N}_{i,P}}{\dot{N}_{i,F}} \quad (5.4) \qquad Y_{j,i} = \left|\frac{\nu_i}{\nu_j}\right| \frac{\dot{N}_{j,R} + \dot{N}_{j,P}}{\dot{N}_{i,F}} \quad (5.5)$$

The key task of the membrane is selective H_2O removal and the membrane is assessed accordingly. The degree of H_2O removal R_{H2O} relates the amount of H_2O recovered in the permeate to the total amount of H_2O produced by reaction:

$$R_{H2O} = \frac{\dot{N}_{H2O,P}}{\dot{N}_{H2O,P} + \dot{N}_{H2O,R}} = \frac{\dot{N}_{H2O,P}}{\dot{N}_{H2O,tot}} \quad (5.6)$$

The total amount of H_2O produced through CO_2 hydrogenation can be determined by an oxygen balance neglecting oxygenates in the hydrocarbon product spectrum:

$$\dot{N}_{H2O,tot} = \left(2 \cdot X_{CO2} - Y_{CO,CO2}\right) \cdot \dot{N}_{CO2,F} \quad (5.7)$$

An accurate determination of R_{H2O} was difficult as the GC solution was not able to measure H_2O quantitatively. R_{H2O} was calculated either on basis of the relative H_2O peaks determined by GC (5.8) or on basis of H_2O recovered by adsorption on desiccant silica gel (5.9). The problems and accuracy of H_2O measurement are discussed in Chapter 12.3.2.

$$R_{H2O,GC} = \frac{A_{H2O,P}/A_{N2,P}}{A_{H2O,P}/A_{N2,P} + A_{H2O,R}/A_{N2,R}} \quad (5.8) \qquad R_{H2O,trap} = \frac{m_{H2O,P}}{m_{H2O,P} + m_{H2O,R}} \quad (5.9)$$

The loss and the removal of reactants, intermediates and products are quantified as follows:

$$R_i = \frac{\dot{N}_{i,P}}{\dot{N}_{i,P} + \dot{N}_{i,R}} \tag{5.10}$$

$$L_i = \frac{\dot{N}_{i,P} - \dot{N}_{i,S}}{\dot{N}_{i,F}} \tag{5.11}$$

A negative loss L_i indicates co-feeding from the sweep side to the reaction zone. Co-feeding should be avoided e.g. by adjusting the pressure ratio accordingly or by choice of the sweep gas. Unruh (2006) used co-feeding to overcome deficiencies in membrane permselectivity by sweeping with H_2 and CO_2 at a pressure ratio of one.

The transmembrane molar fluxes are calculated on basis of an analysis of the permeate stream:

$$j_{i,tmb} = \frac{\dot{N}_{i,P} - \dot{N}_{i,S}}{A_m} \tag{5.12}$$

The membrane permeances under reactive conditions are calculated on basis of mean partial pressures on feed and sweep side:

$$Q_i = \frac{2 \cdot j_{i,tmb}}{(P_{i,F} + P_{i,R}) - (P_{i,S} + P_{i,P})} \tag{5.13}$$

5.1.3 Fixed-bed (membrane) reactor balance

The discussion of the experimental results should be supported by model-assisted calculations. The reactor model is based on an isothermal, one-dimensional plug-flow reactor model. In the case of the membrane reactor, an additional transport term is added, connecting the feed side of the reactor with the sweep side and enabling selective removal/ dosage of reactants and products along the reactor axis (5.15). The boundary conditions (5.16) indicate that the membrane reactor is operated in co-current sweep mode.

$$\frac{d\dot{N}_{i,FS}}{dz^+} = A_m Q_i P_{FS} (y_{i,FS} - \Phi \cdot y_{i,SS}) + \sum_{k=1}^{2} m_{cat,k} \nu_{ik} r_k \tag{5.14}$$

$$\frac{d\dot{N}_{i,SS}}{dz^+} = -A_m Q_i P_{FS} (y_{i,FS} - \Phi \cdot y_{i,SS}) \qquad \text{with} \quad \Phi = \frac{P_{SS}}{P_{FS}} \tag{5.15}$$

$$\begin{aligned} &\dot{N}_{i,FS}(z^+ = 0) = y_{i,F}\dot{N}_F \qquad i = 1....N \\ &\dot{N}_{i,SS}(z^+ = 0) = y_{i,S}\dot{N}_F \cdot \psi \qquad i = 1....N \end{aligned} \qquad \text{with} \quad \psi = \frac{\dot{N}_S}{\dot{N}_F} \tag{5.16}$$

The membrane reactor model is not as complex as models developed e.g. by Koukou et al. (1997, 1998, 2-dimensional, non-isothermal) or Tuchlenski

Table 5.3. Set of permeances and permselectivities applied in the membrane reactor model to describe experimental data. Permeances were set constant in model; concentration or temperature dependence not taken into account. All permeances derived at 250°C (see Figure 3.7), except methanol and DME at 290°C. Refer to Table 12.19 - Table 12.21.

	H_2O	H_2	CO	CO_2	Ar[b]	MeOH	DME
$Q_i \times 10^{-7}$ / mol/ s Pa m^2	0.59	0.16	0.009	0.016	0	0.046	0.007
	1.05[a]				0	0.046	0.007
$S_{H2O,i}$		3.6	64	36	∞	12	84
					∞	23	150

[a] for one series of DME experiments (Φ = 0.15), the H_2O permeance was corrected as a new, fresh CSP2 membrane was applied (Figure 5.14)
[b] Ar was applied as sweep gas generally at low pressure, permeance was assumed to be zero. Ar could not be detected by GC

(1998, axial dispersion, detailed transport through porous media); however, the key assumptions can be justified and are discussed in detail in Chapter 12.7:

- Internal and external mass transfer limitations do not occur; therefore, a reduction to a pseudo-homogeneous model can be validated.
- Axial temperature and pressure gradients are negligible; the energy and impulse balance are omitted.
- Axial dispersion is negligible according to residence-time distribution (RTD) measurements (see Chapter 0).
- Radial concentration gradients do not develop in the lab-scale membrane reactor (for CO_2 hydrogenation) and the membrane transport resistance is located in the functional layer; the transport resistance through the porous support layer is negligible, yielding a one-dimensional model.
- Membrane transport can be modelled according to the solution-diffusion model as individual species permeate independently through the membrane. Pressure and concentration dependencies are weak. Only H_2O, methanol and ethanol show a more complex behaviour (Chapter 3.3).

The reactor model requires input with regard to the reaction kinetics and to the permeation characteristics of the membrane. The kinetics and the rate parameters of the individual reactions were determined in independent, dedicated experiments without membrane. The transport coefficients of the membrane were determined as well in independent, dedicated experiments in a permeation cell without chemical reaction. Though the permeances vary with gas composition and temperature, only one representative set of transport coefficients was chosen. The set of membrane permeances used in the model-supported calculations for CO_2 hydrogenation and DME synthesis are listed in Table 5.3. All supportive membrane reactor model calculations within this chapter are solely based on the independently determined reaction and transport kinetics; model parameters were not adjusted retroactively unless indicated.

5.2 *In-situ* H_2O removal during CO_2 hydrogenation by hydrophilic membranes

Three Fe-based FT catalysts were tested in the catalyst benchmarking, yielding the following ranking with regard to their activity: Fe-5K > Fe-GTLX >> Fe-GTL3. The Fe-GTLX and the Fe-GTL3 were applied in membrane reactor experiments together with the new ceramic-supported polymer membrane CSP2. Experimental data for a Fe-5K catalyst/ amorphous silica membrane system from Unruh (2006) is available as reference. The three FT catalyst/ membrane combinations differ significantly in (a) catalyst activity and (b) membrane permeability and permselectivity. Figure 5.3 compares the CO_2 hydrogenation results of the Fe-GTL3/CSP2 and Fe-GTLX/CSP2 membrane reactor configurations to the conversion and yield data obtained in the reference case (Unruh 2006):

- *Fe-5K/silica membrane (reference case, Unruh 2006):* This combination of an active FT catalyst and a less selective membrane is the reference case and was described by Unruh (2006). The Fe-5K catalyst showed in catalyst benchmarking the highest activity for CO_2 hydrogenation. For temperatures above 150-200°C, H_2 permeated faster through the hydrophilic silica membrane than H_2O due to lack of hydrothermal stability (Unruh 2006). As consequence, the use of Ar as sweep gas resulted in a performance worse than without integrated membrane. A severe reactant loss could be prevented only by the use of H_2 or H_2/CO_2 (3/1) as sweep gas. The obtained increases in CO_2 conversion and yield are due to *in-situ* H_2O removal and not a result of H_2 and CO_2 co-feeding (Unruh 2006).
- *Fe-GTLX/CSP2 membrane:* Fe-GTLX catalyst under CO_2 hydrogenation conditions exhibits a somewhat lower activity than the reference Fe-5K catalyst. The CSP2 membrane has significantly higher permselectivities towards H_2O than the silica membrane, whereby the beneficial effects of *in-situ* H_2O removal on conversion and yield can be demonstrated with Ar and H_2 as sweep gas (Figure 5.3). This allowed a reduction of the pressure on the sweep side to Φ = 0.7 and below. The loss of reactants was not determining in this case. Compared to the reference case, the relative increases in conversion and yield observed are significantly higher for the Fe-GTLX/CSP2 membrane reactor.

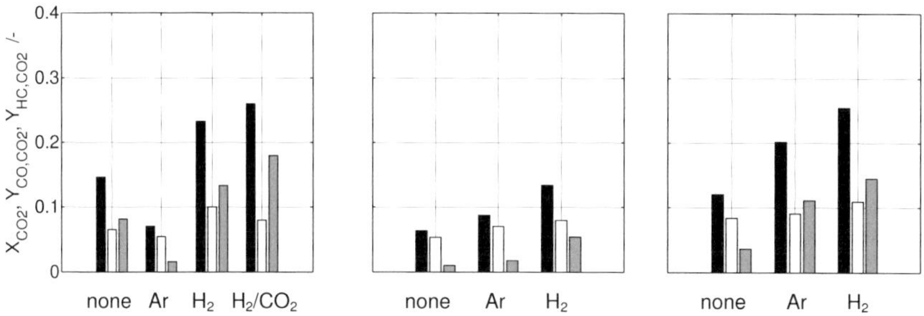

Fe-5K/ silica membrane (Unruh 2006) T = 225°C, $\tau_{mod,n}$ = 4000 kg s/m^3, m_{cat} = $2\cdot10^{-3}$ kg, P_F = 1 MPa, Ψ = 3.3, Φ = 1

Fe-GTL3/ CSP2 membrane T = 255°C, $\tau_{mod,n}$ = 6455 kg s/m^3, m_{cat} = $15\cdot10^{-3}$ kg, P_F = 1 MPa, Ψ = 3.3, Φ = 1

Fe-GTLX/ CSP2 membrane T= 270°C, $\tau_{mod,n}$ = 2000 kg s/m^3, m_{cat} = $4\cdot10^{-3}$ kg, P_F = 1 MPa, Ψ = 3.3, Φ = 0.7

Figure 5.3. Effect of *in-situ* H_2O removal as a function of the sweep gas on CO_2 conversion (■) and CO (□) and hydrocarbon ( ) yield, measured in different FT catalyst/ membrane combinations.

- *Fe-GTL3/CSP2 membrane:* An active catalyst is a prerequisite for successful integration of a catalyst/ membrane system. Fe-GTL3 had the lowest activity under CO_2 hydrogenation conditions of all the tested catalysts. Therefore, the catalyst showed only a weak response to *in-situ* H_2O removal in combination with the selective CSP2 membrane.

The results of the experimental demonstrations in Figure 5.3 indicate that the CO_2 conversion and hydrocarbon yield can be increased significantly by the application of a permselective hydrophilic membrane in combination with an active FT catalyst. The increase in conversion and yield is based on the assumption that H_2O is selectively removed from the reaction zone via the membrane. Lower H_2O partial pressures yield increased rates of the CO_2/CO shift and the FT reaction, which both accelerate CO_2 conversion.

Variation of sweep ratio

The membrane reactors were operated in co-current sweep mode. A variation of the sweep flow rate should have resulted in a shift of the conversion and yield levels. As shown in Figure 5.4, CO_2 conversion increased with increasing sweep flow ratio Ψ in all three membrane reactor systems. In the case of the Fe-GTLX/CSP2 membrane reactor, the yield of the intermediate CO increased much slower compared to the CO_2 conversion. A slowly growing CO yield indicates that the FT reaction is able to convert CO to hydrocarbons and furthermore, it indicates that the membrane is permselective towards CO. CO lost to the permeate side cannot be converted anymore to FT products, however it is still taken into account in the yield calculation; therefore, Unruh

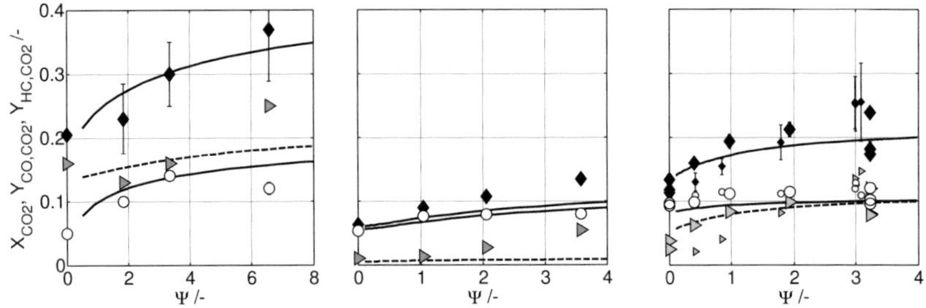

Fe-5K/ silica membrane (Unruh 2006) T= 250°C, P_F = 1 MPa, $\tau_{mod,n}$ = 4000 kg s/m³, $(H_2/CO_2)_F$ = 3; Φ = 1, $(H_2/CO_2)_S$ = 3

Fe-GTL3/ CSP2 membrane T= 255°C, P_F = 1 MPa, $\tau_{mod,n}$ = 6455 kg s/m³, $(H_2/CO_2)_F$ = 3; Φ = 1, $(H_2)_S$

Fe-GTLX/ CSP2 membrane T= 270°C, P_F = 1 MPa, $\tau_{mod,n}$ = 2000 kg s/m³, $(H_2/CO_2)_F$ = 3; Φ = 0.7, $(H_2)_S$

Figure 5.4. Effect of sweep ratio Ψ on CO_2 conversion (◆) and CO (○) and hydrocarbon (▶) yield through *in-situ* H_2O removal, measured in different FT catalyst/ membrane combinations, curves: calculated with membrane reactor model. Left figure: data and model parameters taken from Unruh (2006) & Rohde et al. (2005). Right figure: small symbols repetition with new membrane.

observed a steeper increase in CO yield due to deficiencies in membrane selectivity, though the Fe-5K catalyst was more active.

This effect of a membrane with a higher or lower permselectivity is also reflected in the model-supported calculations when e.g. the results of two membrane reactors are compared. For both cases, the membrane reactor model is able to describe the conversion and yield data. While the membrane parameters of the Fe-5K/silica membrane configuration were derived from the reactive experiments (Unruh 2006), model input for the Fe-GTLX/CSP2 reactor was determined in separate experiments. The model follows the observed trends, but generally underestimates CO_2 conversion data.

The next paragraphs will discuss the Fe-GTLX/CSP2 membrane reactor configuration in more detail. The experimental data – in particular the CO_2 and H_2O data – scattered as data was collected in two separate runs (large and small symbols) using fresh CSP2 membranes and catalysts samples. Membrane performance and degradation has affected consistency of experimental data.

As seen in Figure 5.4 (right), the CO_2 conversion rises initially, but flattens out fairly quickly at higher sweep flow ratios. This can be explained by having a closer look at the partial pressure profiles on the feed and sweep side (Figure 5.6). At high sweep flow rates, the maximum driving force for partial pressure driven transport across the membrane is reached. The membrane was swept with H_2 at a sweep pressure of 0.7 MPa; the partial pressure profiles indicate that H_2 fraction on the sweep side approaches unity, while molar fractions of

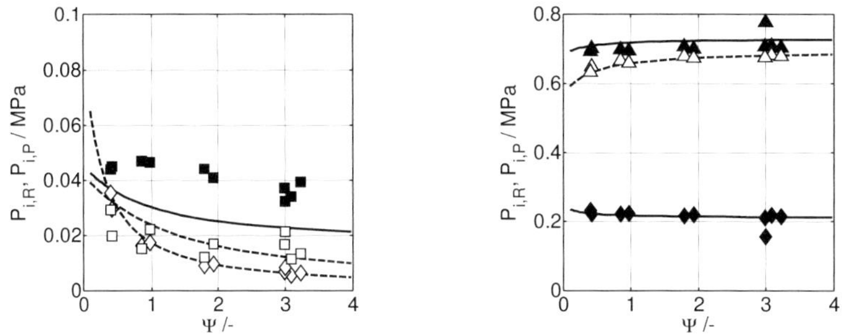

Figure 5.6. Partial pressures of H_2O (■□), CO_2 (◆◇) and H_2 (▲△) as function of the sweep ratio Ψ on the retentate (■◆▲—) and permeate side (□◇△--). Curves: calculated with membrane reactor model.
* Fe-GTLX/ CSP2: T= 270 ℃, P_F = 1 MPa, $\tau_{mod,n}$ = 2000 kg s/m^3, $(H_2/CO_2)_F = 3$; Ψ = var., Φ = 0.7, $(H_2)_S$

H_2O and CO_2 tend to zero. Therefore, a further increase of the sweep flow rate will not increase the conversion and yield any further.

From Figure 5.6, one recognizes that the H_2O partial pressures on the retentate side and therefore the driving partial pressure differences were not very high in general; even at very low sweep ratios, 0.5 bar H_2O partial pressure (at a total pressure of 10 bar) was not exceeded as conversion towards H_2O was low. The model-assisted calculations indicate that the H_2O fraction could be halved by *in-situ* H_2O removal.

The transmembrane molar fluxes, the recoveries and losses of products and reactants are plotted in Figure 5.5. The H_2O flux increases with the sweep ratio Ψ. The obtained H_2O fluxes of about $0.7 \cdot 10^{-3}$ mol/s m^2 (about 0.045 kg/h m^2) are very low due to the low partial pressure differences. For reference, during the membrane tests in the permeation cell (Chapter 3), up to 0.7 kg H_2O /h m^2

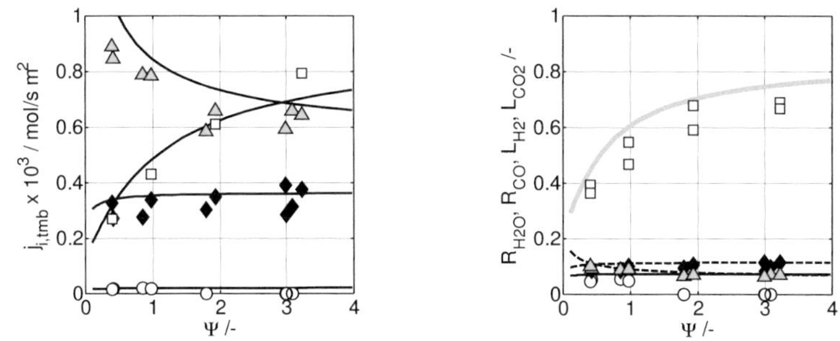

Figure 5.5. Effect of sweep ratio Ψ on transmembrane molar fluxes (left) of (▲) H_2, (□) H_2O, (◆) CO_2 and (○) CO and the resulting recoveries and losses (right) of H_2O (□—) and CO (○—), CO_2 (◆--) and H_2 (▲--). Curves: calculated with membrane reactor model.
* Fe-GTLX/ CSP2: T= 270 ℃, P_F = 1 MPa, $\tau_{mod,n}$ = 2000 kg s/m^3, $(H_2/CO_2)_F = 3$; Φ =0.7, $(H_2)_S$

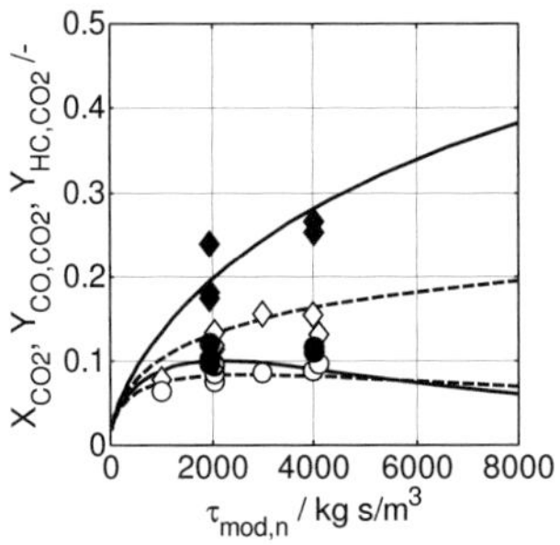

Figure 5.7. CO_2 conversion (◆◇) and CO yield (●○) as function of the modified residence time $\tau_{mod,n}$ for an isothermal fixed bed reactor with (◆●—) and without (◇○--) integrated hydrophilic membrane. Refer also to Figure 4.6.

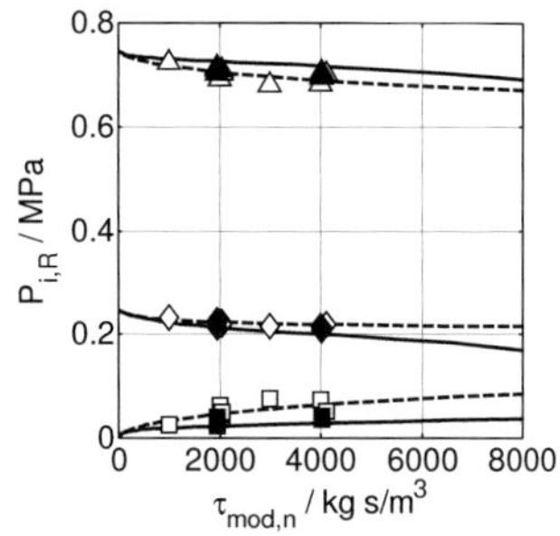

Figure 5.8. Partial pressures of H_2O (■□), CO_2 (◆◇) and H_2 (▲△) as function of the modified residence time $\tau_{mod,n}$ for an isothermal fixed bed reactor with (■◆▲—) and without (□◇△--) CSP2 membrane.

Curves: calculated with membrane reactor model

* Fe-GTLX/ CSP2: T= 270°C, P_F = 1 MPa, $\tau_{mod,n}$ = var., $(H_2/CO_2)_F$ = 3; Ψ = 3.3 (const), Φ = 0.7, $(H_2)_S$

were removed through the CSP2 membrane. During the CO_2 hydrogenation experiments, up to 70% of the H_2O produced by shift and FT reaction was removed (see also Figure 12.36).

At low sweep ratios, the H_2 flux from the reaction zone to the sweep side was considerably larger than the H_2O flux. At higher sweep ratios, however, the H_2 flux levelled out at a lower value as the driving partial pressure difference was reduced (see Figure 5.6). The loss of H_2 could be limited to 8-10% of the total amount fed to the reaction zone by choosing H_2 as sweep gas at 0.7 MPa. The H_2 pressure on the feed side is with 0.75 MPa slightly higher. The positive sign of the H_2 flux indicates that H_2 is removed and not co-fed to the reaction zone; therefore it can be ruled out that co-feeding led to an improvement of reactor performance.

The transmembrane fluxes of CO_2 and CO are more or less invariant with sweep ratio. Both components have a significant lower permeance than H_2O and H_2, and thus, the rate of permeation becomes determining. About 12% CO_2 and maximum 8% of the CO are lost to the permeate side (see also Figure 12.37).

Variation of the residence time

The residence time in the membrane reactor was not varied over a wide range systematically, nevertheless by means of Figure 5.7 and Figure 5.8, a comparison of conversion, yield and partial pressure profiles in a fixed-bed reactor with and without hydrophilic membrane is possible. *In-situ* H_2O removal is able to decrease the relative low H_2O partial pressure even further. This

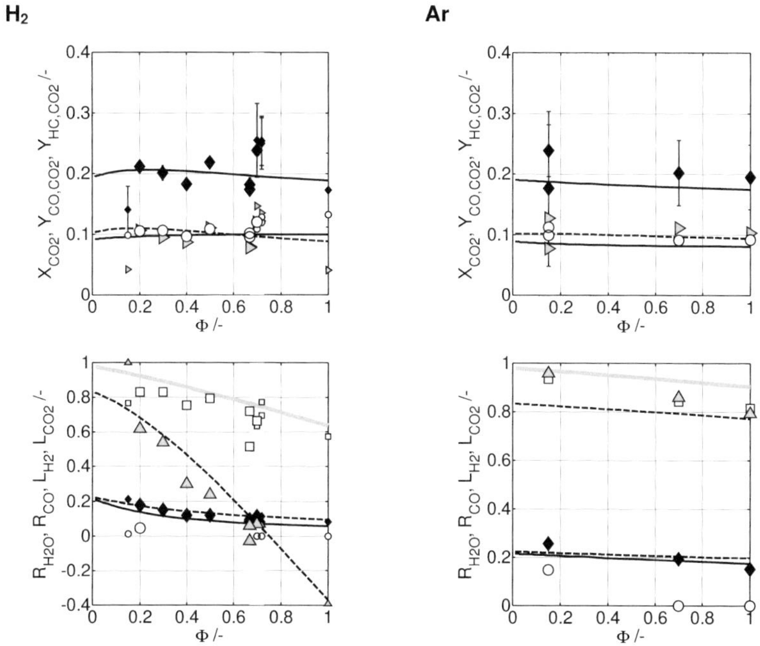

Figure 5.9. Top row: Effect of pressure ratio Φ on CO_2 conversion (◆) and CO (○) and hydrocarbon (▶) yield with H_2 (left) and Ar (right) as sweep gas. Bottom row: Effect of pressure ratio Φ on removal of the products H_2O (□—) and CO (○—) and loss of educts CO_2 (◆--) and H_2 (▲--) to the permeate side with H_2 (left) and Ar (right) as sweep gas. Curves: calculated with membrane reactor model.
* Fe-GTLX/ CSP2: T= 270°C, P_F = 1 MPa, $\tau_{mod,n}$ = 2000 kg s/m³, $(H_2/CO_2)_F = 3$; Ψ = 3.3, Φ = var., $(Ar)_S$

decrease leads to a noteworthy increase of the reaction rates and a significant increase in conversion and hydrocarbon yield.

Variation of pressure ratio and sweep gas

The pressure ratio is an effective measure to increase the driving force across the membrane. In the case of selective membranes, one would use a low pressure on the sweep side rather than a large sweep flow rate (to avoid compression costs). Additionally, it is preferable to sweep with an abundant inert gas (such as low pressure N_2 in the xTL plants) than with an expensive reactant as H_2. The results of pressure ratio variations and use of argon as sweep gas are compared in Figure 5.9.

The reference point is the case (highlighted as grey bar in Figure 5.9) using H_2 as sweep gas with a sweep ratio Ψ of 3.3 and a pressure ratio Φ of 0.7. The results suggest – supported by the model-assisted calculations - that a variation of the pressure ratio has only a small effect on conversion and yield

as a result of two counter-acting effects. By the choice of a lower sweep pressure, significant amounts of H_2 were lost. However, this led to an increase of CO_2 and CO partial pressures in the reaction zone (Figure 12.38). Additionally, a higher fraction of H_2O was removed to the permeate side, and a low H_2O level could be maintained on the feed side. Due to these compensating effects, the overall CO_2 conversion and hydrocarbon yield remained unaltered.
With argon as sweep gas, the H_2 loss exceeded 80%. The conversion level remained high, though model-assisted calculations predict a drop. The membrane reactor with Ar as sweep gas achieved higher CO_2 conversions than the conventional fixed bed reactor without membrane. For comparison, Unruh (2006) observed a slump of the CO_2 conversion by using an inert gas as sweep gas due to significant loss of reactants across the low permselective silica membrane (Figure 5.3). The operation with H_2 as sweep gas at low pressure ratios or with Ar as inert sweep gas yielded similar results.

Membrane permeances under reactive conditions

The permeances under reactive FT conditions were determined for three fresh and for two fresh CSP2 membranes under non-reactive conditions. The summary plot in Figure 5.10 confirms that H_2O is the fastest permeating component – even under reactive conditions –, followed by H_2, CO_2 and CO. The H_2O permeance varies between $2 \cdot 10^{-8}$ and $2 \cdot 10^{-7}$ mol/(s m^2 Pa). The average H_2O permeance is around $6 \cdot 10^{-8}$ mol/(s m^2 Pa). H_2O permeances significant larger than $1 \cdot 10^{-7}$ mol/(s m^2 Pa) can only be obtained with a further modification of the CSP membrane (e.g. thinner polymer layer) or by using a different membrane type (e.g. on zeolite basis).
The H_2O permeance data scatters over a much wider range than the H_2 and CO_2 data. This resulted from various factors such as difficulties to measure H_2O accurately, superimposed membrane degradation (see also Figure 5.21), variation of process conditions and variation in membrane quality.
The permeance data determined under non-reactive conditions (grey) scatters around the permeance data measured in the permeation test cell under ideal conditions (white) (Figure 5.10). The permeances with the exception of H_2O are more or less invariant to variations in sweep and pressure ratio (Figure 5.11). With regard to the model-assisted calculations, the use of a constant set of permeances in the membrane reactor model, which was derived from the experiments in the permeation test cell, is justified.

The CSP2 membrane retained the majority of the hydrocarbons. Only small fractions of hydrocarbons up to heptane were found on the sweep side (Figure 12.11). However, it is assumed that heavier hydrocarbons produced by the FT

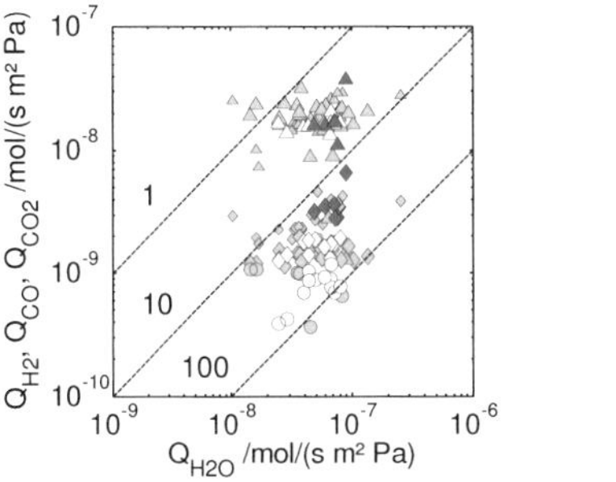

Figure 5.10. Summary of experimental data on permeances Q_i of H_2O, H_2 (▲ ▲ △), CO_2 (◆ ◆ ◇) and CO (● ○), measured under reactive and non-reactive conditions at elevated temperatures T> 250°C. Diagonal lines (- -): indicate lines of constant H_2O permselectivities $S_{H2O,i} = Q_{H2O}/Q_i$.
(▲ ● ◆): Fe-GTL3/Fe-GTLX/CSP2
(△○◇): CSP2 (permeation tests), (▲◆) CSP2 (permeation tests, Vente 2006)

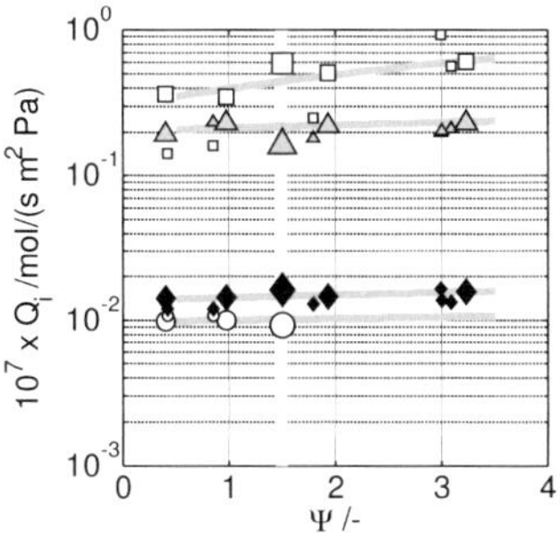

Figure 5.11. Permeances Q_i of H_2O (□), H_2 (▲), CO_2, (◆) and CO (○) determined under reactive conditions in Fe-GTLX/ CSP2 experiments as function of the sweep ratio Ψ. Large symbols at Ψ = 1.5 (grey bar): permeances used in membrane reactor model (Table 5.3).
* Fe-GTLX/ CSP2: T= 270°C, P_F = 1 MPa, $\tau_{mod,n}$ = 2000 kg s/m³, $(H_2/CO_2)_F$ = 3; Φ = 0.7, $(H_2)_S$

reaction contributed to a relative fast decay of the selective membrane transport properties. An operation with a typical FT synthesis gas of $H_2/CO = 2$ was not possible due to rapid blockage of the membrane by heavy hydrocarbons. Therefore, the combination of a polymer-based CSP2 membrane and a Co-based FT catalyst in such a membrane reactor configuration was not possible.

Reactor and catalyst performance

Through the application of a hydrophilic membrane, reactor performance could be improved compared to the conventional fixed-bed reactor as a significantly higher CO_2 conversion and a higher hydrocarbon yield could be achieved. The product distribution remained unaltered by *in-situ* H_2O removal (Figure 12.27).

The model-assisted calculations indicate that both the CO_2/CO shift and FT reaction rate were increased by *in-situ* H_2O removal. The rates at reactor exit ($\tau_{mod,n}$ = 2000 kg s/ m³) were more than twice as high as in the case without membrane, primarily caused by a reduced H_2O inhibition for the FT and in particular for the CO/CO_2 shift reaction (Figure 12.39).

Figure 5.12 shows the conversion-selectivity diagram for CO_2 hydrogenation of a feed gas with $H_2/CO_2 = 3$. The CO_2 conversion and the reaction pathways are limited through the thermodynamic equilibrium of the CO_2/CO shift reaction (recall Figure 4.5). The thermodynamic boundary can be pushed to the right to higher conversion levels by higher temperatures (acting on equilibrium

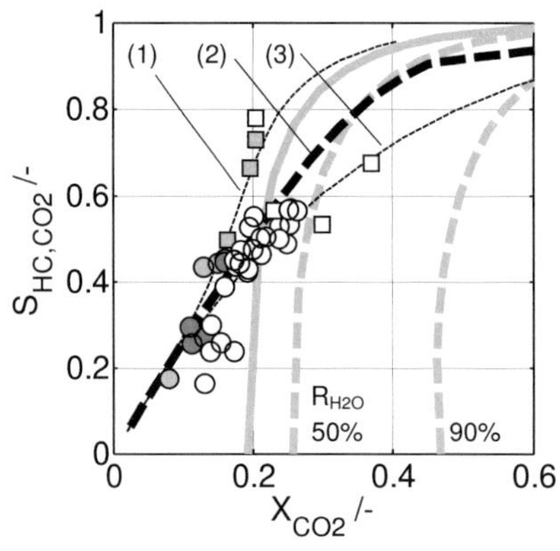

Figure 5.12. Conversion-selectivity diagram for CO_2 hydrogenation of a feed gas with H_2/CO_2 = 3 in membrane reactors and conventional fixed-bed reactors (see also Figure 4.5): (●/○) Fe-GTLX without/ with CSP2 membrane, (■/□) Fe-5K without / with silica membrane (Unruh 2003), (●) Fe-GTLX without membrane (Mena 2009). Curves calculated: Thermodynamic upper boundary of CO_2/CO shift equilibrium calculated at 270 °C for:
(—) 0%, (- -) 50% and 90% H_2O removal
Reaction paths of a fixed-bed reactor with Fe-GTLX catalyst:
(1, --) without membrane
(2, **--**) with CSP2 membrane (real)
(3, --) with CSP2 membrane (ideal)

constant). Or this boundary can be shifted by *in-situ* H_2O removal as indicated in Figure 5.12 (acting on the equilibrium composition). Three reaction pathways are plotted; trajectory (1) describes the Fe-GTLX catalyst employed in a conventional fixed-bed reactor without H_2O removal, case (2) represents the trajectory of the Fe-GTLX/CSP2 membrane configuration as tested in the experiments and (3) the Fe-GTLX catalyst in a membrane reactor with a membrane ideally permselective towards H_2O.

All the data points collected in experiments without H_2O removal are on the left side of the thermodynamic boundary (see also Figure 4.5). However, the data from membrane reactor experiments moves with increasing H_2O removal along the trajectory across the boundary. Due to membrane and catalyst limitations, a maximum CO_2 conversion of only 26% was achieved in the experiments. Higher conversions were obtained by Unruh (2006); the data points follow the trajectory as well, but scattered strongly.

5.3 *In-situ* H_2O removal during DME/DEE synthesis by hydrophilic membranes

The CO_2 hydrogenation experiments in the membrane reactor illustrated the potential, but also proved that selective H_2O removal from synthesis gas is a challenging task and that the loss of reactants is detrimental to reactor and process performance. The etherification of methanol and ethanol to DME and DEE are two example reactions, which distinguish from the CO_2 hydrogenation in two points: first, preceding permeation experiments with a CSP2 membrane under non-reactive conditions suggested that H_2O permeates about 10 times faster than methanol and about 100 times faster than ethanol (Figure 3.9). Therefore, expected permselectivities are much higher. Secondly, H_2O partial

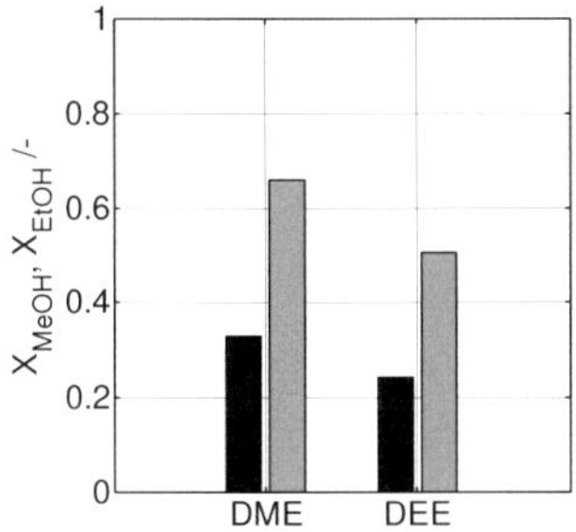

Figure 5.13. Methanol conversion to DME and ethanol conversion to DME in a fixed bed reactor without (■) and with (■) integrated hydrophilic membrane.
* γ-Al_2O_3/ CSP2, T = 290°C, Ψ = 3.3, Φ = 0.15, $(Ar)_S$
* DME: $P_{MeOH,F}$ = 0.67 MPa, $P_{Ar,F}$ = 0.33 MPa, $m_{cat} = 8 \cdot 10^{-3}$ kg, $\tau_{mod,n,MeOH}$ = 5200 kg s/ m^3
* DEE: $P_{EtOH,F}$ = 0.7 MPa, $P_{Ar,F}$ = 0.3 MPa, $m_{cat} = 8 \cdot 10^{-3}$ kg, $\tau_{mod,n,EtOH}$ = 7700 kg s/ m^3

pressures are much higher due to stoichiometry and higher conversion levels. The experiments in the membrane reactor were planned with the aim to boost the rate of the etherification reaction which is strongly inhibited by H_2O.

Methanol or ethanol evaporated into an argon stream was fed to a membrane reactor loaded with acidic γ-Al_2O_3. The CSP2 membrane was swept in co-current mode with argon at a sweep ratio Ψ of 3.3. The pressure on the feed side was kept at 1 MPa, and the sweep side pressure was reduced down to 0.15 MPa. The results of these experiments show that the methanol and ethanol conversion levels to the respective diethers could be increased considerably in the membrane reactor compared to the conversion level achieved in a conventional fixed-bed reactor without a hydrophilic membrane (Figure 5.13 and Figure 5.14).

For the DME case, the equilibrium conversion was not reached or exceeded under the applied conditions as the chosen catalyst loading and therefore the residence time had been too low (Figure 5.14); the upper reactor temperature was determined by membrane limitations. The conversion level was pushed higher by decreasing the pressure ratio Φ.

The reactor model – applying the kinetic and permeance data from independent experiments – is able to describe the experimental results, however, at low pressure ratio (Φ = 0.15), the model-assisted calculations deliver too low conversion levels. This could be explained by a rather high sensitivity of the results on H_2O permeance. The experimental runs at Φ = 0.15 and Φ = 1 were carried out each with a new, fresh membranes, as the first membrane degraded relatively fast. Any variation in membrane quality and in H_2O permeation characteristics will have strong effect on the outcome (Figure 5.14, grey broken curve). From the analysis of the results, it was found that the H_2O and methanol partial pressures were lowered significantly (Figure 5.15), resulting in an accelerated reaction rate (Figure 12.42). The product DME becomes (besides the carrier gas argon) the dominating species in the gas phase.

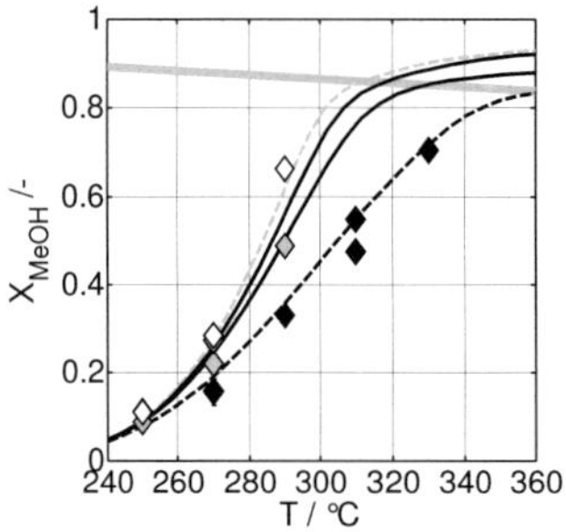

Figure 5.14. Methanol conversion X_{MeOH} as function of the temperature for a fixed bed reactor without (◆) and with integrated CSP2 membrane for two pressure ratios: $\Phi = 1$ (◆), $\Phi = 0.15$ (◇).

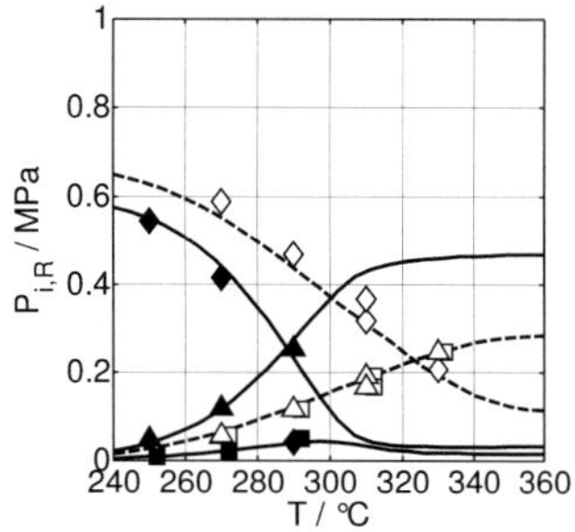

Figure 5.15. Retentate partial pressures of methanol (◇◆), H_2O (□■, estimated) and DME (△▲) as function of the temperature, with CSP2 membrane (◆■▲), without membrane (◇□△) at $\Phi = 0.15$.

Curves: calculated with reactor model with (–) and without (--) CSP2 membrane, grey broken line (--) calculated with 1.5 x Q_{H2O}, (—) thermodynamic equilibrium conversion $X_{MeOH,eq}$. * γ-Al_2O_3/ CSP2, T = var., $P_{MeOH,F} = 0.67$ MPa, $P_{Ar,F} = 0.33$ MPa, $\tau_{mod,n,MeOH} = 5200$ kg s/ m^3, $\Psi = 3.3$, $(Ar)_S$

These experiments under reactive conditions indicate that H_2O permeated about 10-17 times faster than methanol, and 70-175 faster than DME (Figure 12.44). The permselectivities decreased with increasing temperature, as the methanol and DME permeances increased. This relatively strong temperature dependence was not taken into account in the model calculations.

Though the methanol permeance is significantly lower than the one of H_2O, still about 30% of methanol fed to the reactor is lost to the low pressure sweep side (Figure 5.16). This is a step forward compared to the membrane reactor application for CO_2 hydrogenation, where 80% of the reactant H_2 went to the permeate side under similar operating conditions.

Due to the low methanol conversion at low temperatures, the methanol flux is clearly larger than the H_2O flux. With higher temperatures and conversions, the methanol loss can be reduced; therefore, medium or high conversion levels are mandatory. At 290 °C, the H_2O flux is estimated as 0.2 kg/(h m^2), which is more than four times higher that encountered in the CO_2 hydrogenation experiments.

Membrane degradation was a serious issue under these conditions and is addressed further below in Chapter 5.5.

Though the membrane reactor configuration demonstrated a clear benefit in conversion level compared to conventional reactors, the results reveal the drawback of reactant loss and the demand for membranes with a higher permeability for H_2O and improved stability. Furthermore, the obtained results

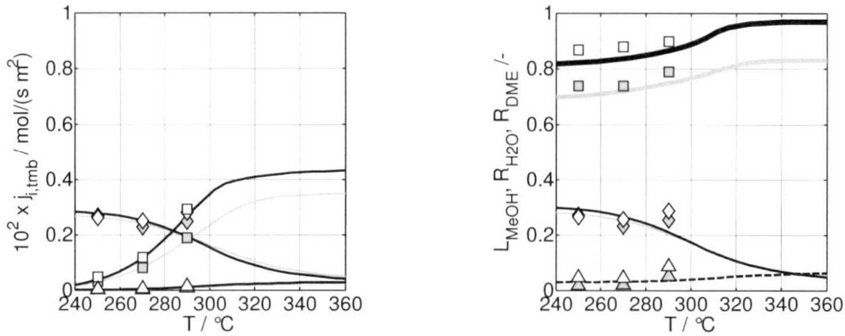

Figure 5.16. Transmembrane molar fluxes of H_2O (□■, estimated), DME (△▲) and methanol (◇◆) as function of the temperature (left) and the resulting recoveries and losses (right) of H_2O (□■, estimated), DME (△▲) and methanol (◇◆) at two pressure ratios $\Phi = 1$ (■▲◆), $\Phi = 0.15$ (□△◇). Curves: calculated with reactor model with CSP2 membrane.
* γ-Al_2O_3/ CSP2, T = var., $P_{MeOH,F}$ = 0.67 MPa, $P_{Ar,F}$ = 0.33 MPa (carrier gas), $\tau_{mod,n,MeOH}$ = 5200 kg s/ m³, Ψ = 3.3, $(Ar)_S$

in the experimental membrane reactor design indicate that the reactor configuration can be improved. Higher residence times achieved by a higher catalyst loading could have driven the reaction beyond the equilibrium conversion already at lower temperatures. A counter-current sweep gas duct and an operation at higher conversion levels appear beneficial to improve H_2O removal and to reduce methanol loss.

5.4 *In-situ* H_2O removal during FT synthesis by chemical reaction

The positive aspects of *in-situ* H_2O removal in a lab-scale membrane reactor for CO_2 hydrogenation and DME/DEE synthesis could be demonstrated in the previous chapters; however, also the limits such as low permselectivities, limiting membrane permeability and membrane stability were identified. Requirements on membrane properties are very high; a failure of membrane jeopardizes process performance.
The use of an additional catalytic function which promotes a H_2O consuming reaction is a totally different approach than applying a physical process as membrane separation. On first sight, a bifunctional catalyst system appears more robust and reactor design less complicated. On the other hand, suitable reactions are rare and applications are very limited. In this case, a CO/CO_2 shift function catalyzed by a low temperature shift catalyst is added to a Co-based FT function.

Figure 5.17 compares the performance of three catalyst systems under the FT conditions, namely (1) the Co-GTL4 catalyst with negligible shift activity, (2) a physical mixture of the Co-GTL4 and the LT shift catalyst and (3) the Fe-GTLX catalyst with a relatively high shift activity.

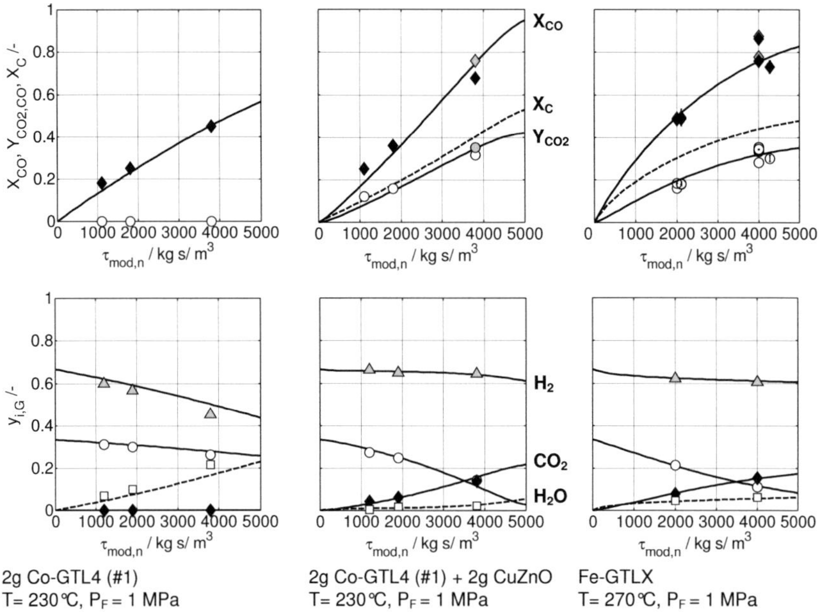

Figure 5.17 Measured CO conversion X_{CO} (◆◆) and CO_2 yield $Y_{CO2,CO}$ (○●) and total carbon conversion X_C and measured molar fractions of H_2 (▲), CO (○), CO_2 (◆) and H_2O (□) for H_2/CO = 2 syngas as function of the modified residence time $\tau_{mod,n}$ for three FT catalyst systems with different CO/CO_2 shift activity. Curves: calculated with the respective kinetic models.†

The Co-GTL4 (#1) operates at a medium conversion level of 40% at a modified residence time of 4000 kg s/ m^3 and the H_2O fraction in the reactor outlet exceeds already 20%. With regard to H_2O partial pressure and H_2O/H_2 partial pressure ratio, this catalyst operates in a safe window, however, a higher operating pressure and conversion level and a reduced H_2/CO feed ratio could change this condition. Due to its relatively high shift activity, the Fe-GTLX catalyst achieves much higher conversion levels; however, only 60% of the CO is converted to hydrocarbons and the other 40% to CO_2. The intrinsic shift function clearly alters the partial pressure profiles: the H_2 partial pressure drops and the H_2O partial pressure rises much slower compared to the Co-based catalyst. The H_2O fraction in the reactor outlet is less than half though

† In the case of the physical mixture of Co-GTL4 (#1) and the LT shift catalyst CuZnO, the rate constant of the FT rate equation (Table 4.6) had to be reduced by 30% to describe the experimental data. There is a high likelihood that the pre-activated catalyst partially reoxidized during the preparation of the physical mixture.

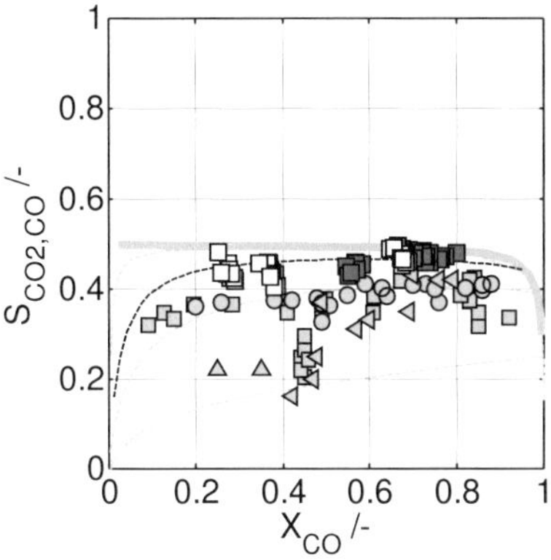

Figure 5.18. Comparison of the CO/CO_2 shift activity of a Co-GTL4/ CuZnO mixture (■□) to various Fe-based catalysts (▲◀●■). For more details refer to Figure 4.5.
Curves calculated:
(—) maximum attainable CO_2 selectivity for a dry feed gas, stoichiometric boundary
(--) reaction pathway calculated with Co-GTL4 (#1)/ CuZnO kinetics
* 2g Co-GTL4 (#1) + 2g CuZnO, T= 230℃, P_F = 1 MPa

both catalysts Fe-GTLX and Co-GTL4 achieve the same hydrocarbon yield or hydrocarbon productivity at a modified residence time of 4000 kg s/ m^3.

The physical mixture of the Co-GTL4 and the LT shift catalyst CuZnO is able to reduce H_2O levels even further (< 5%). The LT shift catalyst demonstrates its ability to operate under FT conditions at low H_2O partial pressures and in presence of hydrocarbons. The dedicated shift catalyst exhibits a much higher shift activity at 230℃ than the Fe-GTLX catalyst at 270℃. With regard to the previous benchmarking study, the Co/CuZnO mixture systems can outperform Fe-based systems with regard to their shift activity as it approaches the theoretical maximum CO_2 selectivity for a dry feed gas much faster than any shift active Fe-based catalyst system (Figure 5.18 and Figure 4.5). Measured CO_2 selectivites larger 45% translate to degrees of H_2O removal > 90%, which are much higher than the one achieved in the membrane reactor experiments.
Catalyst deactivation is a key problem for multifunctional catalyst systems as membrane degradation is for membrane reactor applications. Both catalysts did not show any significant deactivation when they were tested individually. Figure 5.19 documents the time-on-stream behaviour of the combined catalyst system. After a first steep drop, the Co/CuZnO system reached steady-state after about 40 hours. Over the next 300 hours, the activity dropped further. Copper-based LT shift catalysts are known to deactivate in industrial applications because they are prone to easy sintering (Twigg 1989), but as the CO_2 selectivity dropped only slightly over time, the activity loss is contributed mainly to the FT catalyst and not to the shift catalyst. The shift catalysts maintained its activity under FT conditions for more than 400 hours.

Kinetic analysis of the experimental data showed that the FT catalyst lost 30-35% of its original activity. It is not believed that the deactivation of the FT

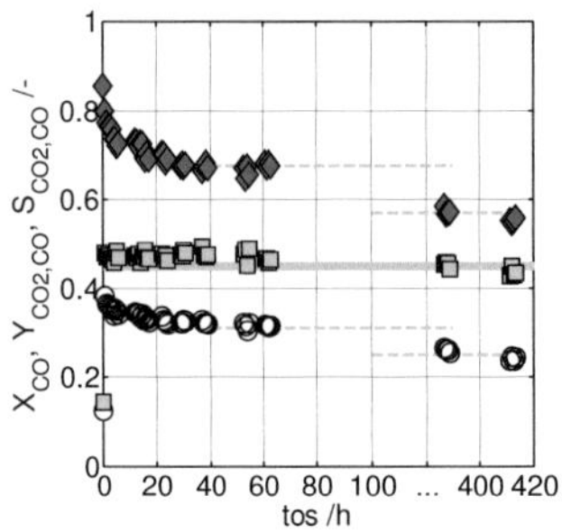

Figure 5.19. CO conversion X_{CO} (◆), CO_2 yield Y_{CO2} (○) and CO2 selectivity $S_{CO2,CO}$ (■) versus time on stream (tos) for Co-GTL4/CuZnO mixture, (—) indicates CO conversion of Co-GTL4(#1) without shift catalyst (reference case).
* 2g Co-GTL4(#1) + 2g CuZnO, $H_2/CO = 2$, T = 230°C, P = 1 MPa, $\tau_{mod,n}$ = 4000 kg s/m^3

catalyst is connected to the presence of the shift catalyst. Chanenchuk, Yates & Satterfield (1991) observed that the FT activity was not altered in the mixture and remained stable for up to 400 hours, while the CO/CO_2 shift activity declined very slowly. The activity loss originates rather from harsh catalyst treatment (pre-activation – potential partial oxidation during mixture preparation – partial re-reduction) and experimental conditions with very high H_2/CO ratios.

The presence of the LT shift catalyst altered the overall carbon distribution slightly. Methane selectivity increased, while the S_{C5+} selectivity dropped. According to Twigg (1989), CuZnO catalysts are not active to methanation reaction, so the selectivity loss may be attributed to much higher H_2/CO ratios. Secondary reactions as hydrogenation of alkenes were accelerated in particular resulting in significantly reduced olefin to paraffin ratios. Additionally, Chanenchuk, Yates & Satterfield (1991) described an increased extent of isomerization, potentially due to acidic sites of the Al_2O_3/ZnO support.

In-situ H_2O removal by chemical reaction can represent an effective measure to maintain low H_2O partial pressures in the reaction zone and to reduce deactivation rates of Co-based FT catalysts. For H_2 balanced syngases ($z_{H2,F}$ = 1) however, the use of shift active FT catalysts is not recommended, as valuable CO is not converted to hydrocarbons, but lost as CO_2. The overall carbon efficiency of the synthesis step drops. As shift active catalysts produce H_2 *in-situ* through conversion of H_2O, shift active FT catalysts are promising candidates in applications with H_2 deficient syngases ($z_{H2,F} < 1$). In particular during the conversion of H_2 deficient feed gases, H_2O/H_2 ratios of 0.6 are easily exceeded.

5.5 Membrane stability

Membrane stability is crucial with regard to future membrane reactor applications; therefore, this chapter is dedicated to the aspects of membrane degradation. Though neither the catalysts in the kinetic experiments nor the

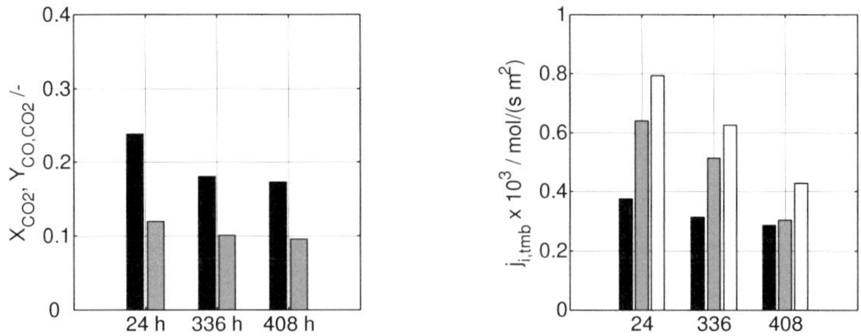

Figure 5.20. Degradation of the CSP2 membrane under reactive FT synthesis conditions over time on stream (24, 336, and 408 hours), (left) CO_2 conversion X_{CO2} (■) and CO yield $Y_{CO,CO2}$ ( ) and (right) transmembrane fluxes of CO_2 (■), H_2 ( ) and H_2O (□).
*Fe-GTLX/ CSP2: T= 270°C, P_F = 1 MPa, $\tau_{mod,n}$ = 2000 kg s/m^3, $(H_2/CO_2)_F$ = 3; Ψ = 3.3, Φ = 0.7, $(H_2)_S$

membranes in the non-reactive permeation experiments showed strong signs of deactivation, the membrane reactor performance was transient. Due to these negative transient effects, the length and the number of *in-situ* H_2O removal experiments were limited. Reproduction experiments with fresh catalyst batches and fresh CSP2 membranes confirmed the measured positive effects of *in-situ* H_2O removal on conversion and yield, but they also showed that the CSP2 membrane has a limited stability under FT and DME/DEE synthesis conditions. Two different types of membrane degradation were observed during the experiments:

Degradation under reactive FT conditions

Experimental data revealed that the CO_2 conversion and CO yield of the reference point declined over time (Figure 5.20) A closer look at the transmembrane fluxes of the key components showed that these dropped significantly, leading to the conclusion that the membrane lost its permeability (Figure 5.20). The plot of the permeances of H_2O, H_2 and CO_2 – extracted from two independent Fe-GTLX/ CSP2 runs – versus time on stream confirms that the membrane does not only lose its overall permeability, but also its beneficial permselectivity towards H_2O (Figure 5.21). Therefore, less H_2O is removed from the reaction zone, and the positive effect of *in-situ* H_2O removal on conversion and yield diminishes.

The presence of long-chain hydrocarbons under reactive conditions may result in enhanced membrane degradation, as the permeation characteristics of the CSP2 membrane were hydrothermally stable for up to four weeks under non-reactive, hydrocarbon-free conditions ($H_2/CO/CO_2/H_2O$, 200-300°C). Operation of the Fe-GTL3/CSP2 membrane reactor under FT conditions with H_2/CO = 2 led to a quick blockage of the membrane, probably due to an increased

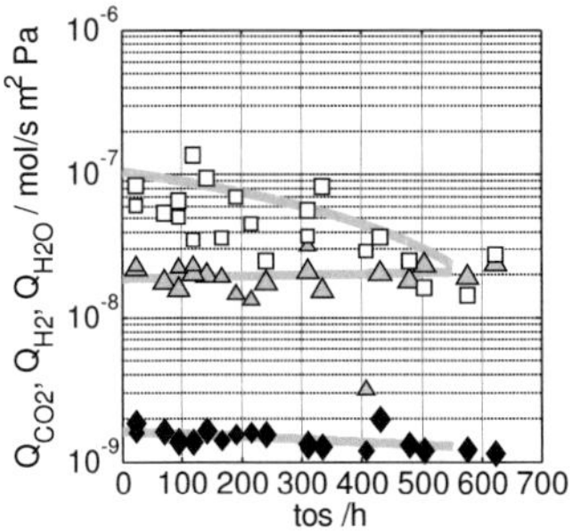

Figure 5.21. Degradation of the CSP2 membrane over time under reactive CO_2 hydrogenation conditions: change of measured permeances Q_i of H_2O (□), H_2 (▲) and CO_2 (◆).
* Fe-GTLX/ CSP2: T= 270°C, P_F = 1 MPa, $\tau_{mod,n}$ = 2000-4000 kg s/m^3, $(H_2/CO_2)_F$ = 3; Ψ = 0.4-3.3, Φ = 0.15-1, $(H_2, Ar)_S$. Curves: trends.

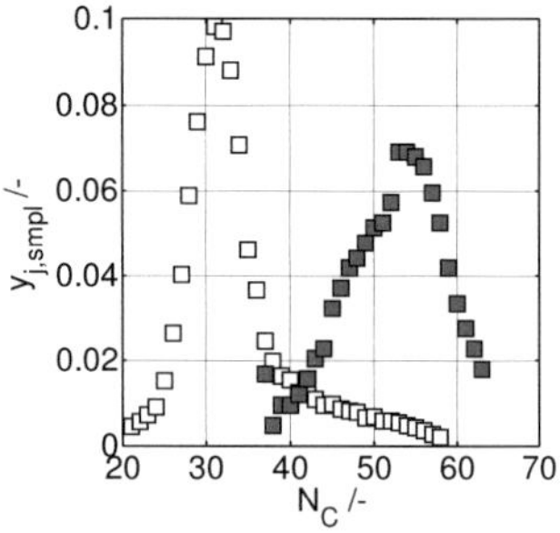

Figure 5.22. Carbon distribution of long-chain hydrocarbons (■) extracted from the CSP2 membrane after experiments under reactive FT conditions; (□) wax sample from Fe-GTLX catalyst as reference (H_2/CO_2 syngas, T = 270°C, P = 1 MPa).

production of waxy hydrocarbons. Attempts to regenerate the membrane by steady inert gas flow through the membrane (stripping) under elevated temperatures did not succeed.

The CSP2 membrane is more or less impermeable for long-chain hydrocarbons, only small fractions of hydrocarbons up to C_6 are found on the permeate side, even at low sweep pressures (Figure 12.11). After the reactive experiments, pieces of a degraded membrane were treated with a solvent (cyclohexane), and the GC analysis detected long-chain hydrocarbons up to C_{63} in the solvent sample (Figure 5.22). This suggests that preferentially heavy hydrocarbons adsorb on or in the functional polymer layer, reducing the permeability.

A closer look at the cross-section of a used CSP2 membrane after reactive FT experiments reveals that the interface between the γ-Al_2O_3 and the coarser α-Al_2O_3 layer was altered (Figure 5.23). It appears that long-chain hydrocarbons penetrated through the polymer and γ-Al_2O_3 layer and covered the α-Al_2O_3 grains. This would be in agreement with the observations. However, another interpretation would be the alteration of the Al_2O_3 layer by a transformation of the instable γ-Al_2O_3 under the aggressive, hydrothermal conditions (pH = 3, organic process water).

Degradation under reactive DME/ DEE conditions

The *in-situ* H_2O removal during the dehydration of methanol to DME and ethanol to DEE had a beneficial effect on the reaction rate and conversion;

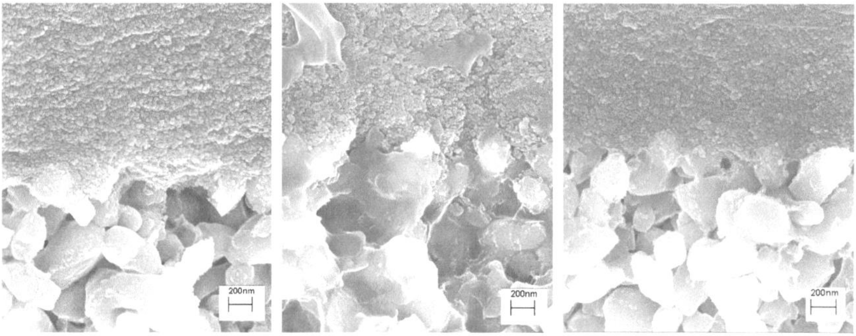

Figure 5.23. Electron microscopic (SEM) pictures of the γ-alumina (top) and α-alumina (bottom) support layers of CSP2 membranes: (left) fresh membrane; (centre) membrane after 16 days under reactive CO_2 hydrogenation conditions; (right) membrane after 10 days under DME- and DEE- synthesis reaction conditions. For reference see also Figure 3.1.

however, it was observed that the conversion declined over time. In contrast to *in-situ* H_2O removal during FTS, the transmembrane fluxes of the key components side increased. In the case of methanol dehydration for example, the loss of methanol to the permeate side more or less doubled in 10 days, leading to a significant decay of the conversion level (Figure 5.24). The increased removal of DME suggests that the CSP2 membrane increased its overall permeability over time. Post-mortem investigation showed that neither the polymer layer nor the support structures of the membrane were altered under reactive DME/DEE conditions (Figure 5.23). However, the presence of oxygenates seems to change the permeation characteristics of the selective polymer layer in a negative way. Such a rapid decay of the membrane was not expected as ECN demonstrated that this type of membrane could be operated stably in a pervaporisation unit for 120 days, dehydrating H_2O/ butanol mixtures at 150℃ (Kreiter et al. 2008).

The hydrophilic CSP2 membrane exhibits a wide temperature operating window and high permselectivities of H_2O with respect to H_2, CO, CO_2 and hydrocarbons and oxygenates. However, in presence of long-chain hydrocarbons and alcohols/ ethers at elevated temperatures (> 270℃), the membrane loses its permselective properties. The most promising application areas are in pervaporisation at lower temperatures - as successfully demonstrated by ECN - or in gas phase permeation at higher temperatures (<300℃) in an environment free of long-chain hydrocarbons and oxygenates. The application of membranes in the reaction zone of a Fischer-Tropsch reactor would rather prefer zeolite than polymer membranes.

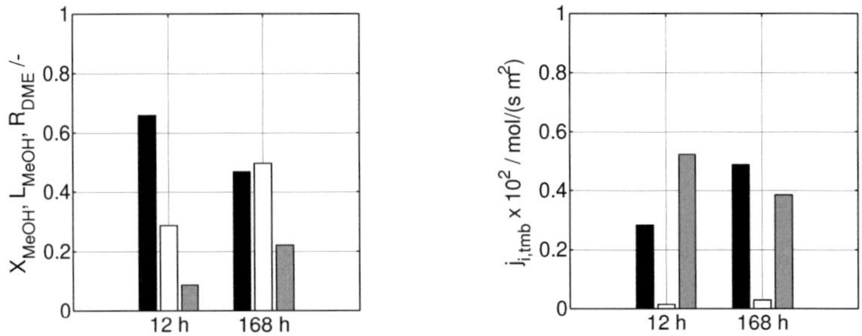

Figure 5.24. Degradation of the CSP2 membrane under reactive DME synthesis conditions over time on stream (12 and 168 hours), (left) methanol conversion X_{MeOH} (■), loss of methanol to the permeate side L_{MeOH} (□) and removal of the product DME to the permeate side R_{DME}( ); (right) transmembrane fluxes of methanol (■), DME (□) and H_2O (, estimated).

* γ-Al_2O_3/ CSP2, T = 290 °C, $P_{MeOH,F}$ = 0.67 MPa, $P_{Ar,F}$ = 0.33 MPa (carrier gas), $\tau_{mod,n,MeOH}$ = 5200 kg s/ m^3, Ψ = 3.3, Φ = 0.15, $(Ar)_S$.

6 Case studies: Potential and limits of in-situ H_2O removal

The final chapter starts with a review of the results from the previous experimental chapters. More general conclusions can be drawn by using the tool of Damköhler-Péclet (Da-Pe) analysis, which allows a simple and comprehensive variation of key parameters. The next step is made from the experimental set-up to applications on industrial scale. In selected case studies, three concepts for H_2O removal are tested against the performance of a state-of-the art reference case, which is represented by a Co-based FT slurry reactor in recycle operation mode. The chapter summarises the findings and provides an outlook on potential applications of in-situ H_2O removal for other reactions beyond the FT synthesis.

6.1 Da-Pe analysis of membrane reactors for *in-situ* H_2O removal

The performance of a membrane reactor in extraction mode is not only determined by the permselectivity of the membrane, but also by the kinetic compatibility of the (by-)product production rate, which has to be removed, and the permeation rate across the membrane. A helpful tool to analyse the kinetic compatibility and the potential and limits of membrane reactors is the so-called Damköhler-Péclet analysis (Bernstein & Lund 1993, Reo, Bernstein & Lund 1997, Battersby et al. 2006). For this purpose, two dimensionless numbers are introduced: the Damköhler number *Da* sets the characteristic times of convection and reaction into relation, the Péclet number *Pe* the characteristic times of permeation and convection; the resulting product *DaPe* represents the ratio of the characteristic times of the permeation and reaction rate:

$$\mathrm{Da}_j = \frac{m_{cat} k_j P_{ref}^{\alpha_j}}{\dot{V}_F c_F} = \frac{\tau}{\tau_R} \tag{6.1}$$

$$\mathrm{Pe}_m = \frac{\dot{V}_F c_F}{A_m Q_{ref} P_F} = \frac{\tau_P}{\tau} \tag{6.2}$$

$$\mathrm{DaPe} = \frac{m_{cat} r_{j,ref}}{A_m Q_{ref} P_F} = \frac{\tau_P}{\tau_R} \tag{6.3}$$

DaPe numbers > 1 indicate that the reaction and therefore the production of the product or byproduct is faster than the membrane transport; *DaPe* numbers < 1 indicate that permeation is faster than the reaction, which is required to decrease the partial pressure of the (by-)product significantly in the reaction zone. For a proper membrane reactor performance, reaction and transport have to be balanced, Bernstein & Lund (1993) give the range 0.1 < *DaPe* < 10 as guideline. According to van Veen et al. (1996), reasonable *Pe*

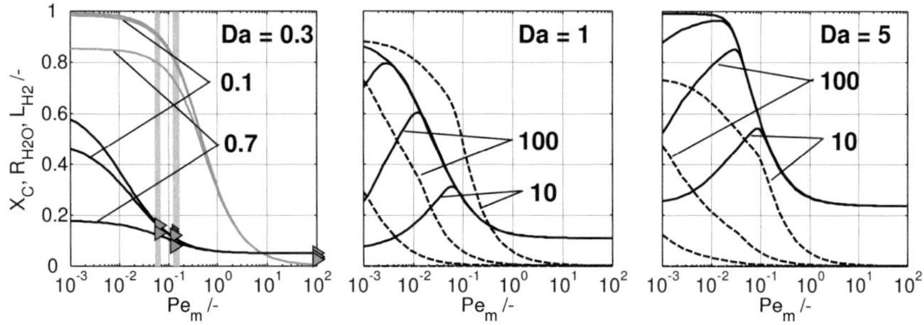

$S_{H2O/H2} = \infty$, $\Phi = 0.7\text{-}0.1\text{-}0.05$, $Da_{FT} = 0.32$, $\kappa = 0.73$

$S_{H2O/H2} = \infty\text{-}1000\text{-}100\text{-}10$, $\Phi = 0.05$, $Da_{FT} = 1$, $\kappa = 0.73$

$S_{H2O/H2} = \infty\text{-}1000\text{-}100\text{-}10$, $\Phi = 0.05$, $Da_{FT} = 5$, $\kappa = 0.73$

Figure 6.1. Effect of *in-situ* H_2O removal during CO_2 hydrogenation on a Fe-based FT catalyst. Total carbon conversion X_C (—), H_2O recovery R_{H2O} (—) and H_2 loss L_{H2} (---) plotted as function of the Péclet Pe_m number. Left: variation of the pressure ratio Φ, centre: variation of permselectivity of H_2O towards to H_2 at a medium Damköhler number Da_{FT}, right: variation of permselectivity of H_2O towards to H_2 at a high Damköhler number Da_{FT}. Experimental data (▶), see Figure 5.5.
* Fe-GTLX kinetics, T = 270 ℃, $(H_2/CO_2)_F = 3$, $\Psi = 3.3$

numbers for technical membrane reactors are between 0.1 and 1. The membrane reactor model (5.14) - (5.16) in Chapter 5.1.3 is rewritten in a dimensionless way, incorporating the Damköhler and the Péclet number:

$$\frac{d\dot{n}^+_{FS,i}}{dz^+} = Da_J \cdot \left[\sum_{j=1}^{J} \nu_{ij} \cdot \kappa_j \cdot R^+_j - \frac{S_{i,ref}}{Da_J Pe_m} \left(\frac{\dot{n}^+_{FS,i}}{\dot{n}^+_{FS}} - \Phi \frac{\dot{n}^+_{SS,i}}{\dot{n}^+_{SS}} \right) \right] \tag{6.4}$$

$$\frac{d\dot{n}^+_{SS,i}}{dz^+} = \frac{Da_J}{\psi} \cdot \frac{S_{i,ref}}{Da_J Pe_m} \left(\frac{\dot{n}^+_{FS,i}}{\dot{n}^+_{FS}} - \Phi \frac{\dot{n}^+_{SS,i}}{\dot{n}^+_{SS}} \right) \quad \text{with} \quad \Phi = \frac{P_{SS}}{P_{FS}} \tag{6.5}$$

$$\begin{aligned} \dot{n}^+_{FS,i}(z^+ = 0) &= \dot{n}^+_{F,i} = y^+_{F,i} \qquad i = 1....N \\ \dot{n}^+_{SS,i}(z^+ = 0) &= \dot{n}^+_{S,i} = y^+_{S,i} \qquad i = 1....N \end{aligned} \quad \text{with} \quad \psi = \frac{\dot{N}_S}{\dot{N}_F} \tag{6.6}$$

with R_j^+ as dimensionless reaction rate, $S_{i,ref}$ as permselectivity Q_i/Q_{ref} with Q_{ref} as reference permeance and κ_j as the ratio of two Damköhler numbers Da_j/Da_J (e.g. of the CO_2/CO shift and the FT reaction).

The results of the Damköhler-Péclet analysis for CO_2 hydrogenation to FT hydrocarbons are shown in Figure 6.1. The graph on the left-hand side reflects the conditions of the experiments described in the previous chapter. Under the experimental and geometrical conditions of the lab-scale membrane reactor, Damköhler numbers of 0.32 for the FT reaction and 0.23 for the CO_2/CO shift reaction were obtained ($\kappa = 0.73$). The Péclet numbers ranged between 0.13 and 0.063. The calculations were carried out for an ideal membrane only permselective towards H_2O.

For Péclet numbers > 10, the effect of the membrane is negligible and the reactor operates as a plug-flow reactor without membrane, which represents the reference case. With decreasing Péclet numbers, an increasing amount of H_2O is removed from the reaction zone, resulting in a rising total carbon conversion or hydrocarbon yield (see Figure 5.4).
Though an ideal permselective membrane was assumed in the Figure 6.1 (left), the predicted total carbon conversion fits well to the experimental data determined with the CSP2 membrane, indicating that a higher permselectivity would have not resulted in a higher conversion level in the experiments.
Degrees of H_2O removal > 80% are key for an enhanced CO_2 conversion; a stepwise decrease of the pressure ratio Φ pushes the H_2O recovery to higher levels, but the response on total carbon conversion will be only significant for Péclet numbers << 0.1 (Figure 6.1, left). The experimental results on pressure ratio variation (Figure 5.9) confirm that fact and highlight that the H_2O permeation across the membrane was limiting. Noteworthy conversion levels would have been achieved e.g. with a membrane with a higher H_2O permeance. Literature reports up to ten times higher H_2O permeances for zeolite membranes than for the CSP2 membrane, indicating a potential to decrease the Péclet numbers significantly at a constant membrane area.

Higher Damköhler numbers – e.g. due to higher catalyst activity, higher catalyst loading or higher residence time – lead to higher overall conversion levels and a more pronounced effect of *in-situ* H_2O removal (Figure 6.1, centre and right). In particular, a higher CO_2/CO shift activity (κ > 0.73) would enhance the hydrocarbon yield.

Figure 6.1 (centre and right) depicts the results of a stepwise reduction of the critical H_2O/H_2 permselectivity from 1000 to 10. For a non-ideal membrane, the total carbon conversion runs through a maximum with decreasing Péclet numbers, followed by a steep decay in conversion and reactor performance, indicating that the loss of hydrogen L_{H2} becomes dominating. The requirements on permselectivity can be formulated in relative terms: if the total carbon conversion level should be doubled, a H_2O/H_2 permselectivity of about 100 is necessary to keep the H_2 loss below 5%. However, if an absolute total carbon conversion level of 60% is targeted, a H_2O/H_2 permselectivity of about 100 results in a H_2 loss of about 6% for the case with the high Damköhler number (Da_{FT} = 5), while the same permselectivity leads to a H_2 loss of about 37% for the case with the low Damköhler number (Da_{FT} = 1).

CO_2 hydrogenation can be enhanced significantly under the following conditions:

- The conversion of the reference case without membrane should be already at medium level, i.e. high Damköhler numbers are preferred. High conversion levels relax the requirements on membrane permeance and permselectivity if a fixed target conversion should be achieved.
- The Péclet numbers Pe_m should be smaller than 0.1. Small Péclet numbers can be obtained through high specific membrane areas and membranes with a high H_2O permeance.
- H_2O/H_2 permselectivities >> 10 are required. H_2O/H_2 permselectivities > 75 are mandatory to keep the loss of H_2 at a reasonable level.

These requirements clearly indicate that the CSP2 membrane, which achieved H_2O/H_2 permselectivities of only up to 5 in permeation experiments, is not a suitable membrane type. One rather has to refer to zeolite membranes, which offer higher H_2O permeances and higher permselectivities.

6.2 Case studies on *in-situ* H_2O removal during FT synthesis

The previous analysis revealed the deficiencies of the tested CSP2 membranes in permeance and permselectivities. However, the question arises if a working reactor concept can be designed which can utilize the potential of a membrane reactor when a suitable membrane is available. Therefore, this chapter should address aspects which were not touched so far such as (a) membrane integration and reactor design, (b) process implications and (c) a comparison of various options for H_2O removal to a state-of-the-art reference case.

On a laboratory scale level, where heat transfer generally plays a minor role, a fixed bed membrane reactor is a straightforward concept as the reactor is simple to construct. The situation becomes more complex for large-scale reactors, when heat and mass transfer areas need to be integrated together into the reaction zone. Packed bed membrane reactors are very limited due to technical and economical drawbacks, and therefore, membrane reactor development tends into the direction of fluidized bed membrane reactors (Deshmukh et al. 2007). The membrane and heat transfer modules are immersed into the fluidized bed; because of the excellent mass and heat transfer properties, smaller specific areas are required compared to packed bed reactors. This concept should be transferable to bubble column reactors, three-phase slurry reactors or stirred tank reactors as well.
Although heat and mass transfer limitations have been overcome, fluidized bed membrane reactors pose additional requirements on membrane design to prevent abrasive destruction of the selective membrane layer e.g. the selective layer could be deposited on the sweep side of the membrane instead of on the

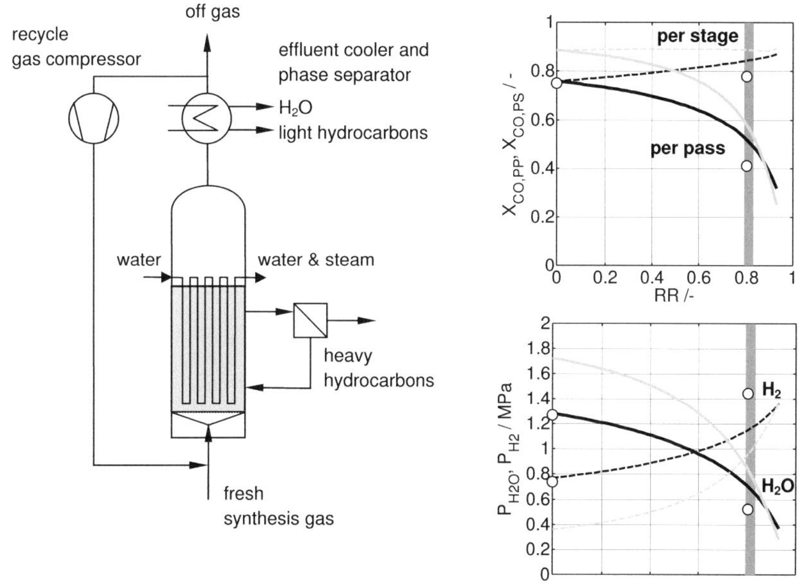

Figure 6.2. Schematic line-up of a three-phase slurry reactor in recycle operation (reference case) and the calculated effect of recycle ratio RR on per pass (—) and per stage (---) CO conversion and on H_2O (—) and H_2 (---) partial pressure in the reactor outlet of a plug-flow reactor (grey) and perfectly mixed reactor (black) and mixed-flow three-phase slurry reactor (○) with a Co-based catalyst.
* Co-GTL4 (#2) kinetics, $(H_2/CO)_{FF}$ = 2, UR = 2.05, P = 3 MPa, T = 230℃

process side; support material could be made from porous metal, such as porous stainless steel (Espinoza et al. 2000). Metal supports would also allow an easier module construction as the metal-ceramic interface falls away. Nonetheless, cheap, high-temperature sealing systems remain still a challenge.The specific membrane areas in membrane modules depend on shape and configuration: tube-shell configurations incorporate up to 250 m^2/m^3 (Saracco & Specchia 1998), a stack of flat membranes up to 30 m^2/m^3, spiral-wound or hollow-fibre configurations up to 1000-10000 m^2/m^3 (Kemmere & Keurentjes 2006).

6.2.1 Reference case

The reactor design of the *reference case* employs three-phase slurry technology in combination with a Co-based FT catalyst today's state-of-the-art, but also the integration of membrane area appears less complex than a multi-tubular reactor design. A schematic line-up of the reactor section is shown in Figure 6.2. The reactor is operated with an internal gas recycle. Light products

Table 6.1. Reactor configuration and operating conditions of the reference case of a three-phase slurry reactor with gas recycle employing Co-based FT catalyst. Estimated reactor production capacity: 17,000 bpd.

reactor configuration					
reactor diameter	/m	8	reaction volume[a]	$/m^3$	1270
dispersion height	/m	30			
operating conditions					
pressure	/MPa	3	conversion per pass	/-	0.41
temperature	/°C	230	conversion per stage	/-	0.78
fresh feed (NTP)	$/m^3/h$	580,000	STY (C_{1+})	$kg/m^3/h$	71
$(P_{H2}/P_{CO})_{FF}$ / UR	/-	2 / 2.05	WTY (C_{1+})	kg/kg/h	0.18
$y_{inert,FF}$	/-	0.02	heat generation	/MW	286
recycle/ fresh feed vol ratio	/-	~1	cooling area[d]	$/m^2$	17650
superficial gas velocity[b]	/m/s	0.44	specific cooling area	$/m^2/kg$	0.035
m_{cat}[c]	/t	507	ε_P	/-	0.3
$\tau_{mod,n}$	$kg\ s/m^3$	1500	$\varepsilon_G/\ \varepsilon_L$	/-	0.47/ 0.53
			P_{H2O} (outlet)	MPa	0.52
			P_{H2O}/P_{H2} (outlet)	/-	0.36

[a] reaction volume represents the volume of the gas/liquid/solid phase without cooling internals under the given operating conditions, [b] superficial velocity at reactor inlet under operating conditions, [c] catalyst particle density assumed as 2500 kg/m^3, [d] cooling area is based on an overall heat transfer coefficient of 1080 W/m^2K and ΔT = 15 K; a cooling tube diameter of 0.05 m yields about 4000 cooling tubes; triangular pitch distance is about 0.12 m; cooling internals require about 15% of the reactor cross-section

and organic process water are separated from the reactor effluent stream in a three-phase separator. Heavy hydrocarbons are removed from the reactor and catalyst particles are retained in filter units. The reaction heat is removed across water-filled cooling tubes immersed in the slurry, generating large amounts of medium-pressure steam.

The reactor should have a capacity of about 17,000 bpd. This requires a diameter of 8 m and a dispersion height of 30 m. The operating temperature is set to 230°C, the operating pressure to 3 MPa. The fresh synthesis gas has a H_2/CO ratio of 2. A guideline for a typical design and operating conditions is given by Maretto & Krishna (1999) and Steynberg et al. (2004). The reaction engineering model from van der Laan (1999) was adopted, which is based on the generalized two-phase model (Krishna, Ellenberger & Sie 1996, de Swart 1996). The model equations are outlined in the appendix in Chapter 12.7. The reaction kinetics of the Co-based catalyst GTL4 (#2) was applied in the reactor model (see Table 12.24 and Figure 12.24 in the appendix) with a rate constant reduced by 40%. The H_2/CO usage ratio UR was assumed as 2.05, the stoichiometric gas phase contraction factor as 0.44.

Following the design guideline of Steynberg et al. (2004), a ratio of the recycle and fresh feed volume flow rates of one is a good starting point to avoid to high per pass conversion and therefore too high H_2O partial pressures and H_2O/H_2 partial pressure ratios.

At a recycle ratio of 0.8, the slurry reactor can be operated up to a per pass conversion of about 40% without exceeding a 6 MPa H_2O partial pressure or P_{H2O}/P_{H2} of 0.6. It is assumed that significantly higher values will lead to increased catalyst deactivation which should be avoided (see chapter 2.1.4). This case is defined as *reference case*; the design parameters and operating conditions are summarized in Table 6.1. Higher operating pressures or lower H_2/CO feed ratios may require lower per pass conversions.

The gas recycle with effluent cooler and phase separator works as external H_2O removal unit. More than 98% H_2O can be removed from the effluent stream per pass. Nonetheless, once-through operation simplifies the line-up of the reactor (Figure 6.3, 1), but leads at similar overall conversion to H_2O partial pressures > 1 MPa and P_{H2O}/P_{H2} ratios >1.5 as illustrated in Figure 6.2. In a slurry reactor, these conditions prevail not only in the reactor exit, but more or less over the entire dispersion height. For plug-flow reactors such as multitubular reactors, only the lower part of the catalyst bed is affected. With reference to Figure 2.10, these conditions will certainly lead to increased catalyst deactivation for particular catalyst systems. Thus, once-through operation demands lower conversion levels. Or a reactor design, that allows a reduction of the H_2O partial pressure in the reaction zone.

6.2.2 Protection of catalyst performance

Figure 6.3 shows three options to protect catalyst stability in once-through operation, namely H_2O removal through hydrophilic membranes, H_2O removal from the saturated hydrocarbon phase and H_2O removal by CO/CO_2 shift reaction.

H_2O removal through hydrophilic membranes

Can a membrane reactor concept replace the recycle gas compressor of the reference case? The membrane area installed should be sized such that the H_2O partial pressure is reduced below 0.6 MPa. The membrane tubes are immersed in the slurry (Figure 6.3, 2). It is assumed that the selective membrane area is deposited on the inside of porous stainless tubes. The gas and liquid phase are in a turbulent, heterogeneous flow regime (see Chapter 12.7) and high mass transfer rates can be expected from the bulk phase to the membrane surface. Individual contributions of liquid phase, small and large bubble classes were not considered in the reactor model due to high mass

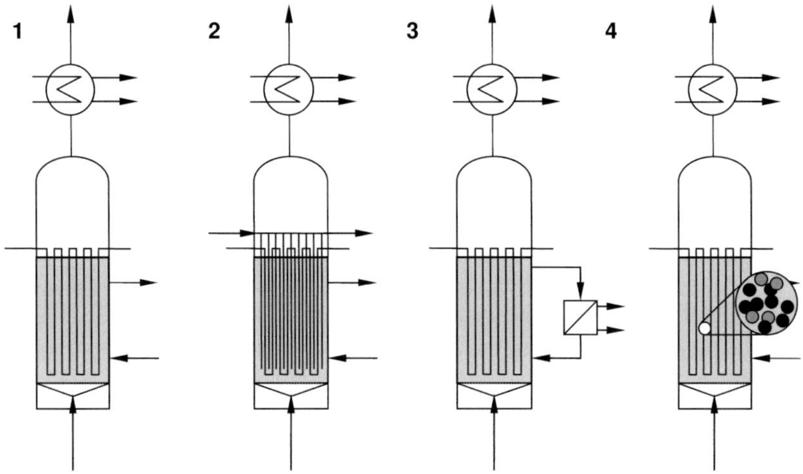

Figure 6.3. Various options of the reference case: Co-based slurry reactor in **1**: once-through operation, **2**: once-through operation with H_2O permselective membrane, **3**: once-through operation with H_2O removal dissolved in liquid phase, **4**: once-through operation with bi-functional FT catalyst.

transfer rates between the phases. The overall permeation properties need to be significantly better than the ones of the tested CSP2 membranes. Therefore, a *future membrane* is envisioned with:

- a H_2O permeance of $3 \cdot 10^{-7}$ mol/s m^2 Pa, which represents an average value of today's state-of-the-art zeolite membranes ($Pe_m = 0.7$),
- permselectivities of H_2O towards H_2 and CO of 60 and 80, which are at or beyond the upper limit of today's membranes.

These membrane properties allow sweeping with N_2 at a low pressure of 0.2 MPa. With 11,000 m^2 installed membrane area (0.023 m^2/ kg catalyst), the H_2O partial pressure and P_{H2O}/P_{H2} ratio are reduced to 0.5 MPa and 0.5, respectively, at a CO conversion of 76%. 74% of the H_2O produced during the reaction is removed across the membrane tubes; about 1% of H_2 and CO is lost. 8 kg H_2O permeate per hour across 1 m^2 membrane; in the CSP2 membrane experiments, the maximum flux achieved was 0.7 kg/ h m^2. 2,500 vertical tubes with length of 30 m and a diameter of 0.05 m are required, decreasing the triangular pitch between cooling and membrane internals to 0.095 m. (Most likely, the membrane tubes have to be arranged in several modules of shorter tube length as 30 m exceeds the maximum possible membrane length.)

As long as the slurry bubble column is operated in the heterogeneous flow regime, lower gas velocities as in once-through operation are advantageous as the liquid hold-up and catalyst mass suspended in the reaction volume increases. And H_2O removal enhances this effect even further. Therefore, the higher liquid hold-up in the membrane slurry reactor outweighs the loss of available reaction volume due to the additional membrane internals. This is an interesting aspect indicating that membrane integration does not necessarily result in larger reactors. However, a 'boost of conversion level' (Zhu et al. 2005) is not observed, as higher CO partial pressures slow down the reaction rate. Overall, the achieved conversion level is just slightly higher compared to the once-through configuration without membrane and even lower than in the reference case with recycle.

Meidersma & Hann (2002) give the price of zeolite membrane modules as 2000 €/m², where only 10-20% are ascribed to the membrane and the rest to the module. This would result in an investment of 22,000,000 € for 11,000 m² membrane area. The power requirement for the recycle gas compressor in the reference case was estimated as 3 MW, the equipment costs excluding installation as 3,000,000 € using the 2002 cost index (Bloch 2006, Peters, Timmerhaus & West 2003). The membrane costs exceed the required investment for a recycle gas compressor by far and therefore, industrial membrane applications on a large scale require significant cost reductions by more than a factor of 10 (Meidersma & Hann 2002). Even if a cost benefit can be identified, the investment decision would most likely be for the recycle compressor configuration due to the risks of membrane failure and membrane decay.

H_2O removal via saturated hydrocarbon streams

H_2O has a much higher solubility than CO or H_2 in hydrocarbon wax (Table 12.34) and large amounts dissolve in the hydrocarbon phase. An alternative concept by Zhang & Espinoza (2003) and Zhang et al. (2006) routes a stream of the H_2O-rich slurry phase from the reactor to an (external) removal unit and recycles the H_2O-deficient hydrocarbon stream back to the reactor (Figure 6.3, 3). Zhang et al. (2006) name a long list of different methods to remove H_2O from the hydrocarbon stream, such as degassing, partial condensation, flash vaporization, extraction, azeotropic distillation, membranes, absorption, H_2 stripping, liquid-liquid extraction, centrifugation, decantation, and more. This concept is similar to the approach of the application of condensable stripping agents for H_2O removal, which was proposed by Battista, Lo & Tatterson (1999) for a different type of reaction.

In this case study, a complete H_2O removal was assumed in the external removal unit, which is not specified further. First, 1500 m^3/h liquid hydrocarbons are sent to the removal unit, which is about fifteen times the volume flow rate of the reactor hydrocarbon production. However, less than 10% of the H_2O produced in the reactor can be removed; the effect on the resulting H_2O partial pressure is very minor. If the hydrocarbon stream is further increased to 7500 m^3/h, about 40% H_2O are removed, but the effect on H_2O partial pressure remains small; the H_2O partial pressure and the H_2O/H_2 partial pressure ratio are reduced to 1.1 MPa and 1.38, respectively. About 1.5 mol% H_2 and CO are lost if they are not recompressed. These numbers indicate that the concept is not feasible to reduce the H_2O partial pressures so significantly that the reactor can be operated in once-through mode at high conversion level unless the hydrocarbon stream is increased even further.

At even higher liquid hydrocarbon streams, the slurry bubble column would operate as a slurry plug flow; heat removal and catalyst separation could take place as well outside of the reaction zone. This concept resembles the concept for monolithic loop reactor with external heat (Güttel & Turek 2009) and H_2O removal.

H_2O removal via chemical reaction

Low H_2O partial pressure even at high per pass conversions can be achieved when a Co-based FT catalysts is combined with a LT shift catalyst as H_2O is consumed *in-situ* by the CO shift reaction. This concept is only promising if it is applied to H_2 deficient synthesis gases – derived from coal or biomass – because additional H_2 is produced *in-situ* by the CO shift reaction. Fe-based catalysts are inherently active in CO/CO_2 shift reaction and therefore they are also suitable for CTL applications (e.g. high and low temperature applications at SASOL). An *in-situ* shift would make the external shift unit unnecessary, which is part of the synthesis gas conditioning step (Figure 2.1).

A synthesis gas stream of 580,000 Nm^3/h is fed to a once-through slurry reactor with a geometrical configuration similar to the reference case. The feed gas composition is $H_2/CO = 0.66$ ($z_{H2,F} = 0.33$) with 2 vol% inerts. This ratio is encountered for coal- or biomass-derived synthesis gases, as indicated in Figure 1.2 (grey line). The slurry reactor is operated at 3 MPa and 230°C. The kinetics of the Co-based FT catalyst Co-GTL4 (#2) and of the CuZnO shift catalyst are used; caution is advised as both kinetics were determined at 1 MPa.

The shift catalyst can be suspended in the slurry as well, or it is introduced as bi-functional catalyst system or it may be fixated on reactor internals. The presence of the shift catalyst boosts the conversion level; H_2O is consumed

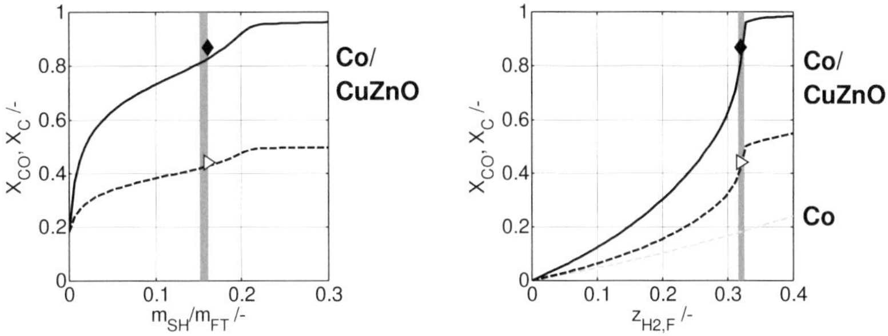

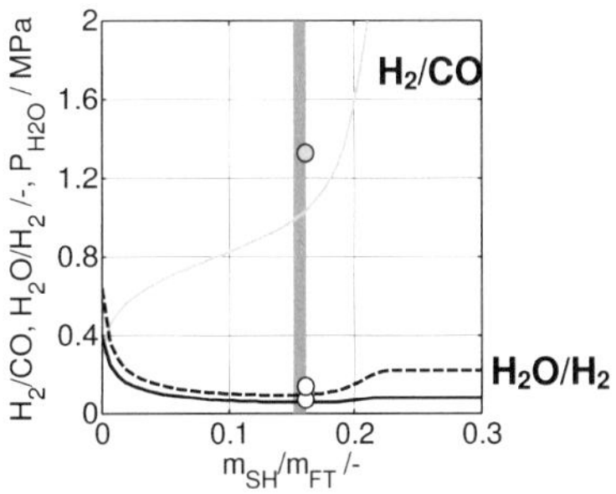

Figure 6.4. Performance of a Co-based FT system in combination with a LT shift catalyst as function of the amount of shift catalyst (left, $z_{H2,F}$ = 0.32) and of the synthesis gas composition $z_{H2,F}$ (right, m_{SH}/m_{FT} = 0.15).
Top row: CO conversion (···) for Co-based FT system, CO conversion (—) and total carbon conversion (---) for bifunctional catalyst system. Bottom left: P_{H2O} (—), P_{H2O}/P_{H2} (---), P_{H2}/P_{CO} (···).
Case study: curves calculated with perfectly mixed reactor model (CSTR), symbols calculated with three-phase slurry model.
* Co-GTL4 (#2) and CuZnO kinetics, T = 230°C, P = 3 MPa, $\tau_{mod,n,FT}$ = 4000 kg s/m^3

and the H_2/CO ratio increases, which is beneficial for the reaction rate (Figure 6.4). The H_2O partial pressure and H_2O/H_2 partial pressure ratio are with 0.07 MPa and 0.14 very low, indicating that H_2O removal by the forward shift reaction is very effective.

As the lower total carbon conversion level is compensated by a much higher CO fraction in the feed gas, this reactor configuration achieves under similar boundary conditions the same production level (17,000 bpd) as the reference case. This fact highlights the potential of this concept as the reactor with the bifunctional catalyst system is operated once-through in safer regime for the FT catalyst; and the recycle gas compressor and an external shift unit could be removed from the line-up.

Relatively small amounts of shift catalyst have to be added, in this case study about 15 wt%. The amount or catalyst activity must not be too high as otherwise the H_2/CO ratio rises steeply above two, which has to be avoided (Figure 6.4). This is also the reason why this concept can only be applied for synthesis gases with H_2/CO ratios < 0.8 ($z_{H2,F}$ < 0.4). While the other process variables as conversion or H_2O partial pressure are less sensitive to the amount or activity of the shift catalyst, the reactor performance becomes very

Table 6.2. Results of case studies. Refer to Figure 6.2 (reference case) and Figure 6.3.

	once-through	recycle operation (reference case)	once-through w/ hydrophilic membrane	once-through w/ liquid phase removal	once-through w/ bifunctional catalyst
$(H_2/CO)_F$	2	2	2	2	0.66
X_{CO} /-	0.75	PS: 0.78 PP: 0.41	0.76	0.76	0.44
Capacity / bpd	16,500	<u>17,700</u>	16,700	16,700	<u>17,700</u>
P_{H2O} /-	<u>1.27</u>	0.52	0.51	<u>1.05</u>	0.07
P_{H2O}/P_{H2} /-	<u>1.72</u>	0.36	0.50	<u>1.38</u>	0.14
R_{H2O} /-	-	> 98% per cycle	74%	36%	96%
comments	+ no recycle compressor	- recycle compressor	- membrane area: 11,000 m^2 - low reliability	- slurry pumps and ext H_2O removal unit	+ no recycle + no shift unit - high inert level

responsive to the H_2/CO ratio in the feed gas, which represents the most significant drawback.

Comparison of case studies

The case studies with regard to the protection of Co-based FT catalyst integrity are summarized in Table 6.2. The reference case with recycle operation presents a working and reliable solution. The application of hydrophilic membranes for catalyst protection is not a viable option. For H_2 deficient synthesis gases, the combination of FT and shift catalysts should be investigated in more detail as it offers a series of incentives, in particular a reduced equipment and unit count. However, stable operation at the operating point appears difficult.

6.2.3 Conversion of CO_2 containing synthesis gases

The last case study relates to the starting point of this thesis, the enhanced conversion of CO_2 containing synthesis gases to FT products. A techno-economical study was carried out in the LTCPO-GTL (2005) project, but this case study is based on a *future membrane* with improved properties. The feed gas –produced from natural gas in a low-temperature catalytic partial oxidation process – gas is characterised by almost equal amounts of CO and CO_2 and a balanced amount of H_2 to convert both CO and CO_2 to hydrocarbons ($z_{CO2,C,F}$ = 0.5 and $z_{H2,F}$ = 1). Due to low temperatures in the gasification step (< 900 °C), the methane conversion is limited and large amounts of the inert slip into the synthesis gas (> 10%) (Rabe, Treong & Vogel 2007).

A synthesis gas stream of 580,000 Nm^3/h is fed to a once-through slurry reactor with a geometrical configuration similar to the reference case. The feed gas composition is $H_2/CO/CO_2$ = 5/1/1 with 10 vol% inerts. The slurry reactor is

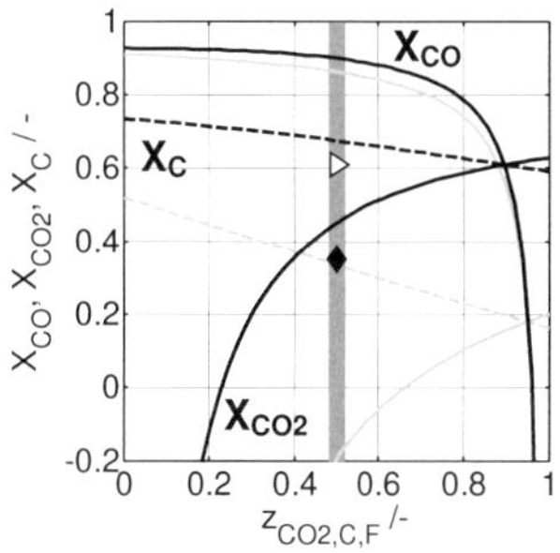

Figure 6.5. CO, CO_2 and total carbon conversion as function of the stoichiometric syngas composition; lines: calculated with perfectly mixed reactor model (CSTR) without membrane (grey) and with membrane (black).

Case study: CO_2 (◆) and total carbon (▷) conversion, calculated with three-phase slurry model with membrane (20,000 m²).
* Fe-5K kinetics (Unruh 2003), T = 250°C, P = 3 MPa, $\tau_{mod,n}$ = 4000 kg s/m³

operated at 3 MPa and 250°C. The higher temperature is due to the application of a Fe-based FT catalyst. The kinetics of Unruh (2006) for a K-promoted Fe-catalyst was applied as representative for an active Fe catalyst.

The CO_2 conversion should be enhanced by H_2O removal through a hydrophilic membrane. This task demands even higher requirements than in the previous Co-based case as the H_2O partial pressure has to be lowered significantly below 0.6 MPa. The future membrane is characterized by:

- a H_2O permeance of $1 \cdot 10^{-6}$ mol/s m² Pa, which represents the highest permeance of today's state-of-the-art zeolite membranes (Pe_m = 0.11),
- permselectivities of H_2O towards H_2 with 60 and towards CO and CO_2 with 80, which are at or beyond the upper limit of today's membranes.

These high permselectivities allow sweeping at low pressure and therefore low sweep flow rates. Figure 6.5 plots the performance of a membrane reactor for different CO_2 containing feed gases in comparison to a reactor without membrane in once-through operation. With about 20,000 m² membrane area installed (0.03 m²/kg catalyst), the total carbon conversion can be increased significantly. The membrane slurry reactor achieves a total carbon conversion of 60% and CO_2 conversion of 35%. The capacity of this reactor stage is about 11,000 bpd (compared to 17,000 bpd of the reference case). 95% of H_2O could be removed, lowering the H_2O partial pressure down to 0.07 MPa. The losses of H_2, CO_2 and CO > 10% are considerable.

Though the membrane permeance is already set to $1 \cdot 10^{-6}$ mol/s m² Pa, still 20 m² membrane area per m³ reaction volume are required. For membrane tubes with 50 mm diameter, cooling and membrane internals close ranks down to 35 mm. This requires the development of the new membrane supports such as resistant fibres, which allow significantly higher volume specific membrane areas.

The techno-economical evaluation within the LTCPO-GTL (2005) project was based on the actual permeation data of the CSP2 membrane. Attributable to the low H_2O permeance, a much higher membrane area was required and the membrane area was installed in external membrane units between a series of multitubular reactor stages. By these means, the membrane was removed from the harsh conditions of the reaction zone and the membrane units could be operated under improved conditions (such as counter-current sweep mode). The low permselectivities were compensated by a sweep gas at high pressure rich in H_2. A large sweep gas loop with H_2O removal by condensation needed to be established, which jeopardized the entire membrane reactor concept. High permeances and permselectivities are required to reduce the equipment count and to simply the process line-up. At the end, the costs of the FT section with membrane were more than twice as high as the costs of the Co-based reference case.

Alternatively, the membrane function and a dedicated shift catalyst can be combined in a membrane shift reactor for reverse shift. The CO_2 containing synthesis gas is converted to a CO rich synthesis gas before it is sent to a Co-based FT reactor. A pre-shift reactor for conversion of CO_2 containing synthesis gas to methanol is also the essential part of the CAMERE process (Joo, Jung & Jung 2004). This reactor needs to be operated at high temperatures (> 500 °C) due to the endothermic character of the reverse shift reaction; additionally, the H_2O produced must be removed from the conditioned synthesis gas stream before it is sent to methanol synthesis unit. The advantages of *in-situ* H_2O removal by membranes are evident as it allows conversion levels beyond the equilibrium conversion at much lower temperatures. However, model-assisted calculations showed that the CO_2 conversion levels could not be improved significantly in the membrane reactor at operating temperatures below 300 °C. As a consequence of the unfavourable equilibrium composition, the H_2O partial pressure is very low already from the beginning. Therefore, extremely large membrane areas and very low sweep pressures are required to reduce it further and reactant loss becomes determining. Higher operating temperatures would facilitate H_2O removal, but suitable hydrophilic membranes are not available for temperature above 250 °C.

6.3 Application to other fuel related reactions

The previous chapters discussed specific examples for *in-situ* H_2O removal. Now general aspects should be outlined by referring to a simple equilibrium reaction:

$$\nu_A A + \nu_B B = \nu_P P + \nu_{H2O} H_2O \qquad K_P = \left(\frac{P_P^{\nu_P} P_{H2O}^{\nu_{H2O}}}{P_A^{\nu_A} P_B^{\nu_B}} \right)_{eq} \tag{6.7}$$

with A and B as educts and P as desired product and K_P as equilibrium constant. With all stoichiometric factors $|\nu_i|$ equal one, equation (6.7) represents reactions such as the CO_2/CO shift reaction or esterification reactions. The following calculations are based on a reaction kinetics with H_2O inhibition:

$$r = k \frac{P_A P_B - P_P P_{H2O} / K_P}{(1 + aP_{H2O})^2} \tag{6.8}$$

Potential

The potential of *in-situ* H_2O removal becomes evident for H_2O recoveries above 50% as shown in Figure 6.6. Though Figure 6.6 was derived for reactive vapour permeation, it should be valid for any type of *in-situ* H_2O removal. Reactions with a different stoichiometry ($\Delta\nu < 0$) should yield similar results. The equilibrium constant K_P allows the division of reactions into three categories:

- *Reactions with low K_P* such as the CO_2/CO shift reaction or the dimethyl carbonate (DMC) synthesis from methanol and CO_2 (van den Broeke 2005). H_2O removal can lead to significant relative conversion increases, but H_2O partial pressures have to be reduced to very low levels.
- *Reactions with medium K_P* such as etherification reactions such as DME and DEE synthesis or esterification reactions such ethanol or butanol with acetic acid to ethyl and butyl acetate (de la Iglesia et al. 2007, Peters, Benes & Keurentjes 2005). H_2O removal through zeolite membranes was successfully applied for esterification reactions in lab-scale pervaporisation units (batch operation) at low temperatures (< 85°C). At higher conversions, H_2O inhibition can start to affect conversion levels.
- *Reactions with high or very high K_P* such as the Claus reaction[‡] with high equilibrium conversions or Co- and Fe-based FT synthesis without equilibrium constraints. Unless H_2O inhibition is not prevailing, effects are relatively small, but they can be obtained already at medium H_2O recoveries (>70%). H_2O removal is an option if conversions above equilibrium level are anticipated (e.g. for the Claus reaction) or if deactivation through high H_2O

[‡] Claus reaction: 2 H_2S + SO_2 = 3/8 S_8 + 2 H_2O. Applied on industrial scale e.g. in oil refineries. Full conversion is target to avoid sulphur emissions.

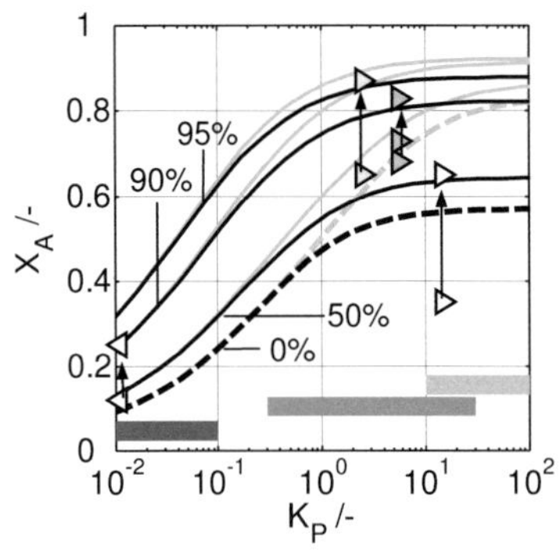

Figure 6.6. Effect of H_2O removal on conversion X_A for equilibrium reactions with (black curves, a = 5 MPa^{-1}) and without (grey curves) H_2O inhibition as function of the equilibrium constant K_P.

Curves calculated with Da-Pe model (6.5)-(6.7) with Da = 10 and varied Pe_m: 1000 (R_{H2O} = 0, ---), 1 ($R_{H2O} \approx 50\%$,), 0.2 ($R_{H2O} \approx 90\%$), 0.1 ($R_{H2O} \approx 95\%$).

Data on *in-situ* H_2O removal:
CO_2 hydrogenation experiment (◁), DME synthesis experiment (▷), ethyl acetate (▶, de la Iglesia et al. 2007), butyl acetate (▶, Peters et al. 2005).

partial pressure has to be avoided (e.g. DME synthesis, DMC synthesis through oxidative carbonylation, Di Muzio 1993).

In-situ H_2O removal could be attractive for a wide range of reactions, and its benefits have been demonstrated for esterification reactions in low temperature lab-scale experiments.

Permselectivity

H_2O removal by vapour permeation requires membranes with high permselectivities Q_{H2O}/Q_i; the previous case studies indicated that permselectivities Q_{H2O}/Q_i need to > 80 to keep reactant or product loss low. Figure 6.7 considers the effect of permselectivity more generally. Permselectivities > 40 have only a small effect on conversion level. Therefore, it is evident that not the impact of reactant loss L_B on conversion level, but its impact on process economy and complexity determines the required permselectivity. If the reactants or products are valuable, loss has to be minimized. Which loss is acceptable – e.g. 5% or < 1% – is an economic decision.

Figure 6.7 discusses a high conversion case (high equilibrium constant K_P) and a low conversion case (low equilibrium constant K_P). High conversion levels lead to lower mean partial pressures of the reactant B on the feed side of the membrane, and therefore less stringent permselectivities are required to keep the reactant loss below a certain level than in a low or medium conversion scenario. Additionally, higher target H_2O recoveries (expressed in a smaller Péclet number) lead to higher demands on permselectivities. For the high conversion case with 90% H_2O recovery, a permselectivity of 40 is sufficient to reduce the reactant loss to 5%, while for the low conversion case with 95% H_2O recovery, permselectivities > 100 are needed.

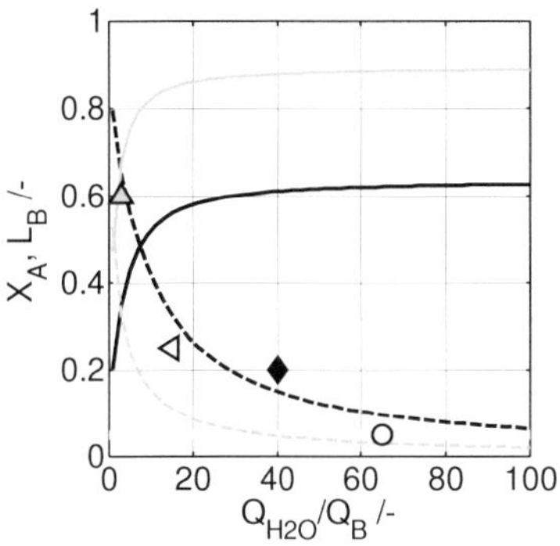

Figure 6.7. Effect of the permselectivity Q_{H2O}/Q_B on conversion X_A (solid line) and reactant loss L_B (broken line) in a membrane reactor.

Curves calculated for two cases:
Reaction with low K_P (black): $K_P = 0.1$, $Pe_m = 0.1$ ($R_{H2O} \approx 95\%$).
Reaction with high K_P (grey): $K_P = 10$, $Pe_m = 0.2$ ($R_{H2O} \approx 90\%$).

Experimental data (CSP2 membrane) on reactant loss:
H_2 (▲), CO_2 (◆), CO (○), methanol (◁)
* Da = 10, a = 0, $\Psi = 1$, $\Phi = 0.01$

Therefore, *reactions with a medium* K_P and operating cases with a high conversion level have the highest potential for membrane applications due to relaxed requirements on H_2O recovery and permselectivity. Low temperature membrane applications (< 150°C) are preferred as permselectivities generally drop with increasing temperature.

Membrane area

For a feasible membrane reactor concept, the H_2O permeation rate has to be *kinetically compatible* to the formation rate of H_2O. The H_2O formation rate of the reactor can be expressed as weight time yield WTY (6.9), the H_2O flux across the membrane as area-time yield ATY (6.10, van de Graaf, Zwiep & Kapteijn 1999). With the assumption of kinetic compatibility, the specific membrane area can be estimated on basis of the WTY/ATY ratio (6.11).

$$WTY = \frac{\dot{N}_{A,F} \cdot Y_{H2O,A} \cdot \tilde{M}_{H2O}}{m_{cat}} \tag{6.9}$$

$$ATY = \frac{J_{H2O,tmb} \cdot \tilde{M}_{H2O}}{A_m} \approx \frac{Q_{H2O} \cdot P_{H2O,R} \cdot \tilde{M}_{H2O}}{A_m} \quad \text{with } P_{H2O,P} = 0 \text{ (maximum driving force)} \tag{6.10}$$

$$a_m = WTY/ATY \tag{6.11}$$

The area time yield ATY depends strongly on the H_2O permeance of the membrane and on the mean H_2O partial pressure, which is determined by the specific purpose of *in-situ* H_2O removal application. The H_2O permeance can stretch from 10^{-7} to 10^{-6} mol/(s m^2 Pa), with polymer and amorphous membranes at the low end and zeolite membranes at the high end.

In-situ H_2O removal applications with the aim to shift equilibrium compositions demand very low target H_2O partial pressures, resulting in very low area time yields. Much higher area time yields can be achieved in applications where

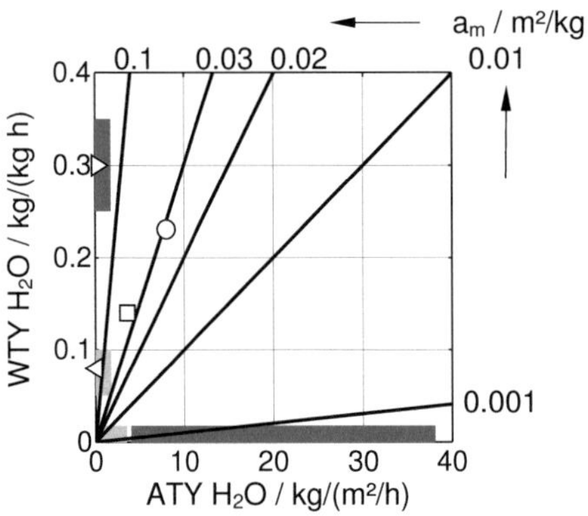

Figure 6.8. Specific membrane area a_m requirement as function of the area time yield (ATY) and weight time yield (WTY), definitions see text.

Data on *in-situ* H_2O removal: CO_2 hydrogenation experiment (◁), DME synthesis experiment (▷), case study on conversion of CO_2 containing syngases (□), case study on catalyst protection for Co-based FTS (○).

Grey bars indicate typical ranges: ■: high H_2O target pressure (0.6 MPa); high specific production ░: low H_2O target pressure (0.06 MPa); low specific production

only a fraction of the produced H_2O needs to be removed. Figure 6.8 visualizes the relationship between ATY and WTY.

On the ATY axis, typical ranges for applications with a high (dark grey bar) and with a low (light grey bar) H_2O target pressure are indicated. On the WTY axis, typical ranges for reactions with a high (dark grey bar) – such as DME or Co-based FT synthesis – and low (light grey bar) productivity are highlighted. The dilemma is obvious: high reactor productivities and low target partial pressures result inevitably in high specific membrane areas (Figure 6.8).

The previous slurry reactor case studies demanded specific membrane areas of about 0.03 m²/kg catalyst though the calculations were already based on membranes with high H_2O permeances. With reaction volume specific catalyst densities ranging from 400 (slurry reactor) to 1000 kg/m³ (fixed bed reactor), this translates to 12 to 30 m^2 membrane area per m^3 reaction volume. For slurry, fluidized bed or fixed bed reactor designs with integrated cooling or heating, these membrane areas already hit the technical limits.
High membrane areas > 0.5 m²/kg can only be obtained in lab-scale membrane reactors or in reactors with catalytic membranes such as catalyst coated or filled hollow fibre membranes. Peters, Benes & Keurentjes (2005) name specific membrane areas of up to 4700 m^2/m^3 or 20 m^2/kg catalyst for catalyst coated hollow fibres. Ho (2004) proposed hollow fibres (3000 m^2/m^3) filled with a CO/CO_2 shift catalyst particles. However, these hollow fibre modules can only be applied for weakly endothermic or exothermic reactions.

The potential application of membrane reactors for *in-situ* H_2O removal appears very limited and a future large-scale application in the fuel industry is very unlikely. *In-situ* H_2O removal by chemical reaction represents in certain cases a better concept. Furthermore, reactive adsorption, where the reaction takes place in presence of an H_2O selective adsorbent, should be considered as a further alternative. E.g. zeolite 3A still can adsorb H_2O at 250 °C (Reßler et al. 2006). The latter concepts avoid the several drawbacks of membrane reactor applications such as reactant and product loss, risk of sudden membrane failure or complex and costly reactor design.

7 Conclusion and outlook

The results of this thesis contributed to the assessment of the working hypothesis whether in-situ H_2O removal during FT synthesis has beneficial effects on reactor performance. This working hypothesis was also addressed by other authors, such as Espinoza et al. (1999), Zhang & Espinoza (2003), Zhu et al. (2005) and Unruh (2006). However, these authors did not draw a decisive conclusion. This thesis assessed various options and gives a clear answer with regard to in-situ H_2O removal in FT synthesis. The main results of this thesis are summarized as follows:

Assessment of the hydrophilic ceramic supported polymer (CSP) membrane

- A new ceramic supported polymer (CSP) membrane (provided by ECN) based on the polyimide P84® was tested in a series of non-reactive permeation experiments. H_2O vapour permeated with a permeance of $7 \cdot 10^{-8}$ mol/(s m^2 Pa) up to 5 times faster than H_2 through the membrane. The membrane revealed high permselectivities towards CO_2 (50…40), CH_4 (60…40), Ar (85…60) and CO (90…60).
- An increase in H_2O partial pressure on the feed side of the membrane did not affect the permeances of the permanent gases. The permeances of the permanent gases increased with temperature, while the H_2O permeance remained constant.
- In the temperature range from 225-275°C, zeolite membranes can achieve higher H_2O permeances and higher H_2O/H_2 permselectivities than the CSP2 membrane. The weakest performance show amorphous microporous membranes.
- The CSP2 membrane properties remained steady over an extended period of time under non-reactive, simulated FT conditions (1 MPa, 225-275°C). From a synthesis gas stream with 25 vol% H_2O, > 90% of H_2O could be removed in co-current operation, while a H_2O flux of 0.7 kg/(h m^2) across the membrane was achieved.
- Under reactive FT conditions, the CSP2 membrane lost its permeability through apparent blockage by long-chain hydrocarbons. Under reactive DME/DEE synthesis conditions, the overall permeability of the CSP2 membrane increased. Both observations highlight the issue of membrane stability in future industrial applications ('initially selective, but not stable over time').

In-situ H_2O removal during CO_2 hydrogenation through membranes

- It was demonstrated experimentally that *in-situ* H_2O removal during CO_2 hydrogenation leads to an increased CO_2 conversion and hydrocarbon yield. With regard to the reference case without membrane, the conversion and yield could be doubled.
- The effect of sweep ratio variations on H_2O recovery could be measured directly and linked to the response of CO_2 conversion. Up to 80% of the H_2O formed during the reaction was removed; a H_2O flux of 0.045 kg/ (h m^2) across the membrane was obtained.
- Due to the high permselectivities of the applied CSP2 membrane, the membrane could be swept with H_2 or Ar at reduced sweep pressure ($\Phi \leq 0.7$) without a significant decay in conversion and yield.
- A membrane reactor model was derived on basis of the species balance of a one-dimensional plug-flow reactor with an ideal solution-diffusion membrane. Model-assisted calculations using parameters derived in independent experiments were able to describe the results obtained in the lab-scale membrane reactor experiments.
- Damköhler-Péclet analysis and case studies with a three-phase slurry bubble column reactor showed that membrane application for *in-situ* H_2O removal in industrial-scale FT synthesis requires H_2O permeances $\geq 5 \cdot 10^{-7}$ mol/(s m^2 Pa) and H_2O permselectivities towards H_2, CO and $CO_2 \geq 80$. With reference to the comprehensive literature study on hydrophilic membranes, these requirements cannot be met by any state-of-the-art membranes.
- Membrane-enhanced FT synthesis can be discarded as an option for hydrocarbon production from CO_2 or CO_2 containing synthesis gases due to unrealistic membrane requirements, very high membrane areas, complex reactor design, risks of membrane failure and high investment.

Feasibility study on H_2O removal during Co-based FT synthesis

- The evaluation of a series of literature publications indicated that H_2O partial pressures above 0.6 MPa and H_2O/H_2 partial pressure ratios above 0.6 can lead to accelerated and irreversible deactivation of Co-based FT catalyst. The susceptibility of the catalyst to deactivation depends on various parameters such as support material, support structure, dispersion of active metal and promoters.
- Case studies with a three-phase slurry bubble column reactor has shown that *in-situ* H_2O removal by membranes can be rejected as an option for

protection of Co-based FT catalyst integrity though less membrane area is required than for the CO_2 hydrogenation case.

- H_2O removal from the saturated liquid phase (Zhang & Espinoza 2003) is not an efficient way of reducing the H_2O partial pressure in a reactor system and can be discarded.
- Today's state-of-the-art design with a gas recycle around the reactor stage with product effluent cooler and liquid removal represents the preferred option for H_2O partial pressure control within the reactor regarding costs and reliability.
- It was experimentally demonstrated that the combination of a low-temperature shift catalyst (CuZnO) with a Co-based FT catalyst resulted in a highly shift-active bifunctional catalyst system. The H_2O partial pressure could be lowered significantly by the forward shift reaction. The bifunctional catalyst system was operated for an extended period without a significant decay of its activity.
- The H_2O removal by the forward CO/CO_2 shift reaction during Co-based FT synthesis was identified as an option with high potential for H_2 deficient synthesis gases, as it combines several incentives: (1) *in-situ* H_2O removal and therefore low H_2O partial pressures, (2) *in-situ* H_2 production and therefore higher per pass conversions, (3) as a result, once-through operation without external shift and therefore reduced equipment count. The largest drawback is the sensitivity to operating and catalyst conditions.

Feasibility study on H_2O removal during fuel-related synthesis reactions

- The experimental results of the DME/DEE synthesis experiments confirm that H_2O removal has a large potential to increase the conversion level of equilibrium limited and/ or H_2O inhibited reactions if suitable membranes are available. Due to higher permselectivities, the membrane reactor could be operated at low sweep pressure. However, the CSP2 membrane lost its permselectivities quickly in presence of oxygenates.
- Model-assisted analysis showed that permselectivities > 40 do not have a strong impact on conversion level. But permselectivities are determined by the tolerable reactant or product loss, and in case of valuable reactants, permselectivities > 80 are required. High permselectivities are essential for membrane reactor operation at a low sweep pressure and low sweep gas flow rates.
- The H_2O permeance of membranes should come close or exceed the H_2O permeance of 10^{-6} mol/(s m² Pa) of today's state-of-the-art membranes to keep specific membrane areas in limits. In case of low target H_2O partial

pressures, the required membrane areas can only be met by hollow fibre modules. This restricts the membrane reactor applications to reactions with a low productivity or low heat of reaction.

Outlook

One outcome of this thesis is that membrane applications for *in-situ* H_2O removal within a FT reactor can be discarded. However, membranes might be considered in a FT plant design in low-temperature applications, such as H_2 removal or CO_2 enrichment in off-gas streams or in the synthesis gas conditioning section.

The combination of a Co-based FT catalyst with a low-temperature CO/CO_2 shift catalyst was identified as an attractive and robust option for H_2O removal for H_2 deficient, coal or biomass derived synthesis gases. A more thorough investigation of the following topics is recommended:

- Combined reduction and activation tests of Co-based FT and LT shift catalyst mixtures
- Experiments with varying H_2/CO feed ratios, temperatures and LT shift catalyst mass fractions with focus on H_2O partial pressures, H_2O/H_2 and H_2/CO partial pressure ratios within the reactor, hydrocarbon yield and product selectivity
- Development of various options for integration of LT shift catalyst into a reactor system
- Model-assisted case studies considering sensitivities and deactivation effects and techno-economical studies comparing external and internal CO-shift and recycle and once-through operation.

With regard to the conversion of CO_2 to hydrocarbons, an adsorption-enhanced reverse shift reactor might be considered as an option.

8 Summary

Introduction and objectives

Economic, political, and environmental constraints have renewed interest in the development of new technologies utilizing unconventional resources such as coal, biomass or CO_2 and providing new types of fuels. Many fuel-related synthesis reactions produce H_2O as the main by-product when 'excess' oxygen is removed as H_2O (or CO_2). Example reactions are the production of hydrocarbons from synthesis gas – a mixture of H_2, CO and CO_2 – either along the Fischer-Tropsch (FT) synthesis or the methanol-to-olefins and oligomerisation route; or the production of the clean fuel dimethyl ether (DME) from methanol or synthesis gas.

In conventional process line-ups, H_2O is generally separated from the products and unconverted reactants after the reaction step. This thesis explored the potential of H_2O removal during the reaction step as *in-situ* H_2O removal can have several incentives as:

- *Accelerated reaction rates.* H_2O inhibits the reaction rate of various synthesis reactions due to strong competitive adsorption on the catalyst surface as reported for example for Fe-based FT synthesis or synthesis of dimethyl and diethyl ether over γ-Al_2O_3 (Bercic & Levic 1992, Butt, Bliss & Walker 1962). Selective H_2O removal during reaction reduces the kinetic inhibition and leads to higher reactant concentrations and higher reaction rates.
- *Improved catalyst lifetime.* Observations of reversible and irreversible catalyst deactivation induced by high H_2O partial pressure have been reported for various reactions. In the case of Co-based FT synthesis, various kinetic studies indicate catalyst deactivation at high H_2O/H_2 partial pressure ratios (> 0.6) and high H_2O absolute pressures (> 0.6 MPa) (e.g. Schanke et al. 1995, 1996, Dry 1990, van Berge et al. 2000, Bartholomew & Farrauto 2005, Saib 2006). In the case of Fe-based FT synthesis, water vapour leads to reoxidation of the catalyst (Dry 1990). Strong catalyst deactivation is also reported for dimethyl ether synthesis from methanol (Kim et al. 2006), for methanol-to-gasoline (MTG) synthesis at H_2O contents above 50 wt% (Aguayo et al. 2001) or for the dimethyl carbonate (DMC) production via oxidative carbonylation at H_2O contents above 10 wt% (Di Muzio 1993). Selective H_2O removal during the reaction could diminish catalyst deactivation, and may allow operation at higher per pass conversions.

- *Conversion levels beyond equilibrium constraints.* Conversion levels of certain reactions are limited by equilibrium constraints. Examples are the synthesis of methanol or FT hydrocarbons from CO_2 containing synthesis gases, dimethyl ether synthesis or the CO_2/CO shift reaction. Selective H_2O removal during reaction shifts the equilibrium composition in favour of the desired product, resulting in conversion levels exceeding the equilibrium constraints.

H_2O can be removed from a chemical process by various methods: condensation and phase separation, adsorption, vapour permeation or by chemical reaction. The general objective of this thesis was to explore the potential of *in-situ* H_2O removal during fuel-related synthesis reactions with focus on *in-situ* H_2O removal by hydrophilic membranes and by chemical reaction. The following three reactions were chosen as examples:

- Membrane-enhanced CO_2 hydrogenation to FT hydrocarbons
- Membrane-enhanced conversion of methanol and ethanol to dimethyl ether (DME) and diethyl ether (DEE)
- H_2O removal by chemical reaction during Co-based FT synthesis

The key questions addressed were whether *in-situ* H_2O removal during Fe- and Co-based FT synthesis represents a feasible option for significantly enhancing reactor performance with regard to conversion and yield levels or catalyst lifetime and if it can be applied to other fuel-related synthesis reactions.

As a part of this thesis was financed and carried out within an EU project related to 'Low Temperature Catalytic Partial Oxidation and Gas-to-Liquids Conversion' (LTCPO-GTL 2005), the conversion of CO_2 containing synthesis gases to FT products in a membrane reactor using a novel hydrophilic membrane forms the major part of this thesis.

Mass transfer in hydrophilic membranes

Fuel-related synthesis reactions such as FT or DME synthesis run at high pressures and temperatures above 200 °C. Therefore, membranes with a high H_2O permeance and high permselectivities are required with a long-lasting hydrothermal stability under these operating conditions. The Centre of Energy Efficiency in Industries of The Netherlands (ECN) provided a new type of ceramic supported polymer (CSP) membranes, whereby a porous tubular support is coated with a 1 µm thin functional layer of the hydrophilic high performance polymer polyimide P84®. The separation performance was tested in non-reactive permeation experiments with single gases, gas mixtures

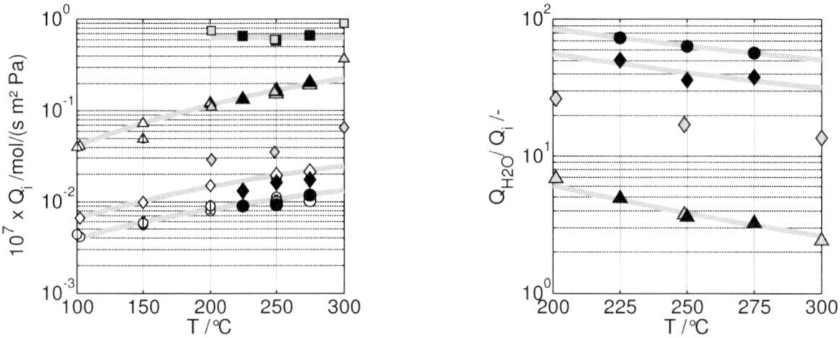

Figure 8.1. Comparison of permeances (left) and permselectivities (right) of the CSP2 membrane determined in (△◇○) transient/ steady-state single gas experiments and permeation experiments with dry gas mixture, (■▲◆●) permeation experiments with gas mixture with 25 vol% H_2O co-fed, and (■▲◆●, grey) permeation experiments with gas mixture with 30 vol% H_2O co-fed (Vente 2006). Curves: trends. Q_i: (■ ■) H_2O, (▲ ▲ △) H_2, (◆ ◆ ◇) CO_2, (● ● ○) CO; Q_{H2O}/Q_i: (▲ ▲) H_2, (◆ ◆) CO_2, (● ●) CO
* Dry gas mixture: $H_2/CO/CO_2$ (50/30/20 vol%), 25 vol% H_2O, P_F = 1 MPa, Ψ = 2, Φ = 0.15, $(Ar)_S$
* ECN: CSP2/dry gas mixture: $H_2/CO/CO_2/CH_4$ (67/0/29/4 vol%), 30 vol% H_2O, ΔP_{tmb} = 0.9 MPa

($H_2/CO/CO_2$) without and with vapours (H_2O, methanol, and ethanol). The following conclusions were drawn from the experimental results (Figure 8.1):

- H_2O is the fastest permeating component with a permeance of $7 \cdot 10^{-8}$ mol/(s m² Pa). The hydrophilic polymer layer favours the selective transport of the polar H_2O molecules over H_2, CO_2 and CO.
- Co-feeding of H_2O (or methanol or ethanol) does not alter the transport of other gases through the membrane. However, higher H_2O fractions in the feed lead to improved H_2O permeance. Due to the weak interactions between the permeating species, the membrane transport can be described by a set of constant permeances at a given temperature.
- The permeance of H_2O remains constant, while the permeances of H_2, CO_2, CH_4, CO and Ar increase with temperature. Consequently, the H_2O permselectivities drop with temperature due to different activation energies.

The CSP2 membrane was ranked amongst transport properties of other hydrophilic membranes, which were treated in a comprehensive literature review (Figure 8.2):

- The H_2O permeance of the CSP2 membrane did not exceed $1 \cdot 10^{-7}$ mol/(s m² Pa) and is in the range of microporous silica or polymer Nafion® membranes. The CSP2 membrane cannot compete with crystalline zeolite membranes which exhibit up to 10 times higher H_2O permeances.

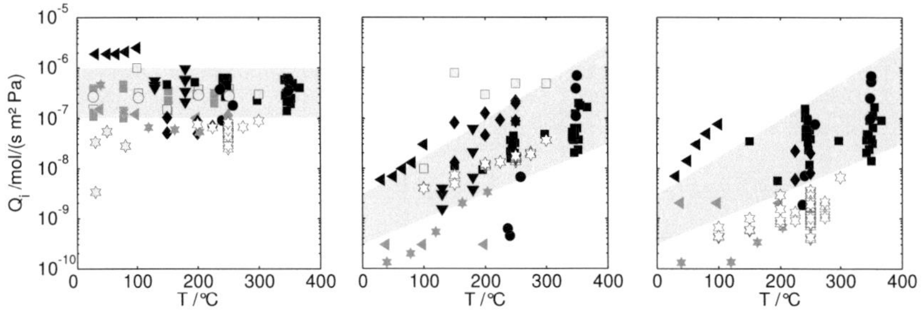

Figure 8.2. Comparison of the CSP2 membrane performance (✩) with data from literature on H_2O permeances (left), H_2 permeances (centre) and CO_2 and CO permeances (right) for different types of membranes: mordenite ●, MFI-type ■, faujasite ▼, zeolite 4A ◀, amorphous silica (TEOS) ◆, polymeric membranes ★. Grey areas: trend.

- The H_2O/H_2 permselectivities found for the CSP2 membrane are higher than for microporous membranes, but lower than for zeolite membranes, in particular, zeolite membranes of the MFI/ZSM-5 type. However, the CSP2 permselectivities for H_2O towards CO_2 and CO (and CH_4 and Ar) are outstanding, also in comparison to zeolite membranes.

In-situ H_2O removal during fuel-related synthesis reaction

Membrane-enhanced CO_2 hydrogenation to FT hydrocarbons

Potassium promoted CO_2/CO-shift active Fe-based FT catalysts are able to convert CO_2 or CO_2 containing synthesis gases to hydrocarbons at elevated temperatures (300-360°C). CO_2 reacts in the reverse water gas shift reaction to CO (8.1) which is then subsequently converted to hydrocarbons in the FT reaction (8.2) (Weatherbee & Bartholomew 1984, Lee, Lee & Chang 1990, Riedel et al. 2003).

$$CO_2 + H_2 \leftrightarrows CO + H_2O \qquad \Delta H_{R}{}^{0}{}_{250°C} = +39.5\ \text{kJ/mol}\ CO_2 \qquad (8.1)$$

$$CO + 2\,H_2 \rightarrow (1/n)\,(C_nH_{2n}) + H_2O \qquad \Delta H_{R}{}^{0}{}_{250°C} = -158\ \text{kJ/mol CO} \qquad (8.2)$$

Unruh, Rohde & Schaub (2002) proposed the application of hydrophilic membranes to keep operating temperatures below 300°C. By the selective removal of H_2O during the FT synthesis, the CO_2/CO shift equilibrium composition is displaced in favour of CO and the CO_2 conversion and hydrocarbon yield are enhanced, which was experimentally demonstrated in a lab-scale membrane reactor by Unruh at 225-250°C (Rohde, Unruh & Schaub 2005). Membrane-enhanced CO_2 hydrogenation was seen as an option for improving the conversion of CO_2 containing synthesis gases in the LTCPO-GTL project.

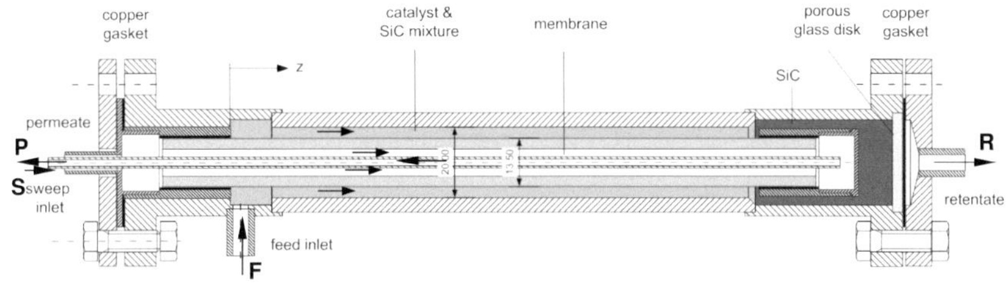

Figure 8.3. Lab-scale fixed-bed membrane reactor for experiments with *in-situ* H_2O removal by hydrophilic membranes; F: feed, S: sweep, R: retentate, P: permeate.

A commercial Fe-based FT catalyst (Fe-GTLX) was combined with the CSP2 membrane in an isothermal lab-scale membrane reactor (Figure 8.3). The catalyst diluted with inert material was filled in the annular gap between the membrane tube and reactor shell. Synthesis gas with H_2/CO_2 ratio of three was fed to reactor inlet; the membrane was swept in co-current mode. In a series of experiments, the sweep gas composition, sweep gas flow rate and pressure were varied.

The experiments yielded the following results: compared to the reference case without membrane, a significantly higher CO_2 conversions and hydrocarbon yields were achieved in the membrane reactor (Figure 8.5). As the CSP2 membrane has much higher permselectivities towards H_2O than the silica membranes applied by Unruh (2006), a performance increase could be observed already even with argon as inert sweep gas though significant amounts of H_2 were still lost to the permeate side. The use of H_2 as sweep gas at a reduced pressure helped to overcome the deficiency in permselectivity

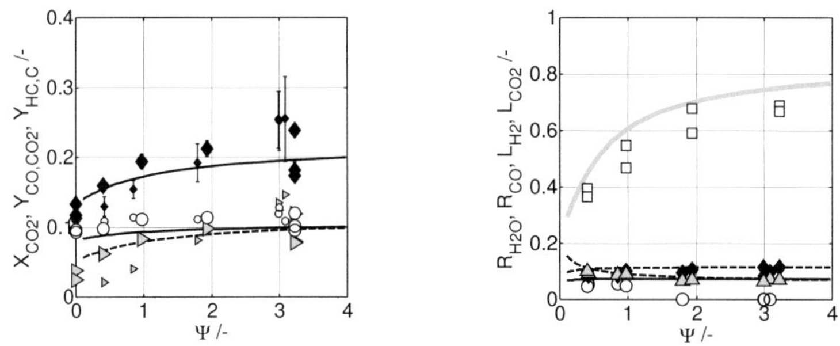

Figure 8.4. Effect of the sweep ratio Ψ on (left) CO_2 conversion (◆) and CO yield (○) and hydrocarbon yield (▶) and on (right) the resulting recoveries and losses of H_2O (□—) and CO (○–), CO_2 (◆--) and H_2 (▲--).Curves: calculated with membrane reactor model.

* Fe-GTLX/ CSP2: T= 270°C, P_F = 1 MPa, $\tau_{mod,n}$ = 2000 kg s/m^3, $(H_2/CO_2)_F$ = 3; Φ = 0.7, $(H_2)_S$

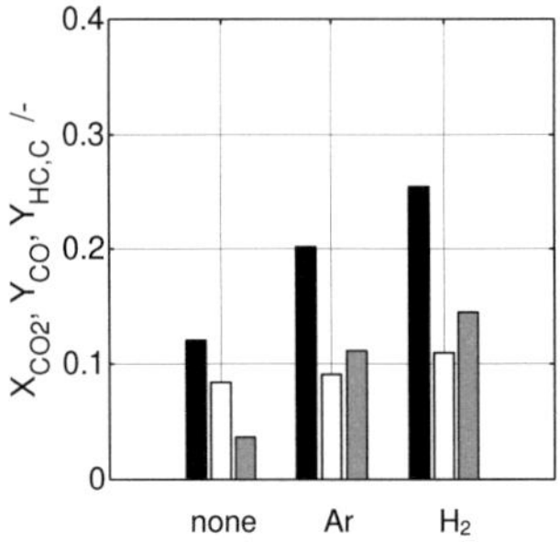

Figure 8.5. CO_2 conversion (■) and CO yield (□) and hydrocarbon yield (■) in a reactor without membrane (none) and a membrane reactor swept with Ar and H_2 as sweep gas.
* Fe-GTLX catalyst/ CSP2 membrane: $(H_2/CO_2)_F = 3$, $\tau_{mod,n} = 2000$ kg s/m^3, $m_{cat} = 4 \cdot 10^{-3}$ kg, T = 270°C, P_F = 1 MPa, sweep ratio $\Psi = 3.3$, pressure ratio $\Phi = 0.7$

and to reduce the reactant loss.

The amount of H_2O removed from the reaction zone to the permeate side – expressed by the degree of H_2O removal R_{H2O} – increased with increasing sweep ratio Ψ. As a result of lower H_2O partial pressures on the feed side, overall higher reaction rates and conversion levels were observed in the membrane reactor (Figure 8.4). Up to 70% of the H_2O produced by the CO_2/CO shift and FT reaction were removed through the membrane, but the H_2O flux of 0.045 kg/ (h m^2) across the membrane was very low. The loss of H_2 could be limited to 8-10% of the total amount fed to the reaction zone by choosing H_2 as sweep gas at 0.7 MPa. Due to their significant lower permeance, only 12% CO_2 and maximum 8% of the CO were lost to the permeate side.

A one-dimensional membrane reactor model, with parameters derived from independent permeation and kinetic experiments, was used to support the experimental results analysis. The trends calculated by the reactor model correlated well with the data collected in membrane reactor experiments (Figure 8.4).

The model-supported calculations indicated that the H_2O permeance of the CSP2 membrane was one of the limiting factors for even higher conversion levels. Besides an improved H_2O permeance, membranes with much higher permselectivities are required. Though the high reactant losses did not have a significant impact on the conversion level, the loss of valuable reactants such as hydrogen has to be kept to a minimum as it jeopardizes the overall membrane reactor concept. Another drawback discovered in the experiments was the decay of the permselective properties over time (Figure 8.6) although the membrane was stable under non-reactive conditions. This suggests that the presence of long-chain hydrocarbons under reactive conditions lead to an accelerated membrane degradation.

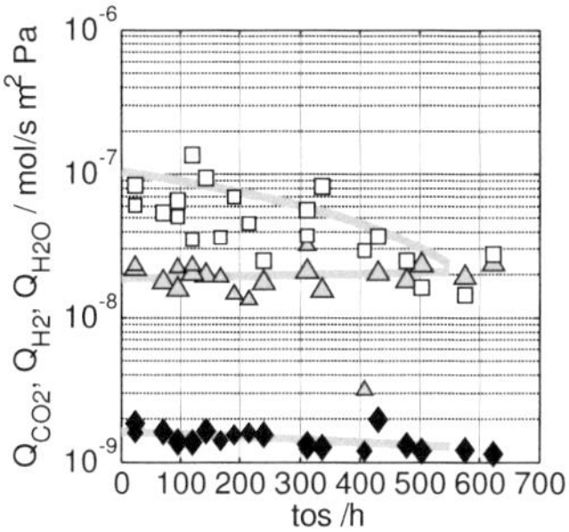

Figure 8.6. Degradation of the CSP2 membrane over time under reactive CO_2 hydrogenation conditions: change of measured permeances Q_i of H_2O (□), H_2 (▲) and CO_2 (◆). Curves: trends.
* Fe-GTLX/ CSP2: T= 270 °C, P_F = 1 MPa, $\tau_{mod,n}$ = 2000-4000 kg s/m^3, $(H_2/CO_2)_F$ = 3; Ψ = 0.4-3.3, Φ = 0.15-1, $(H_2, Ar)_S$

Membrane-enhanced DME/DEE synthesis

Dimethyl ether (DME) is an attractive fuel which has gained increasing public interest in recent years. The etherification of alcohols such as methanol or ethanol takes place on an acidic catalyst and represents a weakly exothermic equilibrium reaction (8.3). The equilibrium conversion of methanol to DME is about 85% at 300 °C. The reaction rate is strongly inhibited by H_2O according to Bercic & Levic (1992). Furthermore, Kim et al. (2006) report that acidic catalysts deactivated in presence of high H_2O partial pressures.

$$2\ \text{MeOH} \leftrightarrows \text{Me-O-Me} + H_2O \qquad \Delta H_{R\ 250°C}^0 = -23\ \text{kJ/mol MeOH} \tag{8.3}$$

Due to the reduced rate inhibition by H_2O, the conversion levels could be raised considerably in the membrane reactor compared to the reactor without CSP2 hydrophilic membrane (Figure 8.7). From the experiments, it was estimated that H_2O permeated about 10-17 times faster than methanol and 70-175 times faster than DME. The H_2O/educt and H_2O/product permselectivities were much higher than those encountered in the CO_2 hydrogenation experiments. Hence, DME synthesis appeared a more suitable to membrane application as the higher permselectivities allow an operation at a low sweep gas pressure. At a sweep pressure of one tenth of the operating pressure, a methanol loss of about 30% was found. However, the exposure of the membrane to oxygenates at high temperatures (290 °C) led to an increase in the overall permeability and fast decay of the permselective properties.

H_2O removal by chemical reaction during Co-based FT synthesis

A potentially more robust alternative to the application of hydrophilic membranes is *in-situ* H_2O removal by chemical reaction. Considering the production of fuel-components or chemicals from synthesis gas, the CO/CO_2 shift reaction is a promising candidate for this task (Ogawa et al. 2003, de Mestier du Bourg 2006, Renk et al. 2006, Post & Sie 1985, Chanenchuk, Yates & Satterfield 1991). Low temperature (LT) shift catalysts operate in a

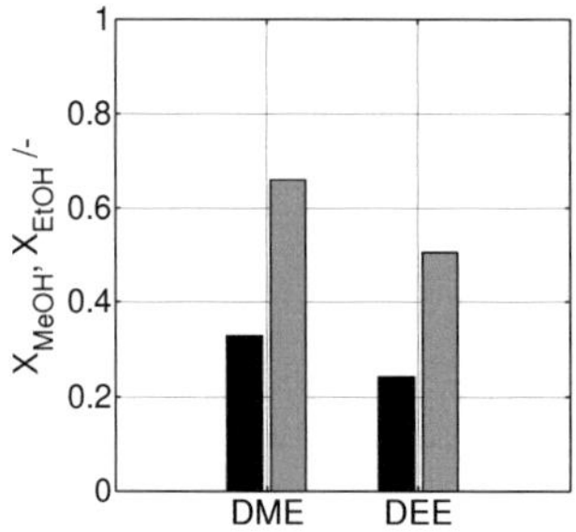

Figure 8.7. Methanol conversion to DME and ethanol conversion to DME in a fixed bed reactor without (■) and with (■) integrated hydrophilic membrane.
* γ-Al_2O_3/ CSP2, T = 290°C, Ψ = 3.3, Φ = 0.15, $(Ar)_S$
* DME: $P_{MeOH,F}$ = 0.67 MPa, $P_{Ar,F}$ = 0.33 MPa,
$m_{cat} = 8 \cdot 10^{-3}$ kg, $\tau_{mod,n,MeOH}$ = 5200 kg s/ m^3
* DEE: $P_{EtOH,F}$ = 0.7 MPa, $P_{Ar,F}$ = 0.3 MPa, $m_{cat} = 8 \cdot 10^{-3}$ kg, $\tau_{mod,n,EtOH}$ = 7700 kg s/ m^3

temperature range from 210 to 270°C (Higman & Supp 2006), which overlaps with the operating window of Co-based FT catalysts. H_2O removal during Co-based FT synthesis is of interest as these catalysts are susceptible to H_2O induced deactivation phenomena.

It was experimentally demonstrated that the combination of a LT shift catalyst (CuZnO) with a Co-based FT catalyst resulted in a highly shift-active bifunctional catalyst system with a higher shift activity than potassium promoted Fe-based catalysts. The LT shift catalyst demonstrated its ability to operate under FT conditions at low H_2O partial pressures and in presence of hydrocarbons for an extended period without a significant decay of activity. The H_2O partial pressure could be reduced to very low levels by the forward shift reaction. Measured CO_2 selectivities larger 45% translate to degrees of H_2O removal > 90%, which are much higher than the ones achieved in the membrane reactor experiments. The ability to reduce H_2O partial pressures effectively and additionally to produce H_2 *in-situ* through conversion of H_2O makes this catalyst combination a promising candidate for applications on H_2 deficient syngases.

Case studies: Potential and limits of *in-situ* H_2O removal

The potential of *in-situ* H_2O removal for equilibrium limited or H_2O inhibited reactions is evident in Figure 8.8. From model-supported analysis, the following general conclusions can be drawn:

- Permselectivities > 40 do not have a strong impact on conversion level. But permselectivities are determined by the tolerable reactant or product loss, and in case of valuable reactants, permselectivities > 80 are required. With regard to permselectivity constraints, membrane reactor applications are more attractive for reaction conditions with medium or high conversion levels.

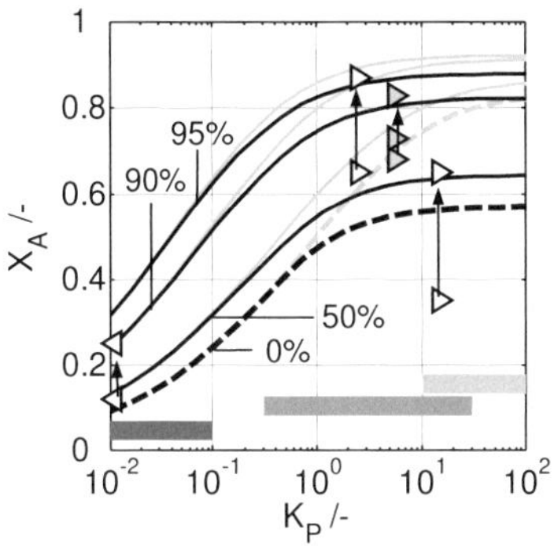

Figure 8.8. Effect of H_2O removal on conversion X_A for a simple equilibrium reaction with (black curves) and without (grey curves) H_2O inhibition as function of the equilibrium constant K_P.

Curves calculated with Da-Pe model (6.5)-(6.7) with Da = 10 and varied Pe_m: 1000 (R_{H2O} = 0, ---), 1 ($R_{H2O} \approx 50\%$), 0.2 ($R_{H2O} \approx 90\%$), 0.1 ($R_{H2O} \approx 95\%$).

Data on *in-situ* H_2O removal:
CO_2 hydrogenation experiment (◁), DME synthesis experiment (▷), ethyl acetate (, de la Iglesia et al. 2007), butyl acetate (▶, Peters et al. 2005).

- The H_2O permeance of membranes should come close or exceed the H_2O permeance of 10^{-6} mol/(s m² Pa) of today's start-of-the-art membranes to keep specific membrane areas in limits. In case of low target H_2O partial pressures, the required membranes areas can only be met by hollow fibre modules. This restricts the membrane reactor applications to reactions with a low productivity or low heat of reaction.

FT related case studies with an industrial-scale three-phase slurry bubble column reactor generated clear results: Membrane-enhanced FT synthesis – either for the conversion of CO_2 containing synthesis gases or for the protection of Co-based FT catalysts – can be discarded due to unrealistic membrane requirements, large membrane areas, risks of membrane failure and high investment. The required permselectivities and permeances cannot be met by today's hydrophilic membranes.

Furthermore, the case studies recommend the application of a bifunctional Co-based FT catalyst as option for H_2O partial pressure control for applications utilizing H_2 deficient synthesis gases.

Conclusions and outlook

This thesis demonstrated that *in-situ* H_2O removal through vapour permeation during CO_2 hydrogenation (270℃) and DME/DEE synthesis (250-290℃) leads to increased conversion and yield levels, which are directly linked to the degree of H_2O recovery. The experimental results are in agreement with the predictions of the applied membrane reactor model and confirm a potential for *in-situ* H_2O removal.

The new ceramic supported polymer membrane (CSP2) represents a progress in membrane development with regard to permselectivities compared to amorphous silica membranes applied by Unruh (2006). However, the membrane is unsuitable for membrane reactor applications due its relative low H_2O permeance and fast degradation of selective permeation properties in presence of hydrocarbons and oxygenates.

Based on a large set of experimental data, various case studies and a comprehensive literature review, this thesis concludes that membrane reactors for H_2O removal will not find an application in FT synthesis or other large-scale synthesis gas based reactions in the foreseeable future. The deficiencies in membrane permeance, permselectivities and stability and the risk of failures and high costs are evident. Today's state-of-the-art design with a gas recycle around the reactor stage with product effluent cooler and liquid removal represent the preferred way for H_2O partial pressure control within the reactor regarding costs and reliability.

Clear incentives for the FT process are seen in the combination of a Co-based FT catalyst with a low-temperature shift catalyst for *in-situ* H_2O removal and H_2 production. This option could lead to a simplified process line-up for coal or biomass based FT plants and should receive further attention.

9 Zusammenfassung

Einführung und Zielsetzung

In der letzten Dekade haben wirtschaftliche, politische und ökologische Zwänge die Erforschung und Entwicklung von Technologien zur Erzeugung von Kraftstoffen aus unkonventionellen Ressourcen wie Kohle, Biomasse oder CO_2 vorangetrieben. Als Beispiele kann man die Produktion von Kohlenwasserstoffen aus Synthesegas – einer Mischung aus H_2, CO und CO_2 – über die Fischer-Tropsch-Synthese (FT) oder über eine komplexere Verfahrenskette mit Methanol als Zwischenprodukt anführen; oder die Produktion sauerstoffhaltiger Kohlenwasserstoffe (Oxygenate) wie Dimethylether (DME) aus Methanol oder direkt aus Synthesegas.

Bei vielen dieser Synthesereaktionen fällt H_2O als Hauptnebenprodukt an, da überschüssiger Sauerstoff in Form von H_2O (oder CO_2) entfernt wird. In konventionellen Prozessketten wird H_2O nach dem Reaktionsschritt abgetrennt. Die vorliegende Arbeit untersucht das Potential der H_2O-Abtrennung während des Reaktionsschritts, da die *in-situ* Entfernung eine Reihe von Vorteilen vorweisen kann:

- *Erhöhte Reaktionsgeschwindigkeiten.* H_2O hemmt die Reaktionsgeschwindigkeiten verschiedener Synthesereaktionen aufgrund konkurrierender Adsorption an der Katalysatoroberfläche wie im Falle der FT-Synthese an Eisenkatalysatoren oder der Dimethyl- und Diethylethersynthese an γ-Al_2O_3 (Bercic & Levic 1992, Butt, Bliss & Walker 1962). Selektive H_2O-Entfernung während des Reaktionsschritts reduziert die kinetische Hemmung, führt zur höheren Konzentrationen der Reaktanden und damit zu höheren Reationsgeschwindigkeiten.
- *Verbesserte Katalysatorstandzeit.* Verschiedene Katalysatorsysteme deaktivieren reversibel oder irreversibel in Gegenwart hoher Wasserdampfpartialdrücke. Verschiedene Untersuchungen zur FT-Synthese an Kobaltkatalysatoren zeigen, dass Deaktivierung bei hohen H_2O/H_2 Partialdruckverhältnissen (> 0.6) und hohen Wasserdampfpartialdrücken (> 0.6 MPa) eintritt (e.g. Schanke et al. 1995, 1996, Dry 1990, van Berge et al. 2000, Bartholomew & Farrauto 2005, Saib 2006). Bei Eisenkatalysatoren führt die Anwesenheit von Wasserdampf zur Reoxidation des Katalysators (Dry 1990). Als weitere Beispiele sind die Dimethylethersynthese aus Methanol (Kim et al. 2006), die Synthese von Ottokraftstoff aus Methanol (methanol-to-gasoline, MTG, Aguayo et al. 2001) oder die Produktion von Dimethylcarbonat (DMC) durch oxidative Karbonylierung (Di Muzio 1993) anzuführen. Durch selektive H_2O-

Entfernung ließe sich die Katalystorlebensdauer verlängern oder der Prozess bei höheren Umsatzgraden betreiben.

- *Umsätze oberhalb der Gleichgewichtslimitierung.* Aufgrund des chemischen Gleichgewichtes sind bei bestimmten Reaktionen die erreichbaren Umsatzgrade begrenzt. Beispiele sind die Methanol- oder FT-Synthese ausgehend von CO_2-haltigen Synthesegasen, die Dimethylethersynthese oder die CO_2/CO-Konvertierungsreaktion. Die selektive H_2O-Entfernung verschiebt die Gleichgewichtszusammensetzung auf die Seite des Zielproduktes, wodurch höhere Umsatzgrade und Produktausbeuten erzielt werden können.

Kondensation mit nachgeschalteter Phasentrennung, Adsorption, Dampfpermeation oder die Kombination mit einer Reaktion mit H_2O als Reaktand stellen einige Möglichkeiten dar, H_2O zu entfernen. Die Zielsetzung der vorliegenden Arbeit war die Untersuchung des technischen Potentials der *in-situ* H_2O-Entfernung mit Bezug zu kraftstoffrelevanten Reaktionen, wobei der Schwerpunkt auf die H_2O-Entfernung durch hydrophile Membranen und durch chemische Reaktion gerichtet war. Drei Beispielreaktionen wurden dazu ausgewählt:

- Membranunterstützte CO_2-Hydrierung zu FT-Kohlenwasserstoffen
- Membranunterstützte Umsetzung von Methanol und Ethanol zu Dimethylether (DME) und Diethylether (DEE).
- H_2O-Entfernung durch chemische Reaktion während der FT-Synthese an einem Kobaltkatalysator

Die Schlüsselfragen waren, ob *in-situ* H_2O-Entfernung während der FT-Synthese an Eisen- und Kobaltkatalysatoren eine realisierbare technische Option darstellt, um die Umsatz- und Ausbeutegrade oder die Katalysatorstandzeit in einem FT-Reaktor signifikant zu erhöhen, und ob dieses Konzept auf andere kraftstoffrelevante Synthesereaktionen angewendet werden kann.

Ein Teil dieser Arbeit wurde im Rahmen eines europäischen Projektes durchgeführt, welches die Entwicklung einer neuen Verfahrenskette zur Erzeugung von Kohlenwasserstoffen aus Erdgas (Low Temperature Catalytic Partial Oxidation and Gas-to-Liquids Conversion, LTCPO-GTL 2005) zum Ziel hatte. Daher nimmt die Untersuchung zur Umsetzung CO_2-haltiger Synthesegase zu FT-Kohlenwasserstoffen in einem Membranreaktor mit einer neuartigen hydrophilen Membran den grössten Teil dieser Arbeit ein.

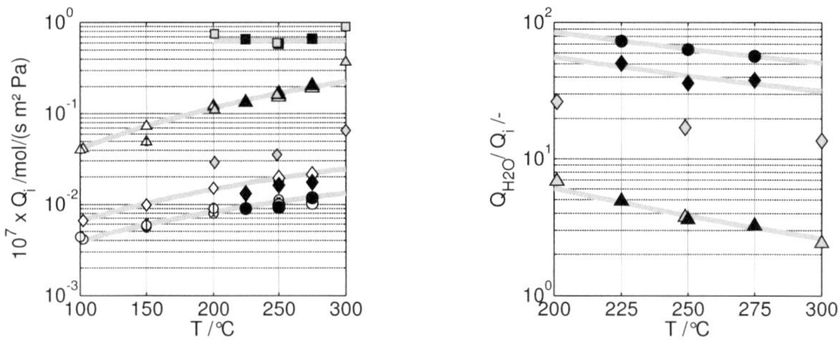

Abbildung 9.1. Vergleich der Permeanzen (links) und Permselektivitäten (rechts) der CSP2-Membran, gemessen in (△◇○) Einzelgasexperimenten und Permeationsexperimenten mit trockenen Gasgemischen, (■▲◆●) Permeationsexperimenten mit Gasgemischen mit 25 vol% H_2O, und (■▲◆●) Permeationsexperimenten mit Gasgemischen mit 30 vol% H_2O (Vente 2006). Q_i: (■ ■) H_2O, (▲ ▲ △) H_2, (◆ ◆ ◇) CO_2, (● ● ○) CO; Q_{H2O}/Q_i: (▲ ▲) H_2, (◆ ◆) CO_2, (● ●) CO

* trockenes Gasgemisch: $H_2/CO/CO_2$ (50/30/20 vol%), 25 vol% H_2O, P_F = 1 MPa, Ψ = 2, Φ = 0.15, $(Ar)_S$ * ECN: CSP2/ trockenes Gasgemisch: $H_2/CO/CO_2/CH_4$ (67/0/29/4 vol%), 30 vol% H_2O, ΔP_{tmb} = 0.9 MPa

Stofftransport in hydrophilen Membranen

Kraftstoffrelevante Synthesereaktionen wie die FT- oder DME-Synthese laufen bei Temperaturen oberhalb von 200°C ab. Daher müssen Membranen nicht nur eine hohe H_2O-Durchlässigkeit und eine hohe Permselektivität, sondern auch eine hohe Stabilität der Membraneigenschaften unter den Prozessbedingungen aufweisen. Das Zentrum für Energieeffizienz in der Industrie in den Niederlanden (ECN) hat zu diesem Zweck eine neuartige Polymermembran bereitgestellt. Bei dieser sogenannten CSP2-Membran (ceramic supported polymer membrane) wurde ein poröses, keramisches Trägerrohr mit einer 1 µm dünnen Funktionsschicht des hydrophilen Polymers Polyimid P84® beschichtet.

Die Trenneigenschaften der Membran wurden in nicht-reaktiven Permeationsexperimenten bestimmt, wobei sowohl Einzelgase als auch Gasmischungen ($H_2/CO/CO_2$) mit und ohne H_2O-, Methanol- und Ethanol-Dämpfen verwendet wurden. Die folgenden Schluss-folgerungen konnten aus den experimentellen Ergebnissen gezogen (Abbildung 9.1):

- H_2O ist mit einer Permanz von $7 \cdot 10^{-8}$ mol/(s m^2 Pa) die am schnellsten permeirende Komponente. Die hydrophile Polymerschicht bevorzugt daher den Transport der polaren H_2O-Moleküle gegenüber H_2, CO_2 und CO.
- Die Zugabe von H_2O (oder Methanol oder Ethanol) verändert den Transport der Permanentgase durch die Membrane nicht. Höhere

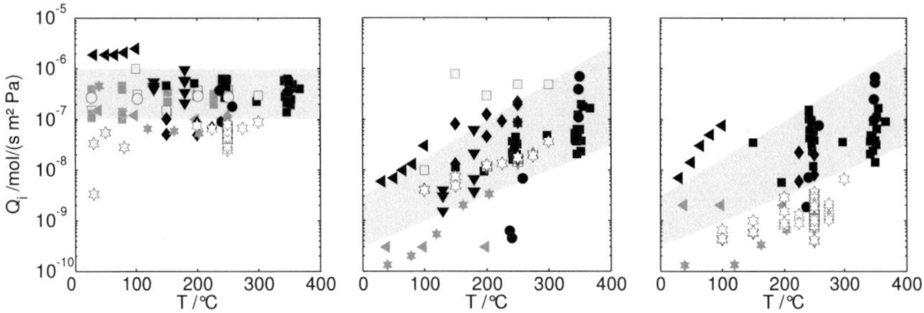

Abbildung 9.2. Vergleich der CSP2-Membran (☆) mit Literaturdaten unterschiedlicher Membran-typen: H_2O-Permeanzen (links), H_2-Permeanzen (mitte) und CO_2- and CO-Permeanzen (rechts): Mordenit ●, MFI-Typ ■, Faujasit ▼, Zeolith-4A ◀, amorphes Silika (TEOS) ◆, Polymermembran ★. Grau hinterlegte Bereiche: Trend.

Wasserdampfanteile im Einsatzstrom führen dagegen zu einer verbesserten H_2O-Permeanz. Aufgrund der schwachen Wechselwirkungen der einzelnen Permeanden untereinander kann der Stofftransport durch die Membran bei einer gegebenen Temperatur mit Hilfe eines Satzes konstanter Permeanzen beschrieben werden.

- Aufgrund unterschiedlicher Aktivierungsenergien bleibt die H_2O-Permeanz nahezu konstant, während die Permeanzen von H_2, CO_2, CH_4, CO und Ar mit der Temperatur ansteigen. Daher verschlechtern sich die selektiven Trenneigenschaften bezüglich H_2O mit zunehmender Temperatur.

Der Vergleich der Transporteigenschaften verschiedener hydrophiler Membranen – zusammengefasst in einer umfassenden Literaturübersicht - mit denen der CSP2 Membran ergab die folgende Einordnung (Abbildung 9.2):

- Die H_2O-Permeanz der CSP2-Membrane überschreitet nicht $1 \cdot 10^{-7}$ mol/(s m^2 Pa) und liegt damit im Bereich von mikroporösen Silikamembranen oder polymeren Nafion®-Membranen. Bei kristallinen Zeolithmembranen wurden bis zu 10-fach höhere H_2O-Permeanzen gemessen.
- Die CSP2-Membran weist höhere H_2O/H_2-Permselektivitäten auf als amorphe mikroporöse Membrane, aber niedrigere als kristalline Zeolithmembrane. Bei diesen zeichnen sich insbesondere Membranen des Typs MFI/ZSM-5 aus. Andererseits hält die CSP2 Membran CO_2 und CO (und CH_4 und Ar) am besten zurück.

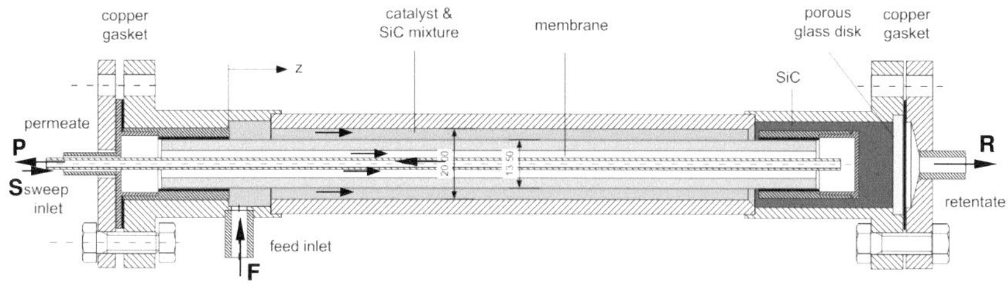

Abbildung 9.3. Laborfestbettreaktor für die Experimente zur *in-situ* H_2O-Entfernung mittels einer hydrophilen Membrane. F: Einsatz, S: Spülgas, R: Retentat, P: Permeat.

In-situ H_2O-Entfernung während kraftstoffrelevanter Synthesereaktionen

Membran unterstützte CO_2-Hydrierung zu FT-Kohlenwasserstoffen

Kaliumpromotierte Eisenkatalysatoren sind sowohl für die FT-Synthese als auch für die CO_2/CO-Konvertierungsreaktion aktiv. Daher können diese Katalysatoren CO_2 oder CO_2-haltige Synthesegase bei höheren Temperaturen (300-360°C) zu Kohlenwasserstoffen umsetzen. CO_2 reagiert mit H_2 in der CO_2/CO-Konvertierungsreaktion zu CO (Gl. 9.1), welches dann im nächsten Schritt in der FT-Reaktion zu Kohlenwasserstoffen umgesetzt wird (Gl. 9.2) (Weatherbee & Bartholomew 1984, Lee, Lee & Chang 1990, Riedel et al. 2003).

$$CO_2 + H_2 \leftrightarrows CO + H_2O \qquad \Delta H_R{}^0{}_{250°C} = +39.5 \text{ kJ/mol } CO_2 \qquad (9.1)$$

$$CO + 2\,H_2 \rightarrow (1/n)\,(C_nH_{2n}) + H_2O \qquad \Delta H_R{}^0{}_{250°C} = -158 \text{ kJ/mol CO} \qquad (9.2)$$

Durch die Anwendung hydrophiler Membranen und der kontinuierlichen, selektiven Entfernung von H_2O aus der Reaktionszone kann der Umsatzgrad erhöht und die erforderliche Reaktionstemperatur abgesenkt werden (Unruh, Rohde & Schaub 2002), da die Gleichgewichtszusammensetzung der CO_2/CO-Konvertierungsreaktion zugunsten zu CO verschoben wird. Dies wurde experimentell von Unruh in einem Laborreaktor bei 225-250°C demonstriert (Rohde, Unruh & Schaub 2005). Die membranunterstüzte CO_2-Hydrierung wurde als Option zur verbesserten Umsetzung CO_2-haltiger Synthesegase im LTCPO-GTL Projekt (LTCPO-GTL 2005) näher untersucht.

Ein kommerzieller FT-Eisenkatalysator (Fe-GTLX) wurde mit einer CSP2 Membran in einem isothermen Laborfestbettreaktor kombiniert (Abbildung 9.3). Der mit Inertmaterial verdünnte Katalysator wurde in den Ringspalt zwischen Membran und Reaktorwand gefüllt. Als Einsatzgas wurde ein stöchiometrisches Synthesegas mit einem H_2/CO_2-Verhältnis von drei gewählt. Die Membran wurde im Gleichstrom gespült, wobei in einer Reihe von

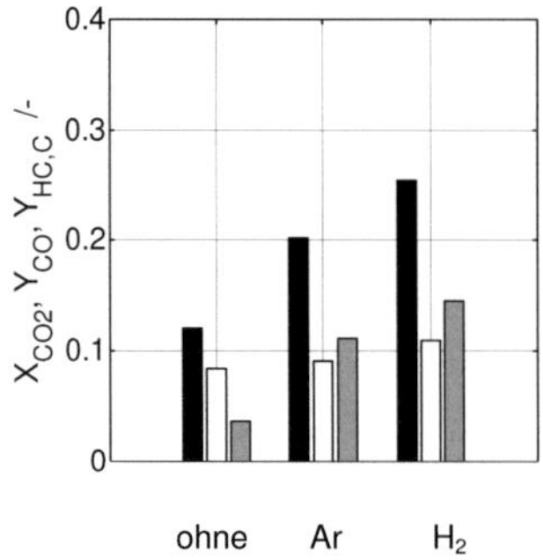

Abbildung 9.4. CO_2-Umsatz (■), CO-Ausbeute (□) und Kohlenwasserstoffausbeute (■) in einem Reaktor ohne Membran und in einem Membranreaktor mit Argon und H_2 als Spülgas. * Fe-GTLX Katalysator/ CSP2-Membran: $(H_2/CO_2)_F = 3$, $\tau_{mod,n} = 2000$ kg s/m^3, $m_{cat} = 4 \cdot 10^{-3}$ kg, T = 270°C, P_F = 1 MPa, Spülgasverhältnis Ψ = 3.3, Druckverhältnis Φ = 0.7

Experimenten die Spülgaszusammensetzung, Spülgasstrom und -druck variiert wurden.

Die Ergebnisse der Experimente zeigen, dass im Membranreaktor signifikant höhere CO_2-Umsätze und Kohlenwasserstoffausbeuten gemessen werden als im Referenzfall ohne integrierter Membran (Abbildung 9.4). Da die CSP2-Membrane höhere Permselektivitäten bezüglich H_2O aufweist als die mikroporösen amorphen Silikamembranen von Unruh (2006), kann eine positive Umsatz- und Ausbeutesteigerung schon mit dem inerten Spülgas Argon festgestellt werden, obwohl mehr als 80% des H_2 über die Membran verloren gingen. Um die zu niedrige H_2O/H_2-Permselektivität zu kompensieren und die H_2-Verluste zu minimieren, wurde in nachfolgenden Experimenten H_2 als Spülgas bei einem reduziertem Spülgasdruck verwendet. Dies führte zu einer weiteren Umsatzsteigerung.

Die Menge an H_2O, die aus der Reaktionszone zur Permeatseite entfernt wird, nimmt mit steigendem Spülgasverhältnis Ψ zu. Als Folge der niedrigeren Wasserpartialdrücke in der Reaktionszone wurden höhere Reaktionsgeschwindig-keiten und Umsatzgrade im Membranreaktor beobachtet (Abbildung 9.5). Bis zu 70% des von der CO_2/CO-Konvertierungs- und FT-Reaktion produzierten H_2O wurden über die Membran entfernt, aber die Flussdichte war mit 0.045 kg H_2O/ (h m^2) sehr niedrig. Der H_2-Verlust konnte durch die Verwendung von H_2 als Spülgas bei einem Druck von 0.7 MPa auf 8-10% limitiert werden. Aufgrund ihrer niedrigeren Permeanzen und Partialdrücke wurden nur 12% des eingesetzten CO_2 und maximal 8% des CO zur Permeatseite verloren.

Die Analyse der experimentellen Ergebnisse erfolgte mit einem eindimensionalen Membranreaktormodell, dessen Parameter in unabhängigen Permeations- und Kinetikexperimenten bestimmt wurden. Die mit dem Reaktormodell berechneten Verläufe korrelieren gut mit den experimentellen Daten aus dem Membranreaktor (Abbildung 9.5). Die modellgestützten

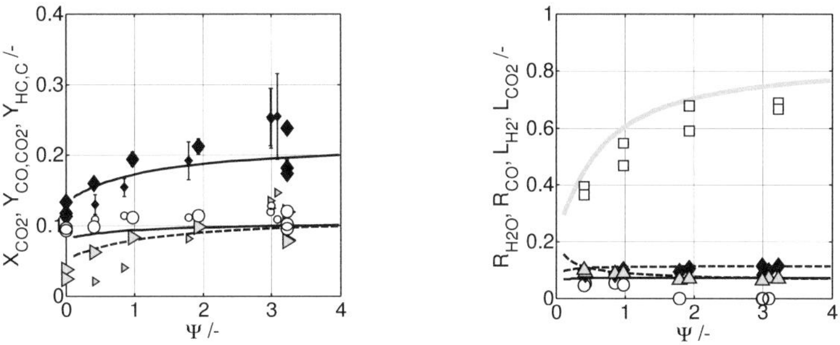

Abbildung 9.5. Effekt des Spülgasverhältnisses Ψ auf (links) den CO_2-Umsatz (◆), CO-Ausbeute (○) und Kohlenwasserstoffausbeute (▶) und auf (rechts) den Grad der H_2O-Entfernung (□—), der CO-Entfernung (○—), des CO_2- (◆--) und H_2-Verlustes (▲--). Kurven: berechnet mit Membranreaktormodell.
* Fe-GTLX/ CSP2: T= 270 °C, P_F = 1 MPa, $\tau_{mod,n}$ = 2000 kg s/m^3, $(H_2/CO_2)_F$ = 3, Φ = 0.7, $(H_2)_S$

Rechnungen zeigten, dass die H_2O-Permeanz der CSP2-Membran einer der limitierenden Faktoren für höhere Umsatzgrade war.

Während der Dauer der Experimente verlor die Membran ihre selektiven Transporteigenschaften (Abbildung 9.6), obwohl diese unter nicht-reaktiven Bedingungen stabil waren. Es wird vermutet, dass die Anwesenheit langkettiger Kohlenwasserstoffe unter reaktiven Bedingungen zu einem beschleunigten Abbau der Membran geführt hat.

Membranunterstützte DME/DEE-Synthese

Dimethylether (DME) hat als Kraftstoff in den letzten Jahren ein wachsendes Interesse erfahren. Alkohole wie Methanol oder Ethanol reagieren an sauren Katalysatoren in einer schwach exothermen Gleichgewichtsreaktion zu den entsprechenden Ethern (8.3). Der Gleichgewichtsumsatz von Methanol zu DME beträgt zirca 85% bei 300 °C. Die Reaktionsgeschwindigkeit wird durch

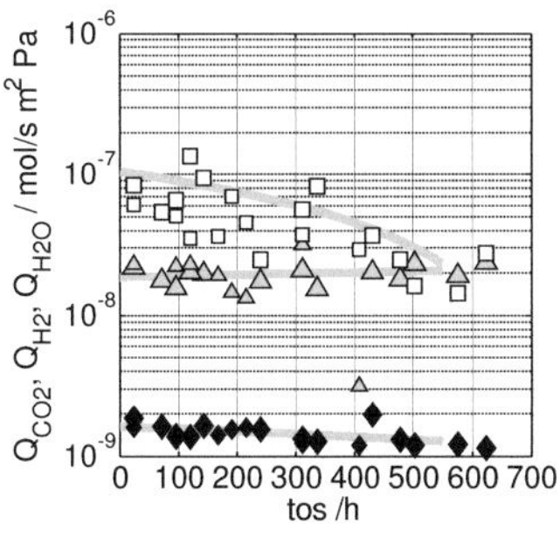

Abbildung 9.6. Abbau der selektiven Transporteigenschaften der CSP2-Membran während der CO_2-Hydrierung zu FT-Kohlenwasserstoffen: Änderung der Permeanzen Q_i von H_2O (□), H_2 (▲) und CO_2 (◆). Kurven: Trends.
* Fe-GTLX/ CSP2: T= 270 °C, P_F = 1 MPa, $\tau_{mod,n}$ = 2000-4000 kg s/m^3, $(H_2/CO_2)_F$ = 3, Ψ = 0.4-3.3, Φ = 0.15-1, $(H_2, Ar)_S$

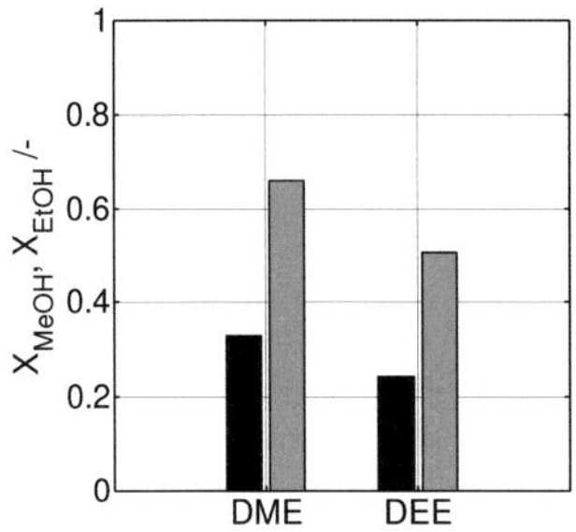

Abbildung 9.7. Methanolumsatz zu DME und Ethanolumsatz zu DEE in einem Festbettreaktor ohne (■) und mit (■) integrierter hydrophiler CSP2-Membran.
* γ-Al_2O_3/ CSP2, T = 290°C, $\Psi = 3.3$, $\Phi = 0.15$, $(Ar)_S$
* DME: $P_{MeOH,F} = 0.67$ MPa, $P_{Ar,F} = 0.33$ MPa,
$m_{cat} = 8 \cdot 10^{-3}$ kg, $\tau_{mod,n,MeOH} = 5200$ kg s/ m^3
* DEE: $P_{EtOH,F} = 0.7$ MPa, $P_{Ar,F} = 0.3$ MPa,
$m_{cat} = 8 \cdot 10^{-3}$ kg, $\tau_{mod,n,EtOH} = 7700$ kg s/ m^3

H_2O stark gehemmt (Bercic & Levic 1992). Desweiteren berichten Kim et al. (2006), dass saure Katalysatoren bei hohen Wasserdampfpartialdrücken deaktivieren.

$$2\ MeOH \leftrightarrows Me\text{-}O\text{-}Me + H_2O \qquad \Delta H_R{}^0{}_{250°C} = -23\ kJ/mol\ MeOH \qquad (9.3)$$

Die Experimente zeigten, dass die Umsatzgrade im Membranreaktor durch die reduzierte H_2O-Hemmung im Vergleich zum Reaktor ohne CSP2-Membran gesteigert werden konnten (Abbildung 9.7). Es wurde abgeschätzt, dass H_2O ungefähr 10-17 Mal schneller als Methanol und 70-175 schneller als DME durch CSP Membrane permeiert. Damit waren die beobachteten H_2O/Edukt- und H_2O/Produkt-Permselektivitäten höher als die im Fall der CO_2-Hydrierung und es war daher möglich, den Spülgasdruck auf ein Zehntel des Drucks in der Reaktionszone abzusenken. Der Verlust von Methanol zur Permeatseite betrug ca. 30%.

Die Anwesenheit von Oxygenaten bei hohen Temperaturen (290°C) führte aber zu einem schnellen Anstieg der Membrandurchlässigkeit und damit zu einem schnellen Abbau der selektiven Transporteigenschaften.

H_2O-Entfernung durch chemische Reaktion während der FT-Synthese an einem Kobaltkatalysator

Eine Variante, die möglicherweise robuster ist als die Anwendung hydrophiler Membranen, ist die *in-situ* H_2O-Entfernung durch eine chemische Reaktion. Betrachtet man die Produktion von Kraftstoffen oder Chemikalien aus Synthesegas, dann ist eine Kombination mit einem CO/CO_2-Konvertierungskatalysator naheliegend (Post & Sie 1985, Chanenchuk, Yates & Satterfield 1991, Ogawa et al. 2003, de Mestier du Bourg 2006, Renk et al. 2006). Tieftemperatur-Konvertierungskatalysatoren können in einem Temperaturfenster von 210 bis 270°C betrieben werden (Higman & Supp 2006), welches mit dem der FT-Synthese an Kobaltkatalysatoren überlappt. Nach heutigem Stand der Literatur deaktivieren Kobaltkatalysatoren bei hohen

H_2O-Partialdrücken und H_2O/H_2-Partialdruckverhältnissen, weshalb die *in-situ* H_2O-Entfernung von Interesse ist.

Es konnte experimentell gezeigt werden, dass die Kombination eines Tieftemperatur-Konvertierungskatalysators (CuZnO) mit einem Kobaltkatalysator in einem bifunktionellen Katalysatorsystem mit einer Konvertierungsaktivität resultiert, die höher ist als bei promotierten FT-Eisenkatalysatoren. Die CO-Konvertierungsreaktion reduzierte den H_2O-Partialdruck auf ein sehr niedriges Niveau. Mehr als 90% des produzierten H_2O wurde zu CO_2 und H_2 umgesetzt. Damit liegt der Wert für den Grad der H_2O-Entfernung höher als in den Membranreaktorexperimenten. Zudem konnte der Konvertierungskatalysator unter FT-Bedingungen bei niedrigen H_2O-Partialdrücken und in Gegenwart von langkettigen Kohlenwasserstoffen ohne signifikanten Verlust der Aktivität betrieben werden. Die Fähigkeit zugleich den H_2O-Partialdruck abzusenken und H_2 *in-situ* zu produzieren macht diese Katalysatorkombination zu einen vielversprechenden Kandidaten für die Umsetzung von H_2-armen Synthesegasen.

Fallstudien: Potenzial und Grenzen der *in-situ* H_2O-Entfernung

Das Potenzial von *in-situ* H_2O-Entfernung bei gleichgewichtslimitierten oder H_2O-gehemmten Reaktionen ist aus Abbildung 9.8 ersichtlich. Basierend auf den Ergebnissen modellgestützter Rechnungen und Fallstudien wurden die folgenden, allgemeinen Schlussfolgerungen gezogen:

- Beim Einsatz von hydrophilen Membranen zur H_2O-Entfernung zeigt sich, dass der positive Einfluss der H_2O-Permselektivität auf den Umsatz bei Werten oberhalb von 40 nur noch klein ist. Die erforderliche Permselektivitität einer Membran wird daher eher durch den tolerierbaren Verlust von Reaktanden oder Produkten bestimmt, und im Falle wertvoller Reaktanden oder Zwischenprodukte sind Permselektivitäten > 80 erforderlich.
- Sollen Membranen zur *in-situ* H_2O-Entfernung eingesetzt werden, dann sollte die H_2O-Permeanz einen Wert von 10^{-6} mol/(s m² Pa) erreichen oder überschreiten, um die spezifischen Membranflächen in Grenzen zu halten. Dieser Wert kann mit heutigen Zeolithmembranen erreicht werden. Im Falle von Anwendungen, in denen der H_2O-Partialdruck auf einen sehr niedrigen Zielwert abgesenkt werden soll, können die erforderlichen Membranflächen nur durch Hohlfasermodule erreicht werden. Dies limitiert die Membrananwendungen auf Reaktionen mit einer niedrigen Produktivität oder mit einer geringen Wärmetönung.

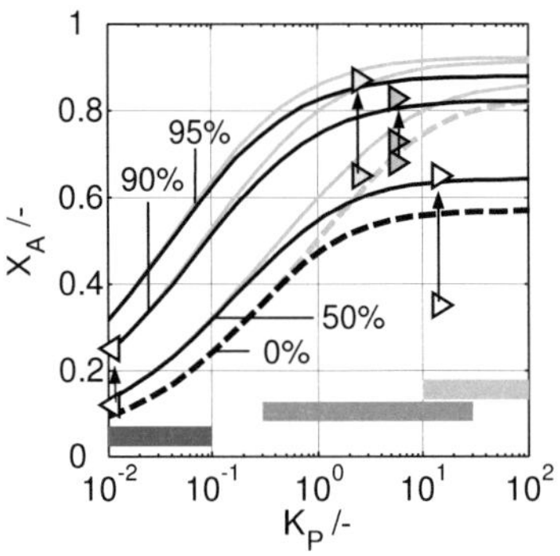

Abbildung 9.8. Effekt von H_2O-Entfernung auf den Umsatz X_A für eine einfache Gleichgewichtsreaktion mit (schwarze Kurven) und ohne (graue Kurven) H_2O-Hemmung als Funktion der Gleichgewichts-konstante K_P.

Kurven berechnet mit Da-Pe-Modell (6.5)-(6.7) mit Da = 10 und verschiedenen Pe_m: 1000 (R_{H2O} = 0), 1 (R_{H2O} ≈ 50%), 0.2 (R_{H2O} ≈ 90%), 0.1 (R_{H2O} ≈ 95%).

Beispiele für *in-situ* H_2O-Entfernung: CO_2-Hydrierung (◁), DME Synthese (▷), Ethylacetatsynthese (▶, de la Iglesia et al. 2007), Butylacetatsynthese (▶, Peters et al. 2005).

Die *in-situ* H_2O-Entfernung in der FT-Synthese wurde anhand von verschiedenen rechnerischen Fallstudien untersucht, wobei ein Modell für einen industriellen Dreiphasenreaktor verwendet wurde:

- Membranunterstützte FT-Synthese – entweder zur verbesserten Umsetzung CO_2-haltiger Synthesegase oder zur Verlängerung der Standzeit von Kobaltkatalysatoren – kann aufgrund unrealistischer Membrananforderungen, hoher Reaktorkomplexität, des hohen Risikos für Membranversagen und zu hoher Kosten als Option zur Prozessintensivierung verworfen werden. Die erforderlichen Permeanzen und Standzeiten können von heutigen Membranen nicht erreicht werden.
- Als eine Option mit hohem Potential zur Kontrolle der Wasserdampfpartialdrücke in einem FT-Reaktor für H_2-arme Synthesegase wurde dagegen die Anwendung eines bifunktionellen, CO/CO_2-konvertierungsaktiven Kobaltkatalysators identifiziert.

Schlussfolgerungen und Ausblick

Die vorliegende Arbeit demonstrierte, dass *in-situ* H_2O-Entfernung durch Dampfpermeation während der CO_2-Hydrierung (270℃) und der DME/DEE-Synthese (250-290℃) zu erhöhten Umsatz- und Ausbeutegraden führt; die Steigerung ist direkt an den Grad der H_2O-Entfernung geknüpft. Die experimentellen Daten stimmen gut mit den Voraussagen des angewendeten Reaktormodells überein und bestätigen ein Potential für *in-situ* H_2O Entfernung.

Die neue keramikunterstütze CSP2-Polymermembran stellt einen Fortschritt in der Membranentwicklung in Bezug auf Permselektivitäten dar, wenn man sie mit den amorphen mikrokristallinen Silikamembranen vergleicht, die von Unruh

(2006) unter FT-Bedingungen getestet hat. Dennoch ist diese Membran nicht geeignet für die vorgeschlagenen Membranreaktoranwendungen aufgrund ihrer relativ niedrigen H_2O-Permeanz und des schnellen Abbaus der selektiven Transporteigenschaften in Gegenwart von Kohlenwasserstoffen und Oxygenaten.

Überträgt man das Konzept der *in-situ* H_2O-Entfernung durch Dampfpermeation auf den industriellen Maßstab, lässt sich erkennen, dass Membranreaktoren zur H_2O-Entfernung keine Anwendung in der FT-Synthese oder in anderen auf Synthesegas basierten Reaktionen finden werden. Die vom Prozess gestellten Anforderungen und die Eigenschaften heutiger Membranen liegen zu weit auseinander; das Risiko eines Membranversagens bei hohen Installationskosten ist zu gross. Die Anwendung eines Gaskreislaufes mit Produktkühler und Wasserabtrennung stellt eine zuverlässige und kostengünstige und damit bevorzugte Option dar, um Wasserpartialdrücke im Reaktor zu regeln.

Andererseits hat die *in-situ* H_2O-Entfernung durch CO/CO_2-Konvertierungsreaktion während der FT-Synthese an einem Kobaltkatalysator vielversprechende Ergebnisse geliefert. Diese Option könnte zu einem vereinfachten Prozessaufbau für auf Kohle oder Biomasse basierte FT-Anlagen führen und sollte weiter verfolgt werden.

10 References

Agar, D.W., Ruppel, W. 1988 'Multifunktionale Reaktoren für die heterogene Katalyse', Chem. Ing. Tech., vol. 60, pp. 731-741.

Agar, D.W. 2005 'The dos and don'ts of adsorptive reactors', In: Sundmacher, K., Kienle, A., Seidel-Morgenstern, A. (eds.) 'Integrated chemical processes – Synthesis, operation, analysis, and control'. Weinheim, Wiley-VCH, pp. 203-232.

Aguayo A., Gayubo A., Castilla M., Olazar M., Bilbao J., 2001, 'MTG process in a fixed-bed reactor: operation and simulation of a pseudoadiabatic experimental unit', Ind. Eng. Chem. Res., vol. 40, no. 26, pp. 6087-6098.

Aoki, K., Kusakabe, K., Morooka, S. 1998 'Gas permeation properties of A-type zeolite membrane formed on porous substrate by hydrothermal synthesis', J. Mem. Sci., vol. 141, pp. 197-205.

Aoki, K., Kusakabe, K., Morooka, S. 2000 'Separation of gases with an A-type zeolite membrane', Ind. Eng. Chem. Res., vol. 39, pp. 2245-2251.

Asfar, K. Al-Rabadi, T. 2003 'Fuel blends in compression ignition engines', Proc. 1st Int. Energy Conversion Eng. Conference, Portsmouth, Virginia.

Baerns, Hofmann & Renken 1987 'Chemische Reaktionstechnik – Lehrbuch der Technischen Chemie Band 1'. Stuttgart, Georg Thieme Verlag.

Baerns, M., Hofmann, H., Renken, A. 1992 'Chemische Reaktionstechnik – Lehrbuch der Technischen Chemie Band 1'. Stuttgart, Georg Thieme Verlag.

Bailey, B., Eberhard, J., Erwin, J. 2006 'Diethyl ether (DEE) as a renewable diesel fuel', http://www.tpub.com/content/altfuels10/dee/dee0001.htm, 01.02.2008.

Baker, R.W. 2002 'Future directions of membrane gas separation technology', Ind. Eng. Chem. Res., vol. 41, pp. 1393-1411.

Baker, R.W. 2004 'Membrane technology and application', Weinheim, Wiley-VCH.

Barbieri, G., Violante, G., di Maio, F.P., Criscuoli, A., Drioli, E. 1997, 'Methane steam reforming analysis in a palladium-based catalytic membrane reactor', Ind. Eng. Chem. Res., vol. 36, pp. 3369-3374.

Barbieri, G., Marigliano, G., Goleme, G., Drioli, E. 2002 'Simulation of CO_2 hydrogenation with CH_3OH removal in a zeolite membrane reactor', Chem. Eng. J., vol. 85, no. 1, pp. 53-59.

Barrer, R.M. 1990 'Porous crystal membranes', J. Chem. Soc. Faraday Trans., vol. 86, no. 7, pp. 1123.

Barrie, J.A. 1968 'Water in polymers', In: Crank, J., Park, G.S. (eds.) 'Diffusion in polymers'. Oxford, Academic Press.

Bartholomew, C.H., Farrauto, R.J. 2005 'Fundamentals of industrial catalytic processes', New York, John Wiley & Sons.

Battersby, S., Teixeira, P.W., Beltramini, J., Duke, M.C., Rudolph, V., Diniz da Costa, J.C. 2006 'An analysis of the Péclet and Damköhler numbers for dehydrogenation reactions using molecular sieve silica (MSS) membrane reactors', Cat. Today, vol. 116, pp. 12-17.

Battista, R.A., Lo, F.S., Tatterson, R.L. 1999 'Process of isolation and removal of unwanted water from chemical reaction', US patent 5,917,078.

Beer, G.L. 1999 'Extended catalyst life two stage hydrocarbon synthesis process', US patent 6,169,120.

Bercic, G., Levic, J. 1992 'Intrinsic and global reaction rate of methanol dehydration over γ-Al_2O_3 pellets', Ind. Eng. Chem. Res., vol. 31, pp. 1035-1040.

Bernal, M.P., Piera, E., Coronas, J., Menéndez, M., Santamaría, J. 2000 'Mordenite and ZSM-5 hydrophilic tubular membranes for the separation of gas phase mixtures', Catal. Today, vol. 56, pp. 221-227.

Bernstein, L.A., Lund, C.R.F. 1993 'Membrane reactors for catalytic series and series-parallel ractions', J. Membr. Sci., vol. 77, pp. 155-164.

Bertole, C.J., Mims, C.A., Kiss, G. 2002 'The effect of water on the cobalt-catalyzed Fischer-Tropsch synthesis', J. Catal., vol. 210, pp. 84-96.

Bhanage, B.M., Fujita, S.I., Ikushima, Y., Arai, M. 2001 'Synthesis of dimethyl carbonate and glycols from carbon dioxide, epoxides, and methanol using heterogeneous basic metal oxide catalysts with high activity and selectivity', Appl. Catal. A, vol. 219, no. 1-2, pp. 259-266.

Bloch, H.P. 2006 'A practical guide to compressor technology'. Hoboken, Wiley-Interscience.

Bonekamp, B.C. 1996 'Preparation of asymmetric ceramic membrane supports by dip-coating', in: A.J. Burggraaf, L. Cot (Eds.) 'Fundamentals of Inorganic Membrane Science and Technology'. Elsevier, Amsterdam.

Bonekamp, B.C., van Horssen, A., Correia, L.A., Vente, J.F., Haije, W.G. 2006 'Macroporous support coatings for molecular separation membranes having a minimum defect density', J. Membr. Sci., vol. 278, pp. 349-356.

Borg, O., Storsaeter, S., Eiri, S., Wigum, H., Rytter, E., Holmen, A. 2006 'The effect of water on the activity and selectivity for γ-alumina supported cobalt Fischer-Tropsch catalysts with different pore sizes', Catal. Lett., vol. 107, no. 1-2, pp. 95-102.

Borodko, Y., Somorjai, G.A. 1999 'Catalytic hydrogenation of carbon oxides – a 10 year perspective', Appl. Catal. A, vol. 186, pp. 355-362.

Botes, F. G., Böhringer, W. 2004 'The addition of HZSM-5 to Fischer-Tropsch process for improved gasoline production', Appl. Cat. A, vol. 267, pp. 217-225.

Bowen, T.C., Noble, R.D., Falconer, J.L. 2004 'Fundamentals and applications of pervaporation through zeolite membranes', J. Membr. Sci., vol. 245, pp. 1-33.

Bracht, M., Bos, A., Pex, P.P.A.C., van Veen, H.M., Alderliesten, P.T. 1995, 'Water gas shift membrane reactor', Proc. Euromembrane 95, September 18-20, University of Bath.

Bradford, M.C.J., Te, M., Pollack, A. 2005 'Monolith loop catalytic membrane reactor for Fischer-Tropsch synthesis', Appl. Catal. A, vol. 283, pp. 39-46.

Butt, J.B., Bliss, H., Walker, C.A. 1962 'Rates of reaction in a recycling system – dehydration of ethanol and diethylether over alumina', AIChE J., vol. 8, no. 1, pp. 42-47.

Burggraaf, A., Cot, L. 1996 'Fundamentals of inorganic membrane science and technology'. Amsterdam, Elsevier Science.

Campbell, J.S 1970 'Influences of catalyst formulation and poisoning on the activity and die-off of low temperature shift catalysts', Ind. Eng. Chem. Process Des. Develop., vol. 9, no. 4, pp. 588-595.

Caro, J., Noack, M., Kölsch, P., Schäfer, R. 2000 'Zeolite membranes – state of their development and perspective', Microporous Mesoporous Mater., vol. 38, pp. 3-24.

Carvill, B.T., Hufton, R.J., Anand, M., Sircar, S. 1996 'Sorption-enhanced reaction process', AIChE J., vol. 42 , no. 10, pp. 259-268.

Centi, G., Perathoner, S. 2004 'Heterogenous catalytic reactions with CO_2: status and perspectives'. In: Park, S.E, Chang, J.S., Lee. K.W. (eds.) 'Carbon dioxide utilization for global sustainability'. Amsterdam, Elsevier, Stud. Surf. Sci. Catal., vol. 153, pp. 1-8.

Chanenchuk, C.A., Yates, I.C., Satterfield, C.N. 1991 'The Fischer-Tropsch synthesis with a mechanical mixture of a cobalt catalyst and a copper-based water gas shift catalyst', Energy Fuels , vol. 5, pp. 847-855.

Chen, J.G., Xiang, H.W., Gao, H.Y., Sun, Y.H. 2001 'Study on deactivation of $Co/ZrO_2/SiO_2$ catalyst for Fischer-Tropsch synthesis', React. Kinet. Catal. Lett, vol. 73, no. 1, pp. 169-177.

Choi, P.H., Jun. K.W., Lee, S.J., Choi, M.J., Lee, K.W. 1996 'Hydrogenation of carbon dioxide over alumina supported Fe-K catalysts', Catal.Lett., vol. 40., pp. 115-118.

Claeys, M. 1997 'Selektivität, Elementarschritte und kinetische Modellierung bei der Fischer-Tropsch Synthese'. Universität Karlsruhe, dissertation.

Claeys, M., van Steen, E. 2002 'On the effect of water during Fischer-Tropsch synthesis with a ruthenium catalyst', Catal. Today, vol. 71, pp. 419-427.

Clark, G.L., Walker, D.G. 2000 'Multiple reactor system and method for Fischer-Tropsch synthesis, US patent 6,156,809.

Coronas, J., Santamaría, J. 1999 'Separations using zeolite membranes', Sep. Purifi. Methods, vol. 28, no. 2, pp. 127-177.

Coronas, J., Santamaría, J. 2004 'State-of-the-art in zeolite membrane reactors', Top. Catal., vol. 29, no. 1–2, pp. 29-44.

Dalai, A.K., Das, T.K., Chaudhari, K.V., Jacobs, G., Davis, B.H. 2005 'Fischer–Tropsch synthesis: water effects on Co supported on narrow and wide-pore silica', Appl. Catal. A, vol. 289, pp. 135-142.

Das, T.K, Conner, W.A., Li, J., Jacobs, G., Dry, M.E., Davis, B.H. 2005 'Fischer-Tropsch synthesis: kinetics and effect of water for a Co/SiO_2 catalyst', Energy Fuels, vol. 19, pp. 1430-1439. OK

Davis, B.H. 2002 'Overview of reactors for liquid phase Fischer-Tropsch synthesis', Catal. Today, vol 71, pp. 249-300.

Davis, B.H. 2003 'Fischer-Tropsch synthesis: relationship between iron catalyst composition and process variables', Catal. Today, vol. 84, pp. 83-98.

de Deugd, R.M., Kapteijn, F., Moulijn, J.A. 2003 'Trends in Fischer-Tropsch reactor technology – opportunities for structured reactors', Top. Catal., vol. 26, no. 1-4, pp. 29-39.

de Graaf, W., Schrauwen, F. 2002 'World scale GTL', Hydrocarbon processing, vol. 5.

de la Iglesia, O., Mallada, R., Menendéz, M., Coronas, J. 2007 'Continuous zeolite membrane reactor for esterification of ethanol and acetic acid ', Chem. Eng. J., vol 131, pp. 35-39.

DENA 2006, 'Biomass-to-liquids – BTL Realisierungstudie', Deutsche Energie Agentur.

de Mestier du Bourg, 2006 'Future prospective of DME', Proc. 23rd World Gas Conference 2006,Amsterdam, http://www.methanol.org/pdfFrame.cfm?pdf=futureprospectiveof DME(dme-org).pdf , 01.02.2008.

Deshmukh, S.A.R.K. , Heinrich, S., Mörl, L., van Sint Annaland, M., Kuipers, J. A.M. 2007 'Membrane assisted fluidized bed reactors: Potentials and hurdles', Chem. Eng. Sci., vol. 62, no. 1-2, pp. 416-436.

Detallante, V., Langevin, D., Chappey, C., Métayer, M., Mercier, R., Pinéri, M. 2001 ' Water apour sorption in naphthalenic sulfonated polyimide membranes', J. Mem. Sci., vol. 190, pp. 227-241.

Di Muzio, N., Fusi, C., Rivetti, F., Sasselli, G. 1993 'Process for producing dimethyl carbonate', US patent 5,210,269.

DIN EN 14214 'Kraftstoffe für Kraftfahrzeuge – Fettsäure-Methylester (FAME) für Dieselmotoren – Anforderungen und Prüfverfahren', 2003-11.

Dittmeyer, R. 2001 'Membranreaktorkonzepte in der heterogenen Katalyse', Universität Erlangen-Nürnberg, Habilitationsschrift.

Dittmeyer, R., Svajda, K., Reif, M. 2004 ' A review of catalytic membrane layers for gas/liquid reactions', Top. Catal., vol. 29, no. 1–2, pp. 3-27.

Dry, M.E. 1981 'The Fischer-Tropsch synthesis'. In: Anderson, J.R., Boudart, M. 'Catalysis Science and Technology'. Berlin, Springer-Verlag.

Dry, M.E. 1983 'The Sasol Fischer-Tropsch processes', In: Leach, B.E. (ed.) 'Applied Industrial Catalysis', vol. 2, pp. 167-213.

Dry, M.E. 1990 'Fischer-Tropsch synthesis over iron catalysts', Catal. Letters, vol. 7, pp. 241-252.

Dry, M.E, Steynberg, A.P. 2004 'Commercial FT process application'. In: Steynberg, A.P., Drey, M.E. (eds.) 'Fischer-Tropsch Technology'. Amsterdam, Elsevier, Stud. Surf. Sci. Catal., vol. 152, pp. 406-481.

Duvenhage, D.J., Coville, N.J. 2006 'Deactivation of a precipitated iron Fischer-Tropsch catalyst – a pilot plant study', Appl. Catal. A, vol. 298, pp. 211-216.

Eilers, J., Posthuma, S.A., Sie, S.T. 1990 'The Shell Middle Distillate Synthesis process', Catal. Lett., vol. 7, pp. 253-270.

Elsner, M.P. 2004 'Experimentelle und modellbasierte Studien zur Bewertung des adsorptiven Reaktorkonzeptes am Beispiel der Claus-Reaktion'. Universität Dortmund, dissertation.

Espinoza R., Santamaria J., Menendez M., Coronas J., Irusta S., 1999a 'Production of hydrocarbons', International patent WO 1999/64380.

Espinoza R., Steinberg A., Jager B., Vosloo A., 1999b 'Low temperature Fischer-Tropsch synthesis from a Sasol perspective', Appl. Catal. A, vol. 186, pp. 13-26.

Espinoza R., du Toit E., Santamaria J., Menendez M., Coronas J., Irusta S. 2000 'Use of membranes in Fischer-Tropsch reactors', Stud. Surf. Sci. Catal., vol. 30, no. 130, pp. 389-394.

Espinoza R., Santamaria J., Menendez M., Coronas J., Irusta S. 2002 'Production of hydrocarbons', US patent 6,403,660.

ExxonMobil 2008 'The outlook for energy: a view to 2030', http://www.exxonmobil.com.

Fleisch, T.H., Sills, R.A., Briscoe, M.D. 2002 '2002 – Emergence of the gas-to-liquids industry: a review of global GTL developments', J. Nat. Gas Chem., vol. 11, pp. 1-14.

Fleisch, T.H., Sills, R.A. 2004 'Beyond GTL-FT: Large-scale gas conversion through oxygenates', Proc. 7th Natural Gas Conversion Symposium, June 9, Dalian, China.

FNR 2005 'Basisdaten Biokraftstoffe', Schriftenreihe Nachwachsende Rohstoffe, Fachagentur Nachwachsende Rohstoffe.

FNR 2005a 'Biokraftstoffe – Eine vergleichende Analyse', Schriftenreihe Nachwachsende Rohstoffe, Fachagentur Nachwachsende Rohstoffe.

FNR 2006 'Biofuels ', Schriftenreihe Nachwachsende Rohstoffe, Fachagentur Nachwachsende Rohstoffe.

Fogler, H.S. 1999 'Elements of chemical reaction engineering'. 3rd edition, New Jersey, Prentice Hall.

Font Freide, J.J., Hardy, L. 2006 'Modified catalyst and use of this catalyst for the conversion of synthesis gas to hydrocarbons'. International patent WO 2006/087552A2.

Fougerit, J.M., Gnep, N.S., Guisnet, M. 1998 'Selective transformation of methnaol into light olefins over a mordenite catalyst: reaction scheme and mechanism', Microporous Mesoporous Mat., vol. 29, no. 1-2, pp 79-89.

Galluci, F., Paturzo, L., Basile, A. 2004 'An experimental study of CO_2 hydrogenation into methanol involving a zeolite membrane reactor', Chem. Eng. Proc., vol 43, pp. 1019-1036.

Guillou, L., Léonard, S., Le Courtois, V., Payen, E., Vanhove, D., 2004 'Membrane reactors for Fischer-Tropsch synthesis', Proc. ICCMR-6, July 6-9, Lahnstein.

Güttel, R., Turek, T. 2009 'Mikrostrukturierte Festbettreaktoren für stark exotherme Gasphasen-reaktionen: Eine Machbarkeitsstudie', Chem. Ing. Tech., vol. 81, pp. 495-500.

Hammer, G., Lübcke, T., Kettner, R., Pillarella, M.R., Recknagel, H., Commichau, A., Neumann, H.J., Paczysnska-Lahme, B. 2006, 'Natural gas'. In: Ullmann's Encyclopedia, Weinheim, Wiley-VCH, DOI: 10.1002/14356007.a17_073.pub2.

Harold, M.P., Lee, C., Burggraaf, A.J., Keizer, K., Zaspalis, V.T., de Lange, R.S.A 1994 'Catalysis with inorganic membranes', MRS Bulletin, April, pp. 34-39.

He, J., Yoneyama, Y. Xu, B., Nishiyama, N., Tsubaki, N. 2005 'Designing a capsule catalyst and its application for direct synthesis of middle isoparaffins', Langmuir, vol. 21, pp. 1699-1702.

Henrich, E., Dinjus, E. 2003 'Das FZK Konzept zur Kraftstoffherstellung aus Biomasse', Schriftenreihe Nachwachsende Rohstoffe, Fachagentur Nachwachsende Rohstoffe, vol. 24, pp. 298-337.

Higman, C., van der Burgt, M. 2003 'Gasification', Amsterdam, Elsevier.

Higman, C., Supp, E. 2006 'Carbon monoxide shift conversion'. In: Ullmann's Encyclopedia of Industrial Chemistry, Weinheim, Wiley-VCH, DOI: 10.1002/14356007.a12_169.pub2.

Ho, W.S. 2004 'Development of novel water-gas-shift membrane reactor', DOE report, Ohio state university.

Hofbauer, H., Rauch, R., Foscolo, P., Matera, D. 2000 'Hydrogen rich syngas from biomass steam gasification', 1st World Conference on Biomass for Energy and Industry, 1999-2001.

Hougen, O.A., Watson, K.M. 1943, 'Solid catalysts and reaction rates', Ind. Eng. Chem, vol. 35, no. 5, pp. 529-541.

Hu, J., Wang, Y., Cao, C., Elliott, D., Stevens, D., White, J. 2005 'Conversion of biomass syngas to DME using a microchannel reactor', Ind. Eng. Chem. Res., vol. 44, pp.1722-1727.

Huff, G.A., Satterfield, C.N. 1984 'Intrinsic kinetics of the Fischer-Tropsch synthesis on a reduced fused-magnetite catalyst', Ind. Eng. Chem. Process Des. Dev., vol. 23, pp. 696-705.

Iglesia, E., Reyes, S.C., Madon, R.J., Soled, S.L. 1993 'Selectivity control and catalyst design in Fischer-Tropsch synthesis: Sites, pellets, and reactors', Adv. Catal., vol. 39, pp. 221-301.

Iglesia, E. 1997 'Design, synthesis, and use of cobalt-based Fischer-Tropsch synthesis catalysts', Applied Catal. A, vol. 161, pp. 59-78.

IPCC 2007 'IPCC Fourth assessment report: climate change 2007', http://www.ipcc.ch/ publications_and_data/publications_and_data_reports.htm#1 , 01.02.2008.

Jacobs, G., Patterson, P.M., Zhang, Y., Das, T., Li, J., Davis, B.H. 2002 'Fischer–Tropsch synthesis: deactivation of noble metal-promoted Co/Al_2O_3 catalysts', Appl. Catal. A, vol. 233, pp. 215-226.

Jess, A., Popp, R., Hedden, K. 'Fischer-Tropsch synthesis with nitrogen-rich syngas: Fundamentals and reactor design aspects', Appl.Cataly. A., vol. 186, pp.321-342.

Jongsomjit, B., Panpranot, J., Goodwin, J.G. 2001 'Co-support compound formation in alumina-supported cobalt catalysts', J. Catal., vol. 204, pp. 98-109.

Joo, O.S., Jung, K.D., Jung, Y. 2004 'CAMERE process for methanol synthesis from CO_2 hydrogenation', Stud. Surf. Sci. Catal., vol. 153, pp. 67-72.

Jun K.-W., Shen W.-J., Lee K.-W. 1999 'Co-production of methanol and dimethyl ether from CO2 hydrogenation: investigation of reaction conditions', Bull. Korean Chem. Soc., vol. 20, no. 9, pp. 993-998.

Kaltschmitt, M., Hartmann, H. 2001 'Energie aus Biomasse', Berlin, Springer-Verlag.

Kaya, Y. 2004 'The role of CO_2 fixation in the strategy for mitigating global warming', In: Park, S.E, Chang, J.S., Lee. K.W. (eds.) 'Carbon dioxide utilization for global sustainability'. Amsterdam, Elsevier, Stud. Surf. Sci. Catal., vol. 153, pp. 555-560.

Kemmere, M.F., Keurentjes, J.T.F. 2006 'Industrial membrane reactors'. In: Nunes, S.P., Peinemann, K.V. (eds.) 'Membrane technology in the chemical industry'. Weinheim, Wiley-VCH, pp. 229-258.

Keyser, M.J., Everson, R.C., Espinoza, R.L. 2000 'Fischer-Tropsch kinetic studies with cobalt-manganese oxide catalysts', Ind. Eng. Chem. Res., vol. 39, pp. 49-54.

Khassin, A.A., Sipatrov, A.G., Chermashetseva, G.K., Yurieva, T.M., Parmon, V.N. 2005 'Fischer–Tropsch synthesis using plug-through contactor membranes based on permeable composite monoliths. Selectivity control by porous structure parameters and membrane geometry', Top. Catal., vol. 32 (1-2), pp. 39-46.

Khassin, A.A., Sipatrov, A.G., Chermashetseva, G.K., Rudina, N.A., Parmon, V.N. 2005a 'Performance of a catalytic membrane reactor for the Fischer–Tropsch synthesis', Cat. Today, vol. 105, pp. 362-366.

Kim, H.J., Jung, H., Lee, K.Y. 2001 'Effect of water on liquid phase DME synthesis from syngas over hybrid catalysts composed of $Cu/ZnO/Al_2O_3$ and γ-Al_2O_3', Korean J. Chem. Eng., vol. 18, pp. 838-841.

Kim S.D., Baek S.C., Lee Y.J., Jun K.W., Kim M.J., Yoo, I.S. 2006 'Effect of γ-alumina content on catalytic performance of modified ZSM-5 for dehydration of crude methanol to dimethyl ether', Appl. Catal. A , vol. 309, pp.139-143.

Kimura, S., Hirose, T. 1992a 'Theory of membrane permeation'. In: Toshima, N. (ed.) 'Polymers for gas separation'. New York, VCH, pp.15-50.

Kimura, S., Hirose, T. 1992b 'Theory of membrane permeation', Polym. Gas Sep., pp. 15-49.

Kiss, G., Kliewer, C.E., DeMartin, G.J., Culross, C.C., Baumgartner, J. E. 2003 'Hydrothermal deactivation of silica-supported cobalt catalysts in Fischer-Tropsch synthesis', J. Catal., vol. 217, no. 1, pp. 127-140.

Kölsch, P., Noack, M., Druska, P., Müller, D., Toussaint, P., Caro, J. 1998 'Keramikmembranen mit molekularen Trenneigenschaften – Teil 1: SiO_x-funktionalisierte Membranen', Chemie Ingenieur Technik, vol. 70, no. 7, pp.860-866.

Kölsch, P., Noack, M., Caro, J., Weh, L. 1999 'Anwendungsmöglichkeiten mikroporöser Keramikmembranen mit molekularen Trenneigenschaften', Keramische Zeitschrift, vol. 51, no. 2, pp. 112-117.

Kölsch, P., Sziládi, M., Noack, M., Caro, J., Kotsis, L., Kotsis, I., Sieber, J. 2000 'Keramische Membran zur Wasserabtrennung aus Lösungsmitteln', Chemie Ingenieur Technik, vol. 72, no. 10, pp.1167-1173.

König, L., Gaube, J., Meisel, W., Gütlich, P., Gerhard, W., Plog, C. 1987 'The influence of water and of alkali promoter on the carbon number distribution of Fischer-Tropsch products formed over iron catalysts', Ber. Bunsenges. Phys. Chem., vol 91, pp. 116-121.

Koros, W.J., Chern, R.T. 1987 'Separation of gaseous mixtures using polymer membranes'. In: Rousseau, R.W. (ed.) 'Handbook of Separation process technology'. New York, Wiley Interscience.

Koros, W.J., Ma, Y.H., Shimidzu, T. 1996 'Terminology for Membranes and Membrane Processes', Pure Appl. Chem., vol. 68, no. 7, pp. 1479-1489.

Koukou, M.K., Chaloulou, G., Papayannakos, N., Markatos, N.C. 1997, 'Mathematical modelling of the performance of non-isothermal membrane reactor', Int. J. Heat Mass Transfer, vol. 40, no. 10, pp. 2407-2417.

Koukou, M.K., Papayannakos, N., Markatos, N.C., Bracht, M., Alderliesten, P.T. 1998, 'Simulation tools for the design of industrial-scale membrane reactors', J. Chem. Eng. Res. Des., vol. 76, no. 8, pp. 911-920.

Kreiter, R., Wolfs, D.P., Engelen, C.W.R,van Veen, H.M., Vente, J.F. 2008 'High-temperature pervaporation performance of ceramic-supported polyimide membranes in the dehydration of alcohols', J. Membr. Sci., vol. 319, pp. 126–132.

Kriebel, M. 2006 'Absorption processes'. In: Ullmann's Encyclopedia of Industrial Chemistry, Weinheim, Wiley-VCH, DOI: 10.1002/14356007.a12_169.pub2.

Krishnamoorthy, S., Tu, M., Ojeda, M.P., Pinna, D., Iglesia, E. 2002 'An investigation of the effects of water on rate and selectivity for the Fischer-Tropsch synthesis on cobalt-based catalysts', J. Catal., vol. 211, no. 2, pp. 422-433.

Kuipers, E.W., Vinkenburg, I.H., Oosterbeck, H. 1995 'Chain length dependence of α-olefin readsorption in Fischer-Tropsch synthesis', J. Catal., vol. 152, pp. 137-146.

Kuipers, E.W., Scheper, C., Wilson, J.H., Vinkenburg, I.H., Oosterbeck, H. 1996 'Non-ASF product distributions due to secondary reactions during Fischer-Tropsch synthesis', J. Catal., vol. 158, pp. 228-300.

Ledakowicz, S., Nettelhoff, H., Kokuun, R., Deckwer, W.D. 1985 'Kinetics of the Fischer-Tropsch synthesis in the slurry phase on a potassium-promoted iron catalyst', Top. Catal., vol. 24, p. 1043-1049.

Lee, M.D., Lee, J.F., Chang, C.S. 1990 'Catalytic behaviour and phase composition change of iron catalyst in hydrogenation of carbon dioxide', J. Chem. Eng. Jpn. vol. 23, no. 2, pp. 130-136.

Lee, K.W, Jun, K.W., Hwang, J.S., Riedel, T., Schaub, G. 2000 'Synthesis of hydrocarbons from bio-syngas', Proc. Conf. Greenhouse Gas Control Technologies, August 2000, Cairns.

Lee K.H., Youn, M.Y., Bongkuk S. 2004, 'Method for preparating dimethyl ether using a membrane reactor for separation and reaction'. US patent 6,822,125B2.

Lee, K.H., Youn, M.Y., Sea, B. 2006 'Preparation of hydrophilic ceramic membranes for a dehydration membrane reactor', Desalination, vol. 191, pp. 296-302.

Lee E.Y., Park Y.K,. Jo, O.S., 2006, 'Methanol dehydration to produce dimethyl ether over γ-Al_2O_3', Reac. Kin. Catal. Lett., vol. 89, no. 1, pp.

Lemasle, J. 1984 'Thermochemical processing of biomass – Methanol from wood', London, Butterworths.

Léonard, S., Miachon, S., Vanhove, D. 2003 'Effet de la nature de la membrane sur la synthese de Fischer-Tropsch en reacteur catalytique membranaire a lit fixe', Récents Progres en Génie des Procédés, vol. 89, pp. 226-231.

Li, J., Jacobs, G., Das, T., Davis, B.H. 2002 'Fischer–Tropsch synthesis: effect of water on the catalytic properties of a ruthenium promoted Co/TiO_2 catalyst', Appl. Catal. A, vol. 233, pp. 255-262.

Li, J., Zhan, X., Zhang, Y., Jacobs, G., Das, T., Davis, B.H. 2002a 'Fischer–Tropsch synthesis: effect of water on the deactivation of Pt promoted Co/Al_2O_3 catalysts', Appl. Catal. A, vol. 228, pp. 203-212.

Lindmark, J. 2006 'Tailoring of MFI membranes for enhanced selectivity'. Lulea University of Technology, dissertation.

Löwe, T. 2001 'Abtrennen von Wasserdampf aus Gasströmen mit Hilfe von Polymermembranen', Fortschr.-Ber. VDI Reihe 3, no. 710. Düsseldorf, VDI-Verlag.

Lox, E.S., Froment, G.F. 1993 'Kinetics of the Fischer-Tropsch reaction on a precipitated promoted iron catalyst. 2. Kinetic apour ng', Ind. Eng. Chem. Res, vol. 32, pp. 71-82.

LTCPO-GTL 2005 'New GTL based on low-temperature CPO', 5th framework program, ENK6-CT-2002-00682.

Ma, D., Lund, C.R.F. 2003 'Assessing high-temperature water-gas shift membrane reactors', Ind. Eng. Chem. Res., vol. 42, pp. 711-717.

Maretto, C., Krishna, R. 1999 'Modelling of a bubble column slurry reactor for Fischer Tropsch synthesis ', Catal. Today, vol. 52, pp. 279-289.

Martin, H. 1978 'Low Peclet number particle-to-fluid heat and mass transfer in packed beds, Chem. Eng. Sci., vol. 33, pp. 913-919.

Matsumoto, D.K., Satterfield, C.N. 1989 'Effetcs of temperature and hydrogen/carbon monoxide ratio on carbon number product distribution from iron Fischer-Tropsch catalysts', Energy & Fuels, vol. 3, pp. 249-254.

Mauze, G.R., Stern, S.A. 1982 'The solution and transport of water in poly(acrylonitrile): A re-examination', J. Membr. Sci., vol. 12, no. 99, pp. 51-64.

Mena Subiranas, A. 2009 'Combining Fischer-Tropsch synthesis (FTS) and hydrocarbon reactions in one reactor'. Universität Karlsruhe, dissertation.

Metz, S.J., Potreck, J., Mulder, M.H.V., Wessling, M. 2002 'Water apour and gas transport through a poly(butylenes terephthalate) poly(ethylene oxide) block copolymer', Desalination, vol. 148, pp. 303-307.

Metz, S.J., Ven, W.J.C. van de, Potreck, J., Mulder, M.H.V., Wessling, M. 2005 'Transport of water vapour and inert gas mixtures through highly selective and highly permeable polymer membranes', J. Mem. Sci., vol. 251, pp. 29-41.

Mills, G.A. 1994 'Status and future opportunities for conversion of synthesis gas to liquid fuels', Fuels, vol. 73, no. 8, pp. 1243-1279.

Morooka, S., Yan, S., Kusakabe, K., Akiyama, Y. 1995 'Formation of hydrogen-permselective SiO_2 membrane in macropores of α-alumnina support', J. Mem. Sci., vol. 101, pp. 89-98.

MTR 2008, Membrane Reactor Technology (MTR), membrane reactor for hydrogen generation and purification by steam methane reforming, http://www.membranereactor.com, 01.02.2008.

Mulder, M. 2000 'Basic principles of membrane technology'. 2nd edition, Dordrecht, Kluwer Academic.

Nettelhoff, H., Kokuun, R., Ledakowicz, S., Deckwer, W.D. 1985 'Studies on the kinetics of Fischer-Tropsch synthesis in slurry phase', Ger. Chem. Eng., vol. 8, pp. 177-185.

Newsome, D. 1980 'The water-gas shift reaction', Catal. Rev. Sci. Eng., vol. 21, pp. 275-318.

Nijmeyer, A. 1999 'Hydrogen-selective silica membranes for use in membrane steam reforming'. University of Enschede, Twente, dissertation.

Noack, M., Kölsch, P. Caro, J., Schneider, M., Toussaint, Sieber, I. 2000 'MFI membranes of different Si/Al ratios for pervaporisation and steam permeation', Microporous Mesoporous Mater., vol. 35-36, pp. 253-265.

Nunes, S.P., Peinemann, K.V. 2006 'Membrane materials and membrane preparation'. In: Nunes, S.P., Peinemann, K.V. (eds.) 'Membrane technology in the chemical industry'. Weinheim, Wiley-VCH, pp. 3-92.

Ogawa T., Inoue N., Shikada T. 2003 'Direct dimethyl ether synthesis', J. Nat. Gas Chem., vol.12, pp. 219-227.

Ohlrogge, K., Stürken, K. 2006 'The separation of organic vapours from gas streams by means of membranes'. In: Nunes, S.P., Peinemann, K.V. (eds.) 'Membrane technology in the chemical industry'. Weinheim, Wiley-VCH, pp. 93-117.

Oppenländer, K., Merger, F., Strickler, R., Hovemann, F., Schmidt, H., Starke, K., Stork, K., Vodrazka, W. 1980 'Application of polyethers and acetals based on methanol and/or ethanol as fuels for diesel engines and fuels for diesel engines comprising these compounds'. European patent EP0014992.

Oukaci, R., Singleton, A.H., Goodwin, J.G. 1999 'Comparison of patented Co F–T catalysts using fixed-bed and slurry bubble column reactors', Appl. Catal. A, vol. 186, pp. 129-144.

Park, Y.K., Jeon, J.K., Ihm, S.K. 2004 'Design of copper based hybrid catalysts for CO_2 hydrogenation'. In: Park, S.E, Chang, J.S., Lee. K.W. (eds.) 'Carbon dioxide utilization for global sustainability'. Amsterdam, Elsevier, Stud. Surf. Sci. Catal., vol. 153, pp. 25-32.

Patil, C., van Sint Annaland, M., Kuipers, J.A.M. 2007 'Fluidised bed membrane reactor for ultrapure hydrogen production via methane steam reforming: Experimental demonstration and model validation', Chem. Eng. Sci., vol. 62, no. 11, pp. 2989-3007.

Paturzo, L., Basile, A. 2001 'Methane conversion to syngas in a composite palladium membrane reactor with increasing number of Pd layers', Ind. Eng. Chem. Res., vol. 41, no. 7, pp. 1703-1710.

Patzlaff, J., Liu, Y., Graffmann, C., Gaube, J. 2002 'Interpretation and kinetic apour ng of product distributions of cobalt catalyzed Fischer-Tropsch synthesis', Catal. Today, vol. 71, pp. 381-394.

Paul, D.R., Koros, W.J. 1976 'Effect of partially immobilizing sorption on permeability and the diffusion time lag', J. Polym. Sci. Polym. Phy. Ed., vol. 14, pp. 675-685.

Peters, M.S, Timmerhaus, K.D., West, R.E. 2003 'Plant design and economics for chemical engineers'. New York, McGraw-Hill. http://www.mhhe.com/engcs/ chemical/ peters/ data/, 01.02.2008.

Peters, T.A., Benes, N.E., Keurentjes, J.T.F. 2005 'Zeolite-coated ceramic pervaporisation membranes; pervaporisation-esterification coupling and reactor evaluation', Ind. Eng. Chem. Res., vol. 44, no.25, pp. 9490-9496.

Pichler, H., Schulz, H. 1970 'Neue Erkenntnisse auf dem Gebiet der Synthese von Kohlenwasserstoffen aus CO und H_2', Chem. Ing. Tech., vol. 42, pp.1162-1174.

Piera, E., Salomón, M.A., Coronas, J., Menendéz, M., Santamaría, J. 1998 'Synthesis, characterisation and separation properties of a composite mordenite/ZSM-5/chabazite hydrophilic membrane', J. Mem. Sci., vol. 149., pp. 99-114.

Popp, R. 1996 'Ergebnisse halbtechnischer Untersuchungen zur Fischer-Tropsch-Synthese mit stickstoffreichem Synthesegas'. Universität Karlsruhe, dissertation.

Post, M.F.M., Sie, T. 1985 'Process for the preparation of hydrocarbons'. European patent EP19850200182.

Post, M.F.M., van't Hoog, A.C., Minderhoud, J.K., Sie, S.T. 1989 'Diffusion limitations in Fischer-Tropsch synthesis catalysts', AIChE J., vol. 35, no. 7, pp. 1107-1114.

Qin, Q., Ramkrishna, D. 2004 'Modeling of surface structure of supported catalysts. Application to deactivation of FT reactors', Ind. Eng. Chem. Res., vol. 43, pp. 2912-2921.

Rabe, S., Treong, T. B., Vogel, F. 2007 'Catalytic autothermal reforming of methane: Performance of a kW scale reformer using pure oxygen as oxidant', Appl. Cat. A, vol 318, pp. 54-62.

Reid, R.C., Prausnitz, J.M., Poling, B.E. 1987 'The properties of gases and liquids'. New York McGraw-Hill.

Reimert, R., Schaub, G. 2006 'Gas production from coal, wood and other feedstocks', In: Ullmann's Encyclopedia of Industrial Chemistry, Weinheim, Wiley-VCH, DOI: 10.1002/ 14356007.a12_169.pub2.

Renk C., Mas C., Ederer H., Henrich E., Dinjus E. 2006 'One-step dimethyl ether synthesis from synthesis gas ex biomass', DGMK Fachbereichstagung Energetische Nutzung von Biomassen, Velen, 24.-26. April 2006.

Reo, C.M., Bernstein, L.A., Lund, C.R.F. 1997, 'Cocurrent membrane reactors vs. PFRs for shifting dehydrogenation equilibrium', AIChE J., vol. 42, no. 2, pp. 495-504.

Reßler, S., Agar, D.W. 2005 'Enhancement of the syngas-to-dimethyl ether process by adsorptive water removal', Proc. International Symposium on Multifunctional Reactors (ISMR-4), June 2005, Portoroz.

Reßler, S., Elsner, M.P., Dittrich, C., Agar, D.W., Geisler, S., Hinrichsen, O. 2006, 'Reactive gas adsorption'. In: Schmidt-Traub, H. & Górak, A. (eds.), 'Integrated reaction and separation operations – Modelling and experimental validation'. Berlin Heidelberg, Springer-Verlag, pp. 149-190.

Rezai, S.A.S., Lindmark, J., Andersson, C., Jareman, F., Möller, K., Hedlund, J. 2006 'Water/hydrogen/hexane multicomponent selectivity of thin MFI membranes with different Si/Al ratio', Microporous Mesoporous Mater., vol. 108, no. 1-3., pp.136-142.

Riedel, T., Claeys, M., Schulz, H., Schaub, G., Nam, S.S., Jun, W.W., Choi, M.J., Kishan, G., Lee, K.W. 1999 'Comparitive study of FIscher-Tropsch synthesis with H2/CO and H2/CO2 syngas using Fe- and Co-based catalysts', Appl. Catal. A, vol. 186, pp. 201-213.

Riedel, T., Schaub, G., Jun, K.-W., Lee, K.-W. 2001 'Kinetics of CO_2-hydrogenation on a K-promoted Fe catalyst', Ind. Eng. Chem. Res., vol. 40, pp.1355-1363.

Riedel, T. 2003 'Reaktionen von CO_2 bei der Fischer-Tropsch Synthese – Kinetik und Selektivität', Universität Karlsruhe, dissertation.

Riedel, T., Schaub, G. 2003 'Low-temperature Fischer-Tropsch synthesis on cobalt catalysts – effects of CO_2', Top. Catal., vol .26, no. 1-4, pp. 145-156.

Riedel, T., Schulz, H., Schaub, G., Jun, K.W., Hwang, J.S., Lee, K.W. 2003 'Fischer-Tropsch on iron with H_2/CO and H_2/CO_2 as synthesis gases: the episodes of formation of the Fischer-Tropsch regime and construction of the catalyst', Top. Catal., vol. 26, pp. 41-54.

Roh, N.S., Dunn, B.C., Eyring, E.M., Pugmire, R.J., Meuzelaar, H.L.C 2003, 'Production of diethyl carbonate from ethanol and carbon monoxide over a heterogeneous catalytic flow reactor', Fuel Process. Technol. 83, pp. 27– 38.

Rohde M., Unruh D., Pias P., Lee K., Schaub G. 2004 'Fischer-Tropsch Synthesis with CO_2-containing syngas from biomass – kinetic analysis of fixed bed reactor model experiments'. In: Park, S.E, Chang, J.S., Lee. K.W. (eds.) 'Carbon dioxide utilization for global sustainability'. Amsterdam, Elsevier, Stud. Surf. Sci. Catal., vol. 153, pp. 97-102.

Rohde, M.P., Unruh, D., Schaub, G. 2005 'Membrane application in Fischer-Tropsch synthesis to enhance CO_2 hydrogenation', Ind. Eng. Chem. Res., vol. 44, no.25, pp. 9653-9658.

Rozovskii, A., Lin, G. 2003 'Fundamentals in methanol. Synthesis and decomposition', Topics. Catal., vol. 22, pp. 245-251.

Ryu, J.S., Lee, K.-W., Choi, M.J., Yoo H.-S. 2003 The synthesis of clean fuels from CO_2-rich biosyngas', Korean J.Chem.Eng., vol. 40, pp 13.

Saib, A.M. 2006 'Towards a cobalt Fischer-Tropsch synthesis catalyst with enhanced stability: a combined approach'. Technische Universiteit Eindhoven, dissertation.

Saib, A.M., Borgna, A., van de Loosdrecht, J., van Berge, P.J., Niemantsverdriet, J.W. 2006a 'XANES study of the susceptibility of nano-sized cobalt crystallites to oxidation during realistic Fischer-Tropsch synthesis', Appl. Catal. A, vol. 312, pp. 12-19.

Salomón, M.A., Coronas, J. Menéndez, M., Santamaria, J. 1998 'Synthesis of a mordenite/ZSM-5/chabazite hydrophilic membrane on a tubular support. Application to the separation of a water–propanol mixture', Chem. Commun., pp. 125-126.

Salomón, M.A., Coronas, J. Menéndez, M., Santamaria, J. 2000 'Synthesis of MTBE in zeolite membrane reactors', Appl. Cat. A, vol. 200, no. 1-2, pp. 201-210.

Sanchez Marcano, J.G., Tsotsis, T.T. (eds.) 2002 'Catalytic Membranes and Membrane Reactors', Wiley-VCH, Weinheim.

Sanfilippo, D., Patrini, R., Marchionna, M. 2007 'Use of an oxygenated product as a substitute of gas oil in diesel engines'. US patent 7,235,113.

Saracco, G., Specchia, V. 1998 'Inorganic-membrane reactors'. In: Cybulski,A., Moulijn, J.A. 'Structured catalysts and reactors'. New York, Marcel Dekker, pp. 463-500.

Sato, K., Sugimoto, K., Sekine, Y., Takada, M., Matsukata, M., Nakane, T. 2007 'Application of FAU-type zeolite membranes to apour/gas separation under high pressure and high temperature up to 5 MPa and 180°C', Microporous Mesoporous Mater., vol. 101, no. 1-2, pp. 312-318.

Satterfield, C.N., Hanlon, R.T., Tung, S.E., Zou, Z.M., Papaefthymiou, G.C. 1986 'Effect of water on the iron-catalyzed Fischer-Tropsch synthesis', Ind. Eng. Chem. Prod. Res. Dev., vol. 25, no. 3, pp. 407-414.

Schanke, D., Hilmen, A.M., Bergene, E., Kinnari, K., Rytter, E., Adnanes, E., Holmen, A. 1995 'Study of the deactivation mechanism of Al_2O_3-supported cobalt Fischer-Tropsch catalysts', Catal. Lett., vol. 34, pp. 269-284.

Schanke, D., Hilmen, A.M., Bergene, E., Kinnari, K., Rytter, E., Adnanes, E., Holmen, A. 1996 'Reoxidation and deactivation of supported cobalt Fischer-Tropsch catalysts', Energy Fuels, vol. 10, no. 4, pp. 867-872.

Schaub, G., Unruh, D., Rohde, M.P. 2004 'Synthetic hydrocarbon fuels and CO_2 utilization'.In: Park, S.E, Chang, J.S., Lee. K.W. (eds.) 'Carbon dioxide utilization for global sustainability'. Amsterdam, Elsevier, Stud. Surf. Sci. Catal., vol. 153, pp. 17-24.

Schmidt-Traub, H., Górak, A. 2006 'Integrated Reaction and Separation Operations – Modelling and experimental validation'. Berlin Heidelberg, Springer-Verlag.

Schulz, H., Böhringer, W., Kohl, C., Rahmann, N., Will, A. 1984 'Entwicklung und Anwendung der Kapillar-GC-Gesamtprobentechnik für Gas/Dampf-Vielstoffgemische', DGMK-Fortschritts-bericht, vol. 320, pp. 1-6.

Schulz, H., Beck, E., Erich, E. 1988 'Mechanism of the Fischer-Tropsch process', Stud. Surf. Sci. Catal., vol. 36, pp. 457-471.

Schulz, H., Claeys, M. 1999 'Reactions of α-olefins of different chain length added during Fischer–Tropsch synthesis on a cobalt catalyst in a slurry reactor', Appl. Catal. A, vol. 186, no. 1-2, pp. 71-90.

Schulz, H., Schaub, G., Claeys, M., Riedel, T. 1999 'Transient initial kinetic regimes of Fischer-Tropsch synthesis', Appl. Catal A, vol. 186, pp. 215-227.

Schulz, H., Nie, Z., Usmanov, F. 2002 'Construction of the Fischer-Tropsch regime with cobalt catalyst', Catal. Today, vol. 71, no. 3-4, pp. 351-360.

Schulz, H. 2003 'Major and minor reactions in Fischer-Tropsch synthesis on cobalt catalysts', Top. Catal., vol. 26, no. 1-4, pp. 73-85.

Seidel-Morgenstern, A. 2005 'Analysis and experimental investigation of catalytic membrane reactors'. In: Sundmacher, K., Kienle, A., Seidel-Morgenstern, A. (eds.) 'Integrated chemical processes – Synthesis, operation, analysis, and control'. Weinheim, Wiley-VCH, pp. 359-390.

Shell 2008, 'Shell energy scenarios to 2050', http://www.shell.com/home/content/aboutshell/ our_strategy/shell_global_scenarios/dir_global_scenarios_07112006.html, 01.02.2008.

Sharifinia, S., Mortazavi, Y., Khodadadi, A. 2005 'Enhacement of distillate selectivity in Fischer-Tropsch synthesis on a Co/SiO_2 catalyst by hydrogen distribution along a fixed-bed reactor', Fuel Process. Technol., vol. 86, pp. 1253-1264.

Sie, S.T., Krishna, R. 1999 'Fundamentals and selection of advanced Fischer-Tropsch reactors', Appl. Catal. A, vol. 186, pp.55-70.

Sloot, H.J., Versteeg, G.F., van Swaaij, W.P.M. 1990 'A non-permselective membrane reactor for chemical processes normally requiring strict stoichiometric feed rates of reactants', Chem. Eng. Sci. vol. 45, no. 8, pp. 2415-2421.

Specht, M., Bandi, A., Elser, M., Staiss, F. 1998 'Comparison of CO_2 sources for the synthesis of renewable methanol', Stud. Surf. Sci. Catal., vol. 114, pp. 363-367.

Specht, M. 2003 ‚Einführung in die Thematik', FVS Fachtagung Regenerative Kraftstoffe, 2003, Stuttgart.

Spencer, M.S. 1995 'On the activation energies of the forward and reverse water-gas shift reaction', Catal. Lett., vol. 32, pp. 9-13.

Stankiewicz, A.I., Moulijn, J.A. 2000 'Process Intensification: Transforming Chemical Engineering', Chem. Eng. Prog., vol. 96, no. 1, pp. 22-34.

Stern, S.A. 1972 'Gas permeation process'. In: Lacey, R.E., Loeb, S. (eds.) 'Industrial processing with membranes'. New York, Wiley Interscience.

Stern, S.A., Saxena, V. 1980 'Concentration-dependent transport of gases and vapours in glassy polymers', J. Membr. Sci., vol. 7, no. 47, pp. 47-59.

Stern, N. 2006 'The economics of climate change', http://www.hm-treasury.gov.uk/ sternreview_index.htm, 01.02.2008.

Steynberg, A.P., Vosloo, A.C., van Berge, P. 2003 'Handling of catalyst'. US patent 6,512,017.

Steynberg, A.P, Dry, M.E., Davis, B.H., Breman, B.B. 2004 'Fischer-Tropsch reactors'. In: Steynberg, A.P., Drey, M.E. (eds.) 'Fischer-Tropsch Technology'. Amsterdam, Elsevier, Stud. Surf. Sci. Catal., vol. 152, pp. 64-195.

Steynberg, A.P. 2004 'Hydrocarbon Synthesis'. International patent WO/2004/026994.

Stookey, D.J. 2006 'Gas-separation membrane applications'. In: Nunes, S.P., Peinemann, K.V. (eds.) 'Membrane technology in the chemical industry'. Weinheim, Wiley-VCH, pp. 119-150.

Storsaeter, S. 2005 'Fischer-Tropsch synthesis over cobalt supported catalysts'. Trondheim, Norwegian University of Science and Technology NTNU, dissertation.

Storsaeter, S., Borg, O., Blekkam, E.A., Totdal, B., Holmen, A. 2005 'Study of the effect of water on Fischer-Tropsch synthesis over supported cobalt catalysts', J. Catalysis, vol. 231, pp. 405-419.

Storsaeter, S., Borg, O., Blekkam, E.A., Totdal, B., Holmen, A. 2005a 'Fischer-Tropsch synthesis over Re-promoted Co catalysts supported on Al_2O_3, SiO_2 and TiO_2: Effect of water', Catal. Today, vol. 100, pp. 343-347.

Struis, R.P.W.J., Stucki, S. 2001 'Verification of the membrane reactor concept for the methanol synthesis', Appl. Catal. A, vol. 216, pp. 117-129.

Taylor, R., Krishna, R. 1993 'Multicomponent mass transfer'. New York, John Wiley & Sons.

Thijmensen, M.J.A., Faaij, A.P.C., Hemelinck, C.N., van Hardeveld, M.R.M. 2002 'Exploration of the possibilities for production of Fischer-Tropsch liquids and power via biomass production', Biomass Bioenergy, vol. 23, pp. 129-152.

Thomas, S. 2003 'Kontrollierte Eduktkzufuhr in Membranreaktoren in Membranreaktoren zur Optimierung der Ausbeute gewünschter Produkte in Parallel- und Folgereaktionen'. Otto-von-Guericke Universität Magdeburg, dissertation.

Tsotsis, T.T., Minet, R.G., Champagnie, A.M., Liu, P.K.T. 1993 'Catalytic membrane reactors'. In: Becker, E.R., Pereira, C.J. (eds.) 'Computer-aided design of catalysts'. New York, Marcel Dekker, pp. 471-551.

Tuchlenski, A. 1998, 'Charakterisierung poröser Membranen und ihre Anwendung in Membranreaktoren'. Otto-von-Guericke Universität Magdeburg, dissertation.

Twigg, M.V. 1989 'Catalyst handbook'. London, Wolfe Publishing, pp. 283-338.

Unruh, D., Rohde, M.P., Schaub, G. 2002 'In-situ removal of H_2O during Fischer-Tropsch Synthesis – a modelling Study', DGMK-Tagungsband, 2002-4, pp. 1-6.

Unruh, D., Rohde, M.P., Schaub, G. 2004 'Improving carbon utilization in biomass conversion to synthetic hydrocarbons via Fischer-Tropsch synthesis', Stud. Surf. Sci. Catal., vol. 153, pp. 91-96.

Unruh, D. 2006 'Fischer-Tropsch Synthese mit Synthesegasen aus Biomasse – Verbesserung der Kohlenstoffnutzung durch Anwendung eines Membranreaktors'. Universität Karlsruhe, dissertation.

Ushikoshi, K., Mori, K., Watanabe, T., Takeuchi, M., Saito, M. 1998 'A 50 kg/day class test plant for methanol synthesis from CO_2 and H_2', Stud. Surf. Sci. Catal., vol. 114, pp. 357-362.

van Berge, P.J., van de Loosdrecht, J., Barradas, S., van der Kraan, A.M. 2000 'Oxidation of cobalt based Fischer-Tropsch catalysts as a deactivation mechanism', Catal. Today, vol. 58, pp. 321-334.

van den Broeke, L.J.P. 2005 'Catalytic membrane reactor for the direct synthesis of dimethyl carbonate using high density carbon dioxide', http://www.nwo.nl/projecten.nsf/pages/ 2300129149, 01.02.2008.

van de Graaf, J.M., Kapteijn, F., Moulijn, J.A. 1998 'Zeolitic membranes'. In: Cybulski,A., Moulijn, J.A. 'Structured catalysts and reactors'. New York, Marcel Dekker, pp. 543-573.

van de Graaf, J.M., Zwiep, M., Kapteijn, F. 1999 'Application of a silicalite-1 membrane reactor in metathesis reactions', Appl. Catal. A, vol. 178, p. 225-241.

van der Laan, G. 1999 'Kinetics, selectivity and scale up of Fischer-Tropsch synthesis'. Rijksuniversiteit Groningen, dissertation.

van der Laan, G.P., Beenackers, A.A.C.M., Krishna, R. 1999 'Multicomponent reaction engineering model for Fe-catalyzed Fischer–Tropsch synthesis in commercial scale slurry bubble column reactors', Chem. Eng. Sci., vol. 54, no. 21, pp. 5013-5019.

van Hardeveld, R.M. 'Fischer-Tropsch process'. European patent EP1746143.

van Herwijnen, T., de Jong, W.A. 1980 'Kinetics and mechanism of the CO shift on Cu/ZnO: 1. Kinetics of the forward and reverse CO shift reactions', J. Catal., vol. 63, no. 1, pp. 83-93.

van Steen, E. 1993 'Elementarschritte der Fischer-Tropsch CO-Hydrierung mit Eisen- und Kobaltkatalysatoren'. Universität Karlsruhe, dissertation.

van Steen, E., Schulz, H. 1999 ' Polymerisation kinetics of the Fischer–Tropsch CO hydrogenation using iron and cobalt based catalysts', Appl. Catal. A, vol. 186, no. 1-2, pp. 309-320.

van Steen, E., Claeys, M., Dry, M.E., van de Loosdrecht, J., Viljoen, E.L., Visagie, J.L. 2005 'Stability of nanocrystals: Thermodynamic analysis of oxidation and re-reduction of cobalt in water/hydrogen mixtures', J. Phys. Chem B, vol. 109, pp. 3575-3577.

van Veen, H.M., Bracht, M., Hamoen, E., Alderliesten, P.T 1996 'Feasibility of the application of porous inorganic gas separation membranes in some large-scale chemical processes', in: A.J. Burggraaf, L. Cot (Eds.) 'Fundamentals of Inorganic Membrane Science and Technology'. Elsevier, Amsterdam.

van Veen, H.M., Alderliesten, P.T 1997 'Technical and economical evaluation on the use of membrane reactors for the dehydration of propane to propylene', Proc. 4th workshop in the ESF network on catalytic membrane reactors, May 1997, Oslo.

Vanhove, D., Léonard, S. 2003 'Synthese de Fischer-Tropsch en alimentations separees (IMPBR)', Récents Progres en Génie des Procédés, vol. 89, pp. 194-198. OK

Vente, J.F. 2006, personal communication.

VDI-Wärmeatlas 1994 'VDI-Wärmeatlas: Berechnungsblätter für den Wärmerübergang'. 7th edition, Düsseldorf, VDI-Verlag.

Vielstich, W. (ed.) 'Handbook of fuel cells- fundamentals, technology and applications'. New York, Wiley.

Vos, R.M. de, Verweij, H. 1998, 'Improved performance of silica membranes for gas separation', J. Membr. Sci., vol. 143, pp. 37-51.

Wakatsuki, T., Morita, Y., Okado, H., Inaba, K., Hirayama, H., Shimura, M., Kawazuishi, K., Iwamoto. O., Suzuki, T. 2001 'Development of a high efficiency GTL process based on CO_2/steam reforming of natural gas and slurry phase FT synthesis', Stud. Surf. Sci. Catal., vol. 116, pp. 117-122.

Weatherbee, G.D., Bartholomew, C.H. 1984 'Hydrogenation of CO_2 on group VIII metals', J. Catal., vol. 77, p. 352-362.

Weimer, T., Schaber, K., Specht, M., Bandi, A. 1996 'Methanol from atmospheric carbon dioxide: a liquid zero emission fuel for the future', Energy Convers. Mgmt., vol. 37, no. 6-8, pp 1351-1356.

Weimer, T., Specht, M., Bandi, A., Schaber, K., Maier, C.U. 1997 'CO_2 removal and fixation – solar high temperature syngas generation for fuel synthesis', Energy Convers. Mgmt., vol. 38, suppl., pp 379-384.

Withers, H.P., Eliezer, K.F., Mitchell, J.W. 1990 'Slurry-phase Fischer-Tropsch synthesis and kinetic studies over supported cobalt carbonyl derived catalysts', Ind. Eng. Chem. Res., vol. 29, no. 9, pp. 1807-1814.

Wurzel, T. 2005 'The Lurgi MegaMethanol Technology – Delivering the building blocks for the future fuel and monomer demand', DGMK-Tagungsbericht 2006-4, ISBN 3-936418-57-8, pp. 9-18.

Xu, A., Indala, S., Hertwig, T. 2005 'Development and integration of new processes consuming carbon dioxide in multi-plant chemical production complexes', Clean Technol. Environ. Policy.

Yates, I.C., Satterfield, C.N. 1991 'Intrinsic kinetics of the Fischer-Tropsch synthesis on a cobalt Catalyst ', Energy Fuels, vol. 5, p. 168-173.

Zennaro, R., Tagliabue, M., Bartholomew, C.H. 2000 'Kinetics of Fischer-Tropsch synthesis on titania-supported cobalt', Catal. Today, vol. 58, pp. 309-319.

Zhang, Y., Jacobs, G., Sparks, D.E., Dry, M.E., Davis, B. 2002 'CO and CO_2 hydrogenation study on supported cobalt Fischer-Tropsch synthesis catalysts', Catal. Today, vol. 71, pp. 411-418.

Zhang, L., Espinoza, R.L. 2003 'Method for reducing water concentration in a multiphase column reactor'. US patent 20,030,134,913.

Zhang, J., Wright, H.A., Jiang, Y., Allison, J.D., York, K.M., Mohedas, S.R., Melquist, V.H. 2006 'Water removal in Fischer-Tropsch synthesis'. US patent 7,001,927.

Zhu, W., Gora, L., Berg, A.W.C., Kapteijn, F., Jansen, J.C., Moulijn, J.A. 2005 'Water vapour separation from permanent gases by a zeolite-4A membrane', J. Mem. Sci., vol. 253, pp. 57-66.

Zimmerman, W.H., Bukur, D.B. 1990 'Reaction kinetics over iron catalysts used for the Fischer-Tropsch-Synthesis', Can. J. Chem. Eng., vol. 68, pp. 292-301.

Zürcher, S., Pabst, K., Schaub, G. 2008 'Ceramic Foams as structured catalyst inserts in gas-particle filters for gas reactions – Effect of backmixing', Appl. Cat. A, vol. 357, no. 1, pp. 85-92.

11 Notations

Latin symbols

symbol	denomination	definition	unit
a	inhibition coefficient in rate equations		
a	activity		-
a	thermal diffusivity	$\lambda / (\rho\ c_P)$	m^2/s
a	volume specific area	A / V	m^2 / m^3
a_m	catalyst mass specific membrane area	$\frac{A_m}{m_{cat}}$	m^2/kg
A	area	L^2	m^2
A_{cs}	cross-sectional area perpendicular to flow direction		m^2
A_j	GC: peak area of component j	-	m^2
A_m	membrane area		m^2
ATY	area time yield	$\frac{J_{i,tmb} \cdot \tilde{M}_i}{A_m}$	$kg/(m^2\ h)$
b	inhibition coefficient in rate equations		
B_0	permeability constant for porous material (Hagen-Poiseuille)	$\frac{\varepsilon}{\tau} \frac{d_{pore}^2}{32}$	m^2
Bi	Biot number, external mass transfer	$\frac{\beta_{ext}}{Q_i RT}$	-
Bo_{ax}	Bodenstein number	$\frac{\varepsilon u_G L}{D_{i,ax}}$	-
c	inhibition coefficient in rate equations		
c	concentration	N_i / V	mol/m^3
c_i^*	concentration of component i in the membrane matrix	V_i / V	m^3/m^3
c_P	heat capacity	-	$J/(kg\ K)$
D_{ax}	axial dispersion coefficient	-	m^2/s
D_r	radial dispersion coefficient	-	m^2/s
D_i	diffusion coefficient in membrane matrix		m^2/s
D_i^K	Knudsen diffusion coefficient	$\frac{1}{3} d_{pore} \sqrt{\frac{8RT}{\pi \tilde{M}_i}}$	m^2/s
d_{pore}	pore diameter	L	m
d_P	particle diameter	L	m
Da	Damköhler number	$\frac{m_{cat} k_j P_{ref}^{\alpha_j}}{\dot{V}_F c_F}$	-
E	energy	$m \cdot du/dt \cdot L$	J
E_A	activation energy	-	J

$E_{A,D}$	diffusion activation energy	-	J
$E_{A,P}$	permeation activation energy	-	J
$E(t)$	residence-time distribution function or exit-age distribution function E(t)	$c(t)/\int_0^\infty c(t)\,dt$	-
F	force	$m \cdot du/dt$	kg m/s^2
f_j	GC: calibration factor of component j	-	-
ΔH_R^0	standard heat of reaction	$\sum_i \nu_i \Delta H_{F,i}^0$	J/mol
ΔH_F^0	standard heat of formation	-	J/mol
ΔH_S^0	standard heat of solubility	-	J/mol
$J_{i,tmb}$	molar flow rate across membrane	$A_m \cdot Q_i \cdot \Delta P_i$	mol/s
$j_{i,tmb}$	molar flow flux across membrane	$\dot{N}_i / A_m$	mol/(s m^2)
k	rate constant		mol/(s kg Pa^n)
k	overall heat transfer coefficient	$\frac{1}{\left(\frac{1}{\alpha_{int}}+...+\frac{\Delta s}{\lambda}+\frac{1}{\alpha_{ext}}\right)}$	W/(m^2 K)
k_L	overall mass transfer coefficient	$\frac{1}{\left(\frac{1}{\beta_{int}}+...+\frac{\Delta s}{\delta}+\frac{1}{\beta_{ext}}\right)}$	m/s
k_La	overall volumetric mass transfer coefficient	-	1/s
k_0	frequency factor in Arrhenius equation		mol/(s kg Pa^n)
K_P	equilibrium constant at standard pressure	$\prod_i (P_i/P^0)^{\nu_i}$	-
Kn	Knudsen number	Λ / d_{pore}	-
L	length	basic dimension	m
L	characteristic length	L	m
L_i	loss of component i across membrane	$(\dot{N}_{i,P} - \dot{N}_{i,S})/\dot{N}_{i,F}$	-
m	mass	basic dimension	kg
m_{cat}	catalyst mass		kg
m_i	distribution coefficient between gas and liquid phase	$c_{i,G} / c_{i,L}$	-
$\dot{m}$	mass flow rate	dm / dt	kg/s
M	M-factor for syngas quality	$\frac{\dot{N}_{H2,F} - \dot{N}_{CO2,F}}{\dot{N}_{CO2,F} + \dot{N}_{CO,F}}$	-
$\tilde{M}_i$	molar mass of component i	-	kg/mol
N	number		-
N_C	number of carbon atoms	-	-

$N_{C,j}$	number of carbon atoms in component j	-	-
N_i	number of moles	basic dimension	mol
$\dot{N}_i$	molar flow rate	dN_i / dt	mol/s
$\dot{n}_i$	area specific molar flow rate	$\dot{N}_i / A = c_i u$	mol/(s m^2)
Nu	Nusselt number	$\alpha \cdot L / \lambda$	-
P	pressure	F / A	kg/(m s^2)
P_i	partial pressure	$y_i P$	kg/(m s^2)
ΔP	pressure drop	-	Pa
∇P	pressure gradient	-	Pa/m
Pe	Péclet number	$u \cdot L / a$	-
Pe_m	Péclet number	$\frac{\dot{V}_F c_F}{A_m Q_{ref} P_F}$	-
Pr	Prandtl number	ν / a	-
Q	heat	$d\dot{Q} = Td\dot{S}$	J
$\dot{Q}$	heat flow rate	dQ / dt	J/s
Q_i	permeance	$J_{tmb,i} / (A_m \cdot \Delta P_i)$	mol/(s m^2 Pa)
Q_i^D	permeance for molecular diffusion	$\frac{1}{\Delta s} \frac{\varepsilon}{\tau} \frac{\delta}{RT}$	mol/(s m^2 Pa)
Q_i^K	permeance for Knudsen diffusion	$\frac{1}{\Delta s} \frac{\varepsilon}{\tau} \frac{2}{3} d_{pore} \sqrt{\frac{2}{\pi \tilde{M}_i RT}}$	mol/(s m^2 Pa)
Q_i^{SD}	permeance of a solution-diffusion membrane	$\frac{S_i D_i}{\Delta s}$	mol/(s m^2 Pa)
Q^V	permeance for viscous flow	$\frac{1}{\Delta s} \frac{\varepsilon}{\tau} \frac{d_{pore}^2}{32} \frac{P_{mean}}{RT}$	mol/(s m^2 Pa)
r_j	reaction rate of reaction j	$\frac{1}{m_{cat}} \frac{d\xi_j}{dt}$	mol/(s kg)
r_p	rate of chain propagation	-	1/s
r_t	rate of chain termination	-	1/s
r_i, r_o	inner, outer radius	L	m
Δr	radial diffusion distance	L	m
R	universal gas constant	8.314	J/(mol K)
R_i	degree of removal of component i across membrane	$\dot{N}_{i,P} / (\dot{N}_{i,P} + \dot{N}_{i,R})$	-
Re	Reynolds number	$u \cdot L / \nu$	-
S	entropy		J/K
$\dot{S}$	entropy flow rate	dS / dt	J/(K s)
S_i	gas phase sorption coefficient of component i in the membrane material		
$S_{i,j}$	permselectivity of a membrane	Q_i / Q_j	-
$S_{j,i}$	product selectivity to product j on basis of educt i	$\frac{Y_{j,i}}{X_i} = \left\vert\frac{\nu_i}{\nu_j}\right\vert \frac{\dot{N}_j - \dot{N}_{j,F}}{\dot{N}_{i,F} - \dot{N}_i}$	-

Δs	membrane layer thickness	-	m
Sc	Schmidt number	ν / δ	-
Sh	Sherwood number	$\beta \cdot L / \delta$	-
STY	space time yield	$\frac{\dot{N}_{i,F} \cdot Y_{j,i} \cdot \tilde{M}_j}{V_R}$	$kg/(m^3\ h)$
t	time	basic dimension	s
T	temperature	basic dimension	K
u	velocity	dL / dt	m/s
u_{sf}	superficial velocity	$\dot{V} / A$	m/s
u_G	gas velocity in packed bed	$\dot{V} / (A \cdot \varepsilon)$	m/s
V	volume	L^3	m^3
$\dot{V}$	volume flow rate	dV / dt	m^3/s
V_R	volume reaction zone	L^3	m^3
w_j	weight fraction based on components i = 1…N	$\dot{N}_j \tilde{M}_j / \sum_{i=1}^{N} (\dot{N}_i \tilde{M}_i)$	-
WTY	weight time yield of product j	$\frac{\dot{N}_{i,F} \cdot Y_{j,i} \cdot \tilde{M}_j}{m_{cat}}$	kg/(kg h)
X_i	conversion	$1 - \dot{N}_i / \dot{N}_{i,F}$	-
y_j	molar fraction based on components j = 1…N	$\dot{N}_j / \sum_{i=1}^{N} \dot{N}_i$	-
$y_{j,smpl}$	molar fraction based on number of molecules in the sample	$\dot{N}_j / \sum_{i=1}^{N_{smpl}} \dot{N}_i$	-
$Y_{j,i}$	yield of product j on basis of educt i	$\left\| \frac{\nu_i}{\nu_j} \right\| \frac{\dot{N}_j - \dot{N}_{j,F}}{\dot{N}_{i,F}}$	-
z	axial coordinate	L	m
z^+	dimensionless axial coordinate	z / L	-
$z_{H2,F}$	H_2 fraction in the feed versus H_2 required for CO/ CO_2 conversion	$\frac{\dot{N}_{H2,F}}{3 \cdot \dot{N}_{CO2,F} + 2 \cdot \dot{N}_{CO,F}}$	-
$z_{CO2,C,F}$	CO_2 fraction in the feed versus total carbon in the feed	$\frac{\dot{N}_{CO2,F}}{\dot{N}_{CO2,F} + \dot{N}_{CO,F}}$	-
$z_{n-alkene,lin\ HC}$	fraction of n-alkenes in the class of linear hydrocarbons with N carbon atoms	$\frac{y_{n-alkene,N_c}}{y_{n-alkene,N_c} + y_{n-alkane,N_c}}$	-

Greek symbols

symbol	denomination	definition	unit
α	chain growth propagation probability	$r_p / (r_p + r_t)$	-
α_1, α_2, ξ	chain growth parameters in two-alpha model	see (2.8)	-

α	heat transfer coefficient	$\dot{Q}/(A\Delta T)$	
β, β_{ext}	mass transfer coefficient, external	$\dot{N}/(A\Delta c)$	m/s
δ	binary diffusion coefficient	-	m^2/s
Δ	difference	-	-
ε	void fraction	V_G / V_R	-
ε	porosity	V_G / V_S	-
ε_G	total gas hold-up	V_G / V_R	-
ε_{GS}	small gas bubble hold-up	$V_{GS} /(V_R - V_{GL})$	-
ε_{GL}	large gas bubble hold-up	V_{GL} / V_R	-
ε_L	liquid hold-up	V_L / V_R	-
ε_S	solid (catalyst) hold-up	V_S / V_L	-
Φ	pressure ratio	P_S / P_F	-
η	dynamic viscosity	$\tau/(du / dz)$	kg/(m s)
κ			
λ	thermal conductivity	$-\dot{Q}/(A \cdot dT / dz)$	W/m K
λ_i	stability parameter of the intermediate product i	-	-
Λ	mean free path	-	m
μ	chemical potential	$\mu_0 + RT \ln a_i + \upsilon_i (P - P^0)$	J/mol
$\nabla\mu$	chemical potential gradient		J/(mol m)
ν	kinematic viscosity	η / ρ	m^2/s
ν_{ij}	stoichiometric coefficient	-	-
ρ	density	m / V	kg/m^3
ρ_b	bulk density	ρ_{cat} (1-ε), m_{cat} / V_R	kg/m^3
ρ_P	density of catalyst particles	m_P / V_P	kg/m^3
ρ_G	gas density	m_G / V_G	kg/m^3
τ	shear stress	F / A	$kg/(m\ s^2)$
τ	particle tortuosity	-	-
$\tau_{mod,n}$	modified residence time for norm conditions (NTP)	$m_{cat} / \dot{V}_{F,n}$	$kg\ s/m^3$
υ_i	molar volume	V_i / N_i	m^3/mol
ξ_i	extent of reaction j	$(N_{ij} - N_{ij,F})/ \nu_{ij}$	-
Ψ	sweep ratio	$\dot{N}_S / \dot{N}_F$	-
∇	gradient	$\left(\frac{\partial}{\partial x}, \frac{\partial}{\partial y}, \frac{\partial}{\partial z}\right)$	1/m

Indices

symbol	denomination
0	basis
c	critical temperature
cat	catalyst
CPr	GC: FID reference compound cyclopropane
C_{1+}	fraction of hydrocarbons with $N_C \geq 1$

C_{5+}	fraction of hydrocarbons with $N_C \geq 5$
eff	effective
EtOH	ethanol
eq	equilibrium
exp	experimental/ measured
F	feed (inlet)
FS	feed side
fxn	functional membrane layer
g	glass transition temperature
G	gas phase
GL	gas bubbles, large
GS	gas bubbles, small
HC	Fischer-Tropsch hydrocarbon molecules (including oxygenates)
init	initial
L	liquid phase
m	membrane
MeOH	methanol
model	calculated by model
N_2,ref	GC: TCD reference compound N_2
P	permeate (outlet sweep side SS)
para	parameter
R	reaction volume
R	retentate (outlet feed side FS)
ref	reference
rel	relative
sat	saturated
sf	superficial
smpl	sample (based only on the molecules within ampoule sample)
spprt	membrane support layers
S	solid
S	sweep (inlet)
SS	sweep side
tot	total
trans	transition (from homogeneous to heterogeneous flow regime)

Abbreviations

symbol	denomination
a	year
Ar	argon
Ar/p	argon/ pressure control
ASF	Anderson-Schulz-Flory
BCO	bio crude oil
bpd	barrels (159 L) per day
BTL	biomass-to-liquids
CHA	chabazite (zeolite)

Co	cobalt
CO-SH	CO shift, water gas shift
CO2-SH	CO_2 shift, reverse water gas shift
CPr	cyclopropane
CSTR	continuously stirred tank reactor = perfectly mixed reactor system
CTL	coal-to-liquids
Cu	copper
DEC	diethyl carbonate
DEE	diethyl ether
DMC	dimethyl carbonate
DME	dimethyl ether
DMM_{3-8}	poly-dimethoxy methane
ETBE	ethyl tert-butyl ether
EtOH	ethanol
FAME	fatty acid methyl ester
Fe	iron
FI	flow indicator
FIC	flow indicator controller/ mass flow controller
FID	flame ionization detector
FT	Fischer-Tropsch
FTS	Fischer-Tropsch synthesis
GC	gas chromatograph
GTL	gas-to-liquids
IPCC	Intergovernmental panel on climate change
K	potassium
LPG	liquefied petroleum gas
LT	low temperature
LTA	Linde type-A (zeolite)
MeOH	methanol
MFI	ZSM-5 (zeolite)
Mn	manganese
MOR	mordenite (zeolite)
MTBE	methyl tert-butyl ether
MTG	methanol-to-gasoline
MTHC	methanol-to-hydrocarbon
NTP	norm conditions (273.15 K, $1.013 \cdot 10^5$ Pa)
PFR	plug-flow reactor
PI	pressure indicator
ppb	parts per billion
ppm	parts per million
PIR	pressure indicator data registration
Pt	platinum
QIR	quality indicator data registration
RME	rapeseed methyl ester
Ru	ruthenium
SH	shift

Si	silica
SIL-1	silicalite-1 (zeolite)
SMDS	Shell middle distillate synthesis
SNG	substitute natural gas
TCD	thermal conductivity detector
Ti	titanium
TI	temperature indicator
TIC	temperature indicator control
tos	time on stream
TPR	temperature programmed reduction
WGS	water gas shift
ZnO	zinc oxide
Zr	zirconium
ZSM-5	Zeolite Socony Mobil 5 (zeolite), also known as MFI
Z4A	zeolite 4A

12 Appendix

12.1 Literature

Figure 1.2 is based on the data collected in Table 12.1. Compositions from coal derived synthesis gases are taken from Reimert & Schaub (2006). Refer also to Higman & van der Burgt (2003).

Table 12.1. Composition of dry synthesis gases from natural gas (NG), coal (C) and biomass (B) for different gasification processes, in vol%.

feed stock	process	H_2	CO	CO_2	CH_4	N_2	z_{CO2}	z_{H2}	
NG	POX, > atm, O_2/H_2O	59.8	35.5	3.8	0.5		0.1	0.72	[1]
NG	SR, > atm, O_2/H_2O	62.2	17	13.2	1.1		0.44	0.86	[1]
NG	triforming	49.4	25.3	5.1	6.5		0.17	0.75	[9]
NG	LTCPO	61.4	13.7	11.4	13.2				[11]
C	moving bed, pressurized	38.3	18.5	32	10.6		0.63	0.29	[2]
B	entrained flow, > atm, O_2	30	50	17	0.1		0.25	0.2	[3]
B	entrained flow, atm, O_2	35.2	41.1	22	0.1		0.35	0.24	[4]
B	circulating bed, atm, H_2O	52.4	28.7	16.8	2.1		0.37	0.48	[5]
B	fluidized bed, > atm, O_2/H_2O	31	38.6	27.2	3.1		0.41	0.2	[6]
B	atm, O_2/N_2	12.5	16.3	13.5	4.4	52	0.45	0.17	[7]
B	atm, O_2/H_2O	38.1	28.1	21.2	8.6		0.43	0.32	[7]
B	NEDO	57.4	28.4	14.2			0.33	0.58	[10]

[1] Reimert et al. (2006), [2] Herbert et al. (1956), [3] Henrich et al. (2003), [4] Althapp (2003), [5] Hofbauer et al. (2000), [6], Lemasle (1984), [7] Kaltschmitt et al. (2001), [9] Wakatsuki et al. (2001), [10] Ryu et al. (2003), [11] LTCPO-GTL (2005).

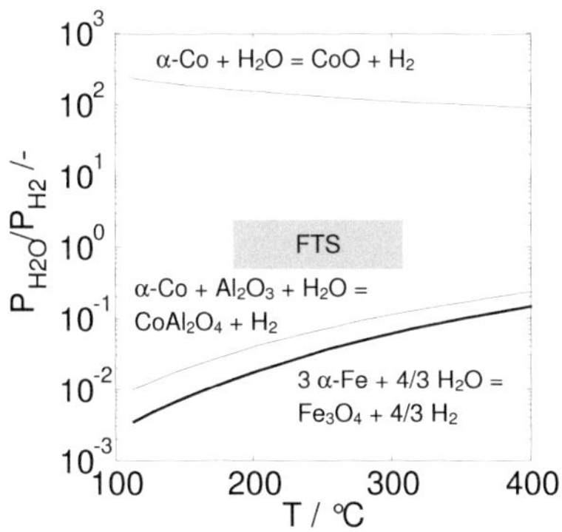

Figure 12.1. Thermodynamic equilibrium constants K = P_{H2O}/P_{H2} for selected Co and Fe oxidation reactions, typical FT conditions (adapted from van Berge et al. 2000, König et al. 1987).

Table 12.2. Examples of kinetic rate equations for the Fischer-Tropsch reaction (formation of organic compounds) for Co-based catalysts.

			T /°C	P / MPa	H_2/CO /-	reference e.g.
1	$r_{FT} = kP_{CO}^{-0.24}P_{H2}^{0.74}$	Co/TiO_2	200	0.8-1.6	1-4	Zennaro et al. (2000)
2	$r_{FT} = k\frac{P_{CO}P_{H2}}{(1+aP_{CO})^2}$	Co/MgO/SiO_2	220-240	0.5-1.5	1.5-3.5	Yates et al. (1991)
3	$r_{FT} = k\frac{P_{CO}^{0.5}P_{H2}^{0.5}}{(1+aP_{CO}^{0.5})^2}$	Co/SiO_2	200	2	1.4-3.4	Keyser et al. (2000)
with H_2O effect						
4	$r_{FT} = k\frac{P_{CO}P_{H2}^{1.5}/P_{H2O}}{(1+aP_{CO}P_{H2}/P_{H2O})^2}$	Co/MgO/ThO_2/SiO_2	190-210	2.7-6.2		van Steen et al. (1999)
5	$r_{FT} = k\frac{P_{H2}}{1+aP_{H2O}/(P_{CO}P_{H2})}$	Co/ZrO_2/SiO_2	240-280	3	1-2	Withers et al. (1990)
6	$r_{FT} = k\frac{P_{CO}^{-0.25}P_{H2}^{0.5}}{1+a(P_{H2O}/P_{H2})}$	Co/SiO_2	210	2.21	1-2.4	Das et al. (2005)

Inhibition parameters: 2: Yates et al. (1991) a = $7.9 \cdot 10^{-6}$ Pa^{-1}; 4: van Steen et al. (1999) a = $4.8 \cdot 10^{-6}$; 5: Withers et al. (1990) a = $1.75 \cdot 10^4$ Pa; Das et al. (2005) a = -0.155.

Table 12.3. Examples of kinetic rate equations for the Fischer-Tropsch reaction (formation of organic compounds) on Fe-based catalysts.

			T /°C	P / MPa	H_2/CO /-	reference e.g.
1	$r_{FT} = k\frac{P_{H2}^2P_{CO}}{P_{CO}P_{H2}+bP_{H2O}}$	fused iron	232-263	0.4-1.5	0.5-1.8	Huff & Satterfield (1984)
2	$r_{FT} = k\frac{P_{CO}P_{H2}}{P_{CO}+cP_{CO2}}$	prec. Fe	220-260 210-270	1 0.5-5.5	0.5-0.6 0.5-3.5	Ledakowicz et al. (1985) Nettelhoff et al. (1985)
3	$r_{FT} = k\frac{P_{CO}P_{H2}}{P_{CO}+bP_{H2O}}$	prec. Fe/Cu/K	250	1.5-3	0.6-1	Zimmerman & Bukur (1990)
4	$r_{FT} = k\frac{P_{CO}P_{H2}}{P_{CO}+bP_{H2O}+cP_{CO2}}$	prec. Fe/Cu/K	250 225-400	1.5-3 1	0.6-1 2-100[a]	Zimmerman & Bukur (1990) Riedel (2003) Unruh (2006)
5	$r_{FT} = k\frac{P_{CO}P_{H2}^{0.5}}{(1+aP_{CO}+cP_{CO2})^2}$	prec. Fe/Cu/K	250	1-2.5	0.25-4	van der Laan (1999), van der Laan et al. (2000)
6	$r_{FT} = k\frac{P_{CO}P_{H2}^{1.5}/P_{H2O}}{(1+aP_{CO}P_{H2}/P_{H2O})^2}$	prec. Fe/Cu/K	205-250	1-1.4	2-100[a]	Riedel (2003)/ van Steen et al. (1999)

Inhibition parameters: 1: Huff et al. (1984) b = $1.15 \cdot 10^6$ Pa; 3: Zimmerman et al. (1990) b = 4.5; 4: Zimmerman et al. (1990) b = 4.8, c = 0.33; Riedel (2003) b = 33, c = 2.7; Unruh (2006) b = 1.12, c = $2.3 \cdot 10^{-5}$, 5: van der Laan (1999) gas-slurry a = $1.185 \cdot 10^{-6}$ Pa^{-1}, c = $0.66 \cdot 10^{-6}$ Pa^{-1}; 6: Riedel (2003): a = $1.55 \cdot 10^{-6}$ Pa^{-1}.

[a] Riedel and Unruh covered synthesis gases from H_2/CO = 2 to H_2/CO_2 = 3 (CO_2 hydrogenation)

Table 12.4. Examples of kinetic rate equations for the CO/CO_2-shift reaction (water gas shift reaction).

			T /°C	P / MPa	H_2/CO /-	reference e.g.
1	$r_{CO\text{-}SH} = k \dfrac{P_{H2O}P_{CO} - P_{CO2}P_{H2}^{0.5} / K_{P,CO\text{-}SH}}{\left(1 + bP_{H2O} / P_{H2}^{0.5}\right)^2}$	prec. Fe/Cu/K	250-350	0.6-2.1	3-6	Lox & Froment (1993)
2	$r_{CO\text{-}SH} = kP_{H2O}P_{CO} - \dfrac{P_{CO2}P_{H2}}{K_{P,CO\text{-}SH}}$	prec. Fe/Cu/K	250	1.5-3	0.6-1	Zimmerman & Bukur (1990)
3	$r_{CO\text{-}SH} = k \dfrac{P_{H2O}P_{CO} - P_{CO2}P_{H2} / K_{P,CO\text{-}SH}}{P_{CO} + bP_{H2O} + cP_{CO2}}$	prec. Fe/Cu/K	250	1.5-3	0.6-1	Zimmerman & Bukur (1990)
4	$r_{CO\text{-}SH} = k \dfrac{P_{H2O}P_{CO} - P_{CO2}P_{H2} / K_{P,CO\text{-}SH}}{\left(P_{CO} + bP_{H2O}\right)^2}$	prec. Fe/Cu/K	250	1-2.5	0.25-4	van der Laan (1999)
5	$r_{CO\text{-}SH} = -k \dfrac{P_{CO2}P_{H2} - P_{H2O}P_{CO}K_{P,CO\text{-}SH}}{P_{CO} + bP_{H2O} + cP_{CO2}}$	prec. Fe/Cu/K	250-400	1	2-100[a]	Riedel (2003) Unruh (2006)

Inhibition parameters: 3: Zimmerman & Bukur (1990) b = 21, c = 0; 4: van der Laan (1999), gas-slurry b = 3.07; 5: Riedel (2003) b = 65, c = 7.4; Unruh (2006) b = 15.01, c = $7.9 \cdot 10^{-6}$.

[a] Riedel and Unruh covered synthesis gases from $H_2/CO = 2$ to $H_2/CO_2 = 3$ (CO_2 hydrogenation)

Table 12.5. List of relevant hydrophilic asymmetric membranes published in literature tested for H_2O permeation under elevated temperatures. Functional layer: CHA: chabazite, FAU: faujasite, MOR: mordenite, SIL-1: silicalite-1, ZSM-5: zeolite socony mobile 5, also known as MFI, Z4A: zeolite Linde type 4A

authors	membrane type	vapours (H_2O vol%) and gases tested	T/ °C	P/ MPa	
zeolite membranes					
Espinoza et al. (1999a, 2000, 2002)	MOR, ZSM-5	mixture: H_2O (10-50), CO, CO_2, CH_4, C_8H_{18}	200-350	1.5-2	●■
Bernal et al. (2000)	MOR, ZSM-5	binary: H_2O (4), He, CH4, C_3H_8, C_4H_{10}	30-230	0.1	●■
Piera et al. (1998)	MOR/ZSM-5/CHA	binary/ ternary: H_2O (2), MeOH, PrOH, O_2	30-255	0.1	
Lindmark (2006), Rezai et al. (2007)	ZSM-5, SIL-1	ternary: H_2O (2), He, H_2, n-C_6H_{18}	30-400	0.1	
Zhu et al. (2005)	Z4A	binary: H_2O (2.2), H_2, CO, CH_4	30-100	0.1	◀
Aoki, Kusakabe & Morooka (2000)	Z4A	binary: H_2O (2), He, H_2, CO_2, O_2, N_2, CH_4, C_2H_6, n-C_4H_{10}, i-C_4H_{10}	30-200	0.1	◀
Sato et al. (2007)	FAU	mixture: H_2O (3-16), H_2, MeOH	130-180	1-5	▼
Noack et al. (2000), Kölsch et al. (1999)	MFI (Si/Al var.)	single: H_2O, H_2, CO, CO_2, CH_4, C_xH_y	105	0.1	

amorphous membranes					
Unruh (2006), Rohde, Unruh & Schaub (2006)	Al_2O_3/SiO_2	single/ binary/ reactive: H_2O (60-80), H_2, CO, CO_2, C_xH_y	100-250	0.1-1	◆
Lee, Youn & Sea (2006)	Al_2O_3/SiO_2	single/ binary: H_2O, CH_3OH, DME	250		◆
Kölsch et al. (1998)	Al_2O_3/SiO_2	single: H_2O, H_2, CO, CO_2, CH_4, C_xH_y	105	0.1	
polymer membranes					
Struis & Stucki (2001)	Nafion (Li)	mixture: H_2O, H_2, CO_2, MeOH, Ar	40-200	0.1	★
Metz et al. (2005)	PEO-PBT binary	binary: H_2O, N_2	20-80	0.1-8	★

12.2 Experimental plan and conditions

12.2.1 Mass transfer in hydrophilic membranes

Table 12.6 summarizes the experiments and experimental conditions to determine the permeances and permselectivities of the CSP2 membrane.

Table 12.6. Determination of permeances and permselectivities of the ceramic supported P84® polymer membrane (CSP2) under non-reactive conditions. Experimental plan and experimental conditions.

	feed gas	T	P_F	$V_{F,n}$ (dry)	sweep gas	Ψ (dry)	Φ
		/°C	/MPa	/ml/min		-	-
single gas, transient ΔP	H_2, CO, CO_2, Ar	100-150-200-250		-			
multi gas, steady state	$H_2/CO/CO_2/H_2O$	225-250-270	1	200	Ar, H_2	2	0.15
multi gas, steady state	$H_2/CO/CO_2/H_2O$	250	0.5-1	200	Ar	2	0.15-1
multi gas, steady state	$H_2/CO/CO_2$/MeOH	200-250-270-290	1	200	Ar	2	0.15
multi gas, steady state	$H_2/CO/CO_2$/EtOH	200-250-270-290	1	200	Ar	2	0.15
binary, dynamic	Ar – H_2	250	1	200			
multi gas, dynamic	Ar – $H_2/CO/CO_2$	250	1	200			

12.2.2 Reaction kinetics

The benchmarking experiments with Fe-5K and Co-GTL1 catalysts are not listed in the following tables. Table 12.7 and Table 12.8 summarize the experiments carried out to determine the kinetics and the effect of H_2O co-feeding on the kinetics and selectivity of the Fe-GTLX catalyst. The number of

kinetic experiments was limited as one could revert to similar experiments carried out by Mena (2009).
Table 12.9 and Table 12.10 summarize experiments carried out to determine the kinetics and effect of H_2O co-feeding on the kinetics and selectivity of the Co-GTL4 catalyst. Three experimental runs were carried out and though the catalyst samples were taken from the same batch, it was found that Co-GTL4/ run #1 had a significant lower activity than Co-GTL4/ run #2a and #2b.

Table 12.11 summarizes the experiments carried out to determine the kinetics of the low temperature $CuZnO/Al_2O_3$ catalyst. The CO/CO_2 shift catalyst was applied in the experiments of *in-situ* H_2O removal during Co-based FT synthesis by chemical reaction.

Table 12.12 summarizes the experiments carried out to determine the kinetics of the dehydration of methanol to DME and ethanol to DEE on γ-Al_2O_3. These two reactions were example reactions for the experiments of *in-situ* H_2O removal during fuel-related synthesis reactions.

Table 12.7. Experiments to determine the kinetics of the Fe-GTLX catalyst: Range of experimental variables. Experiments were carried out in the fixed-bed glass reactor.

	T /°C	P /MPa	$\tau_{mod,n}$ /kg s/m^3	m_{cat} /kg	$z_{CO2,C,F}$ /-	$z_{H2,F}$ /-	$y_{H2O,F}$ /-
Activation	250	1	4000	$4\cdot10^{-3}$	0	1	0
Kinetics CO_2	270	1	2000 4000	$4\cdot10^{-3}$	0-0.25-0.5-0.75-1 0-1	1	0

Table 12.8. H_2O co-feeding experiments to determine the effect of H_2O on the kinetics and selectivity of the Fe-GTLX catalyst: Range of experimental variables. Experiments were carried out in the (membrane) reactor with gas-tight membrane replica.

	T /°C	P /MPa	$\tau_{mod,n}$ /kg s/m^3	m_{cat} /kg	$z_{CO2,C,F}$ /-	$z_{H2,F}$ /-	$y_{H2O,F}$ /-
Activation	250	1	2000	$4\cdot10^{-3}$	0	1	0
Kinetics	270	1	2000 4000	$4\cdot10^{-3}$	0-1 0	1	0
Kinetics H_2O co-feeding	270	1.06 1.09 1.14 1.21	2000	$4\cdot10^{-3}$	0	1	0.06 0.08 0.12 0.17
Kinetics H_2O co-feeding	270	1.06 1.14 1.21	2000	$4\cdot10^{-3}$	1	1	0.06 0.12 0.17

Table 12.9. Experiments to determine the kinetics of the Co-GTL4 (#1) catalyst: Range of experimental variables. Experiments were carried out in the fixed-bed glass reactor.

	T /°C	P /MPa	$\tau_{mod,n}$ /kg s/m^3	m_{cat} /kg	$z_{CO2,C,F}$ /-	$z_{H2,F}$ /-	$y_{H2O,F}$ /-
Kinetics (#1)	220	1	1800 3600	$2\cdot10^{-3}$	0 0	1	0
Kinetics (#1) CO_2 effect	230	1	1200 1800 3600	$2\cdot10^{-3}$	0-0.3-0.5-0.7 0-0.3-0.5-0.7 0-0.3-0.5-0.7	1	0
Kinetics (#1) CO_2 effect	240	1	1200 1800 3600	$2\cdot10^{-3}$	0 0-0.3-0.5-0.7 0	1	0

Table 12.10. Experiments to determine the kinetics and the effect of H_2O co-feeding on the kinetics and selectivity of the Co-GTL4 (#2a/b) catalyst: Range of experimental variables. Experiments were carried out in the (membrane) reactor with gas-tight membrane replica.

	T /°C	P /MPa	$\tau_{mod,n}$ /kg s/m^3	m_{cat} /kg	$z_{CO2,C,F}$ /-	$z_{H2,F}$ /-	$y_{H2O,F}$ /-
Kinetics (#2a)	200 215 230	1	1000-1200-4000	$4\cdot10^{-3}$	0		
Kinetics (#2b)	230	1	1000 2100 4200	$4\cdot10^{-3}$	0	1	0
Kinetics (#2b) H_2/CO ratio	230	0.67 0.67 1.67	1200 1500 3200	$4\cdot10^{-3}$	0	0.5 1 2	0
Kinetics (#2b) H_2O co-feeding	230	1.11 1.22 1.33	1200 1800 3600	$4\cdot10^{-3}$	0	1	0.1 0.2 0.3

Table 12.11. Experiments to determine the kinetics of the CO_2/CO shift reaction on the low temperature CuZnO/Al_2O_3 catalyst: Range of experimental variables. Experiments were carried out in the fixed-bed glass reactor. In these experiments, calcined sand was used inert material. All experiments with respect to the forward CO/CO_2 reaction reached equilibrium conversion and are not listed here.

	T /°C	P /MPa	$\tau_{mod,n}$ /kg s/m^3	m_{cat} /kg	H_2/CO_2
Kinetics	200-210-220- 230-240	0.1	950-1560-2120	$2\cdot10^{-3}$	1
Kinetics	200-210-220- 230-240	0.1	950-1560-2120	$4.5\cdot10^{-3}$	1
Kinetics	230	1	210-330-440	$0.5\cdot10^{-3}$	1-1.6-2.2
Kinetics	230	1	510-800-1060	$1\cdot10^{-3}$	1-1.6-2.2

Table 12.12. Experiments to determine the kinetics of the dehydration of methanol to DME and ethanol to DEE on γ-Al_2O_3 catalyst. Experiments were carried out in the (membrane) reactor with gas-tight membrane replica. Ar was used as carrier gas to evaporate the liquid. [a] $\tau_{mod,n}$ based on methanol/ ethanol feed flow rate.

	T	P	$y_{MeOH,F}$ $y_{EtOH,F}$	$\tau_{mod,n}$ [a]	m_{cat}
	/°C	/MPa	/-	/kg s/m^3	/kg
DME kinetics	270	1	0.67	5200-8500-17000	$8\cdot10^{-3}$
DME kinetics	270-310-330	1	0.67	5200	$8\cdot10^{-3}$
DEE kinetics	270-290-310-330	1	0.7	7700	$8\cdot10^{-3}$

12.2.3 In-situ H_2O removal by membranes and chemical reaction

Three separate experimental runs were carried out with respect to *in-situ* H_2O removal with CSP2 membranes during CO_2 hydrogenation. Table 12.13 summarizes the experiments that were executed with the Fe-GTL3/CPS2 combination within the European project LTCPO-GTL (2005). Due to the higher FT activity, the Chapter 5.2 focuses on two independent experiments with Fe-GTLX/CSP2 combination (Table 12.14, Table 12.15).

Furthermore, experiments were carried out with respect to H_2O removal during other fuel-related reactions, i.e. for DME synthesis (Table 12.16) and DEE synthesis (Table 12.17).

Table 12.18 covers the limited number of experiments with regard to *in-situ* H_2O removal by chemical reaction.

Table 12.13. *In-situ* H_2O removal by hydrophilic membrane: CO_2 hydrogenation with Fe-GTL3 catalyst with integrated CSP2 membrane.

	T	P_F	H_2/CO_2	$\tau_{mod,n}$	m_{cat}	Φ	Ψ	sweep
	/°C	/MPa	/-	/kg s/m^3	/kg	/-	/-	
sweep ratio	255	1	3/1	6455	$15\cdot10^{-3}$	1	1-2-3.3 1-2-3.3	Ar H_2
pressure ratio	255	1	3/1	6455	$15\cdot10^{-3}$	0.4-0.7-0.9-1 0.4-0.7-0.9-1	3.3	Ar H_2

Table 12.14. *In-situ* H_2O removal by hydrophilic membrane: CO_2 hydrogenation with Fe-GTLX catalyst with integrated CSP2 membrane.

	T	P_F	H_2/CO_2	$\tau_{mod,n}$	m_{cat}	Φ	Ψ	sweep
	/°C	/MPa	/-	/kg s/m^3	/kg	/-	/-	
sweep ratio	270	1	3/1	4000	$4\cdot10^{-3}$	0.7	0.5-1-1.5-3.3	H_2
pressure ratio	270	1	3/1	4000	$4\cdot10^{-3}$	0.15-0.7-1 0.15-0.7-1	3.3	Ar H_2

Table 12.15. *In-situ* H_2O removal by hydrophilic membrane: CO_2 hydrogenation with Fe-GTLX catalyst with integrated CSP2 membrane.

	T /°C	P_F /MPa	H_2/CO_2 /-	$\tau_{mod,n}$ /kg s/m^3	m_{cat} /kg	Φ /-	Ψ /-	sweep
sweep ratio	270	1	3/0/1	4000	$4\cdot10^{-3}$	0.7	0.4-1-2-3.3	H_2
pressure ratio	270	1	3/01	4000	$4\cdot10^{-3}$	0.2-0.3-0.4-0.5-0.7	3.3	H_2
$\tau_{mod,n}$	270	1	3/0/1	2000 4000	$4\cdot10^{-3}$	0.7	3.3 3.3-6.6	H_2

Table 12.16. *In-situ* H_2O removal by hydrophilic membrane: DME synthesis on γ-Al_2O_3 catalyst with integrated CSP2 membrane.

	T /°C	P_F /MPa	$y_{MeOH,F}$ /-	$\tau_{mod,n}$ /kg s/m^3	m_{cat} /kg	Φ /-	Ψ /-	sweep
pressure ratio	270	1	0.67	5200	$8\cdot10^{-3}$	0.15-0.7-1	3.3	Ar
temperature	250- 270- 290	1	0.67	5200	$8\cdot10^{-3}$	0.15-1 0.15-1 0.15-1	3.3	Ar
temperature (new CSP2)	250- 270- 290	1	0.67	5200	$8\cdot10^{-3}$	0.15 0.15 0.15	3.3	Ar

Table 12.17. *In-situ* H_2O removal by hydrophilic membrane: DEE synthesis on γ-Al_2O_3 catalyst with integrated CSP2 membrane.

	T /°C	P_F /MPa	$y_{EtOH,F}$ /-	$\tau_{mod,n}$ /kg s/m^3	m_{cat} /kg	Φ /-	Ψ /-	sweep
pressure ratio	270	1	0.7	7700	$8\cdot10^{-3}$	0.15-0.5-1	3.3	Ar
temperature	250- 270- 290	1	0.7	7700	$8\cdot10^{-3}$	0.15 0.15 0.15	3.3	Ar

Table 12.18. *In-situ* H_2O removal by chemical reaction: FT synthesis with a Co-GTL4 (non-shift active FT catalyst) and $CuZnO/Al_2O_3$ (low temperature shift catalyst) catalyst mixture.

	T /°C	P /MPa	$\tau_{mod,n}$ /kg s/m^3	$m_{cat,FTS}$ /kg	$m_{cat,SH}$ /kg	$z_{CO2,C,F}$ /-	$z_{H2,F}$ /-
$\tau_{mod,n}$	230	1	1200 1800 3800	$2\cdot10^{-3}$	$2\cdot10^{-3}$	0-0.3-0.5-0.7 0-0.3-0.5-0.7 0	1

12.3 Analytical methods

12.3.1 Analysis of gas effluent streams by GC

In the experimental set-up (see e.g. Figure 3.3), an accurately metered flow (Bronkhorst mass flow controller) of reference gas (0.5 vol% cyclopropane in N_2, BASI) is added and mixed to the gas stream to be analyzed. A small purge gas stream is then either fed to an on-line gas chromatograph or gas samples are taken by the ampoule method. Ampoules are analyzed later off-line in a dedicated gas chromatograph.

Figure 12.2 shows the cross-section of an ampoule sampler. The product gas stream mixed with reference gas flows through the sampling chamber. The capillary of a preheated and evacuated glass ampoule (sample volume: 2 cm^3) is inserted through the septum into the sampling chamber. By breaking the tip of the capillary mechanically by turning the fork, the evacuated ampoule fills up rapidly with a representative sample of the product gas. Then, the ampoule is sealed by smelting off the capillary with the flame of a propane burner. The sealed ampoule can be stored and the sample can be analyzed later in an off-line gas chromatograph. This method was developed at the Engler-Bunte-Institut by Schulz and is described in detail in Schulz et al. (1984). The advantages of this method are: (a) rapid sampling, i.e. every 20 seconds, which is not possible with on-line gas chromatography and (b) storage of samples, which allows the analysis in an off-line gas chromatograph with extended methods. Within this thesis, the ampoule method was only applied to get detailed information on the product distribution of the FT synthesis. Conversion, yield and the C_1-C_{10} product distribution were determined by on-line gas chromatography.

Figure 12.3 shows the column line-up of the on-line gas chromatograph HP 6890N (Agilent/ JAS). The permanent gases CO_2, H_2, N_2, CO and CH_4 and H_2O are analyzed in the front detector, hydrocarbons in the back detector.

The small purge stream split from gas stream to be analyzed flows with a

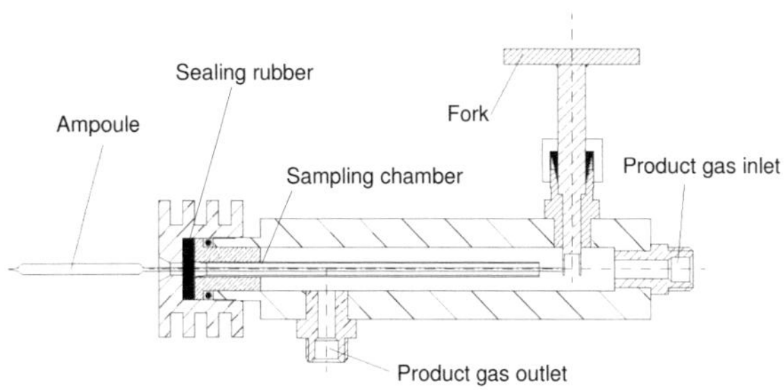

Figure 12.2. Cross-section of an ampoule sampler.

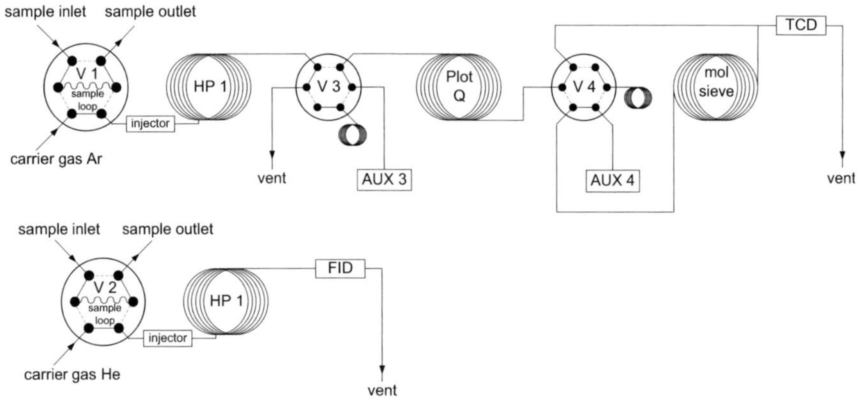

Figure 12.3. Column line-up of the on-line gas chromatograph HP 6890N (Agilent/ JAS), front detector (TCD): product analysis of permanent gases CO_2, H_2, N_2, CH_4 and CO and H_2O, back detector (FID): product analysis of hydrocarbons C_1-C_{15} (Unruh 2006).

constant flow rate through the sample loops of the front and back inlet. By switching the sample valve V1 of the front detector, the carrier gas Ar transports the sample to HP1 pre-column (l/d/s 30 m/ 0.53 mm/ 2.65 µm) where long-chain hydrocarbons are retained. After CO_2, H_2, N_2, CO, H_2O and CH_4 (short-chain hydrocarbons) passed the pre-column, the HP1 is flushed. The polar components CO_2 and H_2O are separated from the gas mixture on the subsequent Plot Q column (l/d/s 30 m/ 0.53 mm/ 40 µm). H_2, N_2, CH_4 and CO are separated on the subsequent mole sieve column (l/d/s 30 m/ 0.53 mm/ 25 µm). Valve V4 is necessary to prevent that H_2O and CO_2 reach the mole sieve column where they would adsorb irreversibly. The separated compounds are detected in a thermal conductivity detector (TCD), in the following order CO_2, H_2, N_2, CH_4, CO, H_2O, methanol, ethanol. The TCD detector is calibrated with a dedicated gas mixture H_2, CO, CO_2, CH_4, N_2 (BASI).

The hydrocarbons are analyzed in the back detector. By switching the sample valve V2, the carrier gas He transports the sample to a HP1 column (l/d/s 50 m/ 0.32 mm/ 0.52 µm). The hydrocarbons are separated into linear and branched paraffins and olefins and oxygenates. Due to a limited temperature program (30 °C – 290 °C), ethane/ ethene and propane/ propene cannot be separated. The components are detected in a flame ionization detector (FID).

Ampoule samples were analyzed in GC (HP 5890), equipped with two ampoule breakers, dedicated to detailed hydrocarbon analysis. The ampoule sample is broken by pneumatic force in an ampoule breaker. The carrier gas H_2 transports the sample to a fused silica column (L/d/s 50 m/ 0.25 mm/ 0.25 µm). The temperature is ramped up from -80 °C to 280 °C according to a

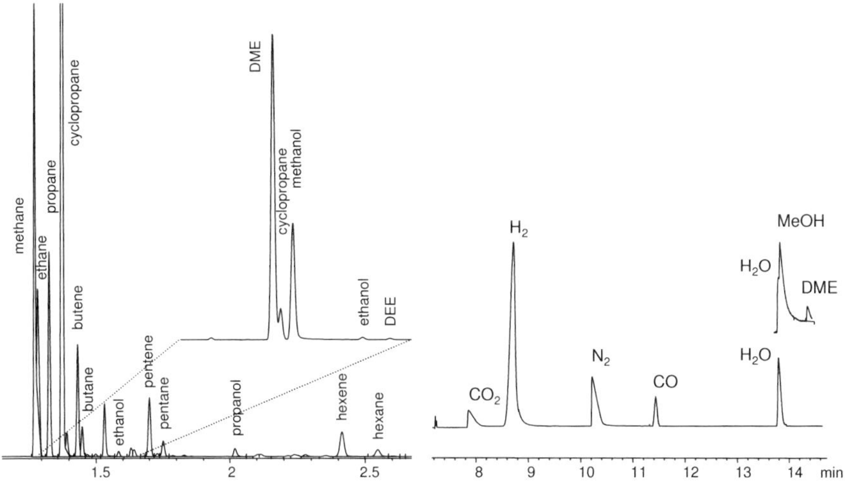

Figure 12.4. Sample chromatograms of HP 6890N (Agilent/ JAS, on-line GC): (left) organic volatile compounds in the FID detector, (right) inorganic gases, methanol and DME in the TCD detector.

temperature program. The components are detected in a flame ionization detector (FID).

Wax samples were analyzed in HP 5890 Series II plus gas chromatograph (offline). A 10 mg wax sample is dissolved in 10 ml cyclohexane in an ultrasound bath at 60-70°C. A 1 µl sample is injected into the gas chromatograph. The carrier gas He transports the sample to a HP-SimDist column (L/d/s 15 m/ 0.53 mm/ 0.15 µm). The temperature is ramped up from 35°C to 350°C according to a temperature program. The components are detected in a flame ionization detector (FID). Typical gas chromatograms of the on-line and off-line GC are shown in Figure 12.4 and Figure 12.5.

Calculation of molar flow rates on basis of the gas chromatographic analysis

The determination of molar flow rates is based on the accurately metered flow rate of the reference gas (0.5 vol% cyclopropane in N_2, BASI). N_2 is the reference component (internal standard) for the TCD, and cyclopropane – a component which is not formed in FT synthesis – the reference component (internal standard) for the FID analysis. The peak area of a specific component in the TCD chromatogram is proportional to its concentration in the carrier gas. With N_2 as internal standard, the molar flow rate of a specific component is calculated as follows:

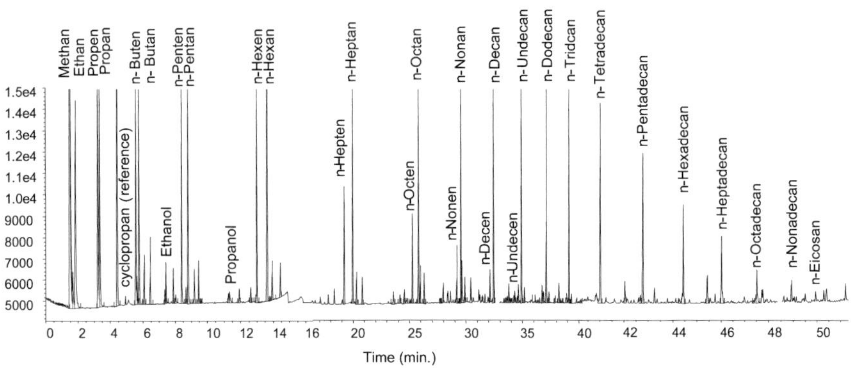

Figure 12.5. Sample chromatograms of HP 5890 (off-line GC, ampoule method): organic volatile compounds in the FID detector.

$$\dot{N}_j = \frac{f_j}{f_{N2}} \cdot \frac{A_j}{A_{N2,ref}} \cdot \dot{N}_{N2,ref} \tag{12.1}$$

with A_j and $A_{N2,ref}$ as the peak areas of species j and the reference component N_2, f_j and $f_{N2,ref}$ as calibration factors of component j and N_2. The component specific calibration factors f_i have to be determined in calibration runs with gas mixtures of known composition (multipoint calibration). H_2O could not be calibrated accurately. Therefore, H_2O has to be determined gravimetrically (see details below). H_2O and methanol could not be separated in the front detector; therefore, methanol and ethanol are analysed quantitatively in the FID. A peak area in the FID chromatogram is directly proportional to the proportion of reduced carbon atoms in the flame of the detector. With cyclopropane (CPr) as internal standard, the molar flow rate of a specific component is calculated as follows:

$$\dot{N}_j = \frac{f_j}{f_{CPr}} \cdot \frac{N_{C,CPr}}{N_{C,j}} \cdot \frac{A_j}{A_{CPr}} \cdot \dot{N}_{CPr} \tag{12.2}$$

with $N_{C,j}$ as the number of carbon atoms in the specific hydrocarbon j, $N_{C,CPr}$ as the number of carbon atoms of cyclopropane, A_j and A_{CPr} as peak areas of hydrocarbon i and the reference component cyclopropane, f_j and f_{CPr} as calibration factors of component j and of cyclopropane. The calibration factors f_i for hydrocarbons equal one. However, these factors differ with 1.3 and 1.2 for methanol and ethanol significantly from one in the case of oxygenates (van Steen 1993, Claeys 1997).

12.3.2 H_2O measurement and oxygen balance

Accurate H_2O measurement in hydrocarbon containing gases appeared to be a difficult task. A GC solution (provided by Joint Analytical Systems) dedicated measuring CO_2, H_2, CO, CH_4 and H_2O did not fulfil the expectations with respect to H_2O measurement. Therefore, gravimetric methods as freezing out in cooling traps and adsorption on desiccant silica gel were applied. During the CO_2 hydrogenation experiments in the membrane reactor, it was found that the permeate effluent does not contain any long-chain hydrocarbons, and therefore, accurate and reliable results could be expected with regard to transmembrane H_2O fluxes. Also the H_2O content in the retentate should be measurable without large errors as the formation of very long-chain hydrocarbons under CO_2 hydrogenation conditions was limited and as silica gel is a selective adsorbent for H_2O (e.g. it is applied in industrial drying of natural gas streams). On the other hand, the application of silica gel traps with Co-based catalysts with H_2/CO syngas was not possible due to the heavier hydrocarbons in the gas phase. The experience on H_2O measurement by GC and gravimetric methods can be summarized as follows:

- A calibration and therefore absolute H_2O flow measurement was not possible with the GC line-up provided and applied though H_2O was separated from CO_2 on a Plot Q column (Figure 12.3). The relative H_2O peak area correlated with the H_2O concentration in the gas stream (Figure 12.4). However, the H_2O peak was always detected, even if no H_2O was present and even after extensive heating of the column system. That H_2O base peak varied in size over time, wherefore absolute water measurement by GC (calibration) was not possible. Figure 12.6 (right) shows the correlation between gravimetrically determined H_2O flows and relative peak

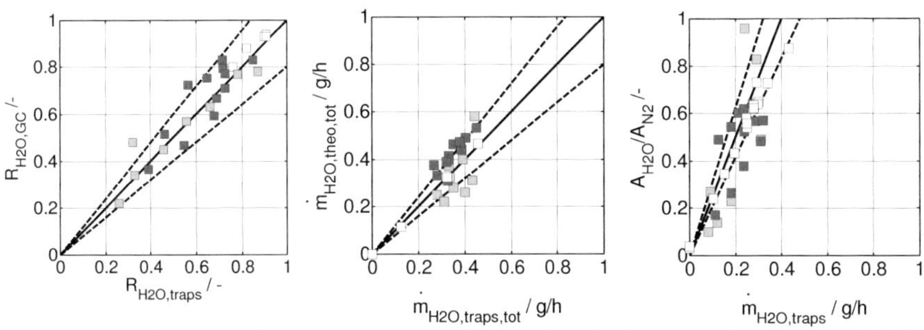

Figure 12.6. Quality of H_2O measurement: (left) parity plot between the degree of H_2O removed determined gravimetrically by H_2O traps and determined by GC (ratio of H_2O peak areas); (centre) parity plot between the total mass flow rate of H_2O determined gravimetrically by H_2O traps and determined on basis of yield and conversion data (GC, C-/O-balance); (right) relationship between mass flow rate of H_2O determined gravimetrically and the specific peak area A_{H2O}/A_{N2} detected by GC.
(□) Experiments in gas permeation cell without catalyst, note that these values are scaled down by a factor of 10 in the figures in the centre and on the right; (■ ■) experiments with Fe-GTLX/ CPS2 membrane. Broken lines: error ± 20%.

areas (with N_2 as reference gas) for three different runs over a time span of several months.

- As the size of the H_2O base peak varied only slowly with time, GC measurements of the H_2O content in retentate and permeate streams relative to each other were possible with a satisfying accuracy. Figure 12.6 (left) correlates the degrees of H_2O removal R_{H2O} measured by GC and by adsorption.
- H_2O flow rates were determined gravimetrically by adsorption on moisture-indicating desiccant silica gel. The silica gels traps were immersed into a cooling bath to ensure high H_2O recovery. Figure 12.6 (centre) compares H_2O flows determined gravimetrically to the total H_2O flows expected according to the carbon and oxygen balance. E.g. in the case of the reactive experiments with membrane, the total amounts of H_2O produced is determined on basis of CO_2 conversion and CO yield data. The results of two different Fe-GTLX/ CSP2 experiments are shown in Figure 12.6 (centre): In the first run (light grey symbols), the data scatters strongly (±20%), while in the second run (dark grey symbols), the results from the traps correlate much stronger with data from the carbon and oxygen balance due to optimized procedures. However, a systematic error is obvious, as either the total H_2O flow rate calculated on basis of the carbon and oxygen balance was too high (i.e. the GC overestimated the hydrocarbon yield) or the traps recovered too little H_2O.

For the FT experiments with integrated membrane, the relative distribution of the H_2O flows on retentate and permeate side, i.e. degree of H_2O removal R_{H2O}, can be determined with satisfying accuracy.
Neither the relative (by GC) nor the absolute (gravimetrically) H_2O flows could be determined for the methanol dehydration experiments, as methanol and H_2O have similar retention times on the Plot Q column and both adsorb on silica gel. H_2O flows are therefore estimated on the basis of model calculations.

12.3.3 Carbon balance

Gas permeation and membrane reactor experiments

Figure 12.7 shows the distribution of CO and CO_2 in the outgoing retentate and permeate streams for a CSP2 permeation experiment under non-reactive conditions. The analysis of several permeation test runs yielded an average apparent CO_2 conversion of -1.4% and an average apparent CO conversion of 1.5% due to inaccuracies in the carbon balance. These low pseudo-conversions found in the non-reactive experiments indicate that the increase in

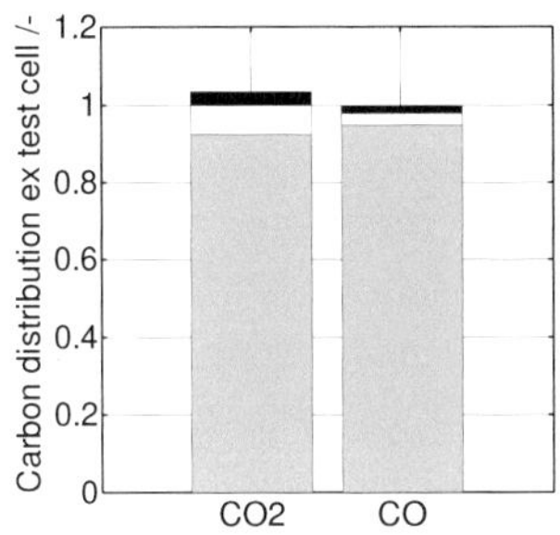

Figure 12.7. Example of a carbon balance: Carbon distribution in effluent streams from the membrane test cell for CO_2 and CO: carbon out in retentate (GC, ■), carbon out in permeate (GC, □), lost/ excess carbon (■).

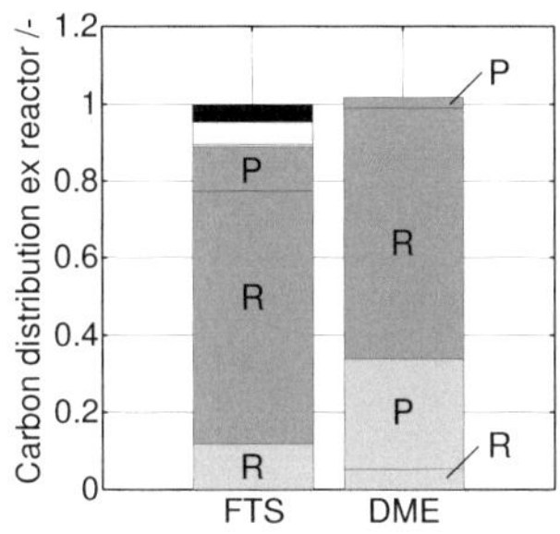

Figure 12.8. Example of a carbon balance: Carbon distribution in effluent streams from the membrane reactor for (a) for FTS/ CSP2: carbon out as CO (GC, ■), as CO_2 (GC, ■), as wax (separator, □), as gaseous/ liquid hydrocarbons (ampoule,); lost carbon (■), R: retentate, P: permeate; and (b) for DME synthesis/CSP2: carbon out as MeOH (GC, ■), as DME (GC, ■), R: retentate, P: permeate.

* CSP2, $(H_2/CO/CO_2)_{dry}$ = 50/20/30 vol%, $y_{H2O,F}$ = 0.25; T = 250°C, P = 1 MPa, , Ψ = 1 (H_2), Φ = 0.5
* Fe-GTLX/ CSP2, H_2/CO_2 = 3, T = 270°C, P = 1 MPa, $\tau_{mod,n}$ = 2000 kg s/m^3, Ψ = 3.3 (H_2), Φ = 0.7.
* DME/ CSP2, T = 290°C, $P_{MeOH,F}$ = 0.67 MPa, $P_{Ar,F}$ = 0.33 MPa (carrier gas), $\tau_{mod,n,MeOH}$ = 5200 kg s/ m^3, Ψ = 3.3., Φ = 0.15, 0.7, $(Ar)_S$

conversion and yield levels in the reactive membrane experiments are not artefacts from an inaccurate carbon balance.

The left bar diagram in Figure 12.9 illustrates the carbon balance, i.e. the distribution of the ingoing carbons atoms in outgoing species as CO, CO_2, gaseous and liquid hydrocarbons and wax, for an Fe-GTLX/ CSP2 membrane reactor experiment. Long-chain hydrocarbons and small amounts of wax were recovered only from retentate side traps. Carbon on the permeate side was found almost only as CO_2; CO and short-chain hydrocarbons were detected only in small amounts (Figure 12.11).

Fischer-Tropsch experiments

Figure 12.9 illustrates the carbon balance, i.e. the distribution of ingoing carbon atoms in outgoing products as CO, CO_2, gaseous and liquid hydrocarbons and wax, for Fe-GTLX and Co-GTL4 catalyst. CO and CO_2 were detected by online GC, the gaseous and liquid hydrocarbons by the ampoule method and offline GC, and the amount of waxy hydrocarbons was determined gravimetrically. Figure 12.9 indicates that the amount of lost carbon – an indicator for the quality of the carbon balance – increases with

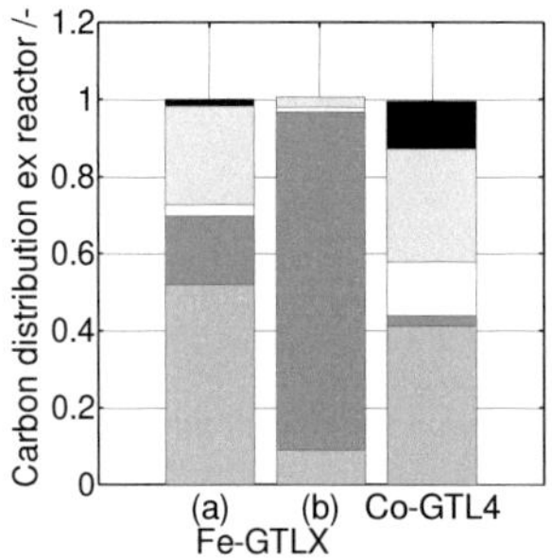

Figure 12.9. Carbon balance: Carbon distribution in effluent streams ex reactor for Fe-GTLX and Co-GTL4 catalyst: carbon out as CO (GC), as CO_2 (GC), as wax (separator □), as gaseous/ liquid hydrocarbons (ampoule), lost carbon (■).
* Fe-GTLX, (a) $H_2/CO = 2$, (b) $H_2/CO_2 = 3$, T = 270 °C, P = 1 MPa, $\tau_{mod,n}$ = 2000 kg s/m³.
* Co-GTL4 (#2b), $H_2/CO = 2$, T = 230 °C, P = 1 MPa, $\tau_{mod,n}$ = 2000 kg s/m³

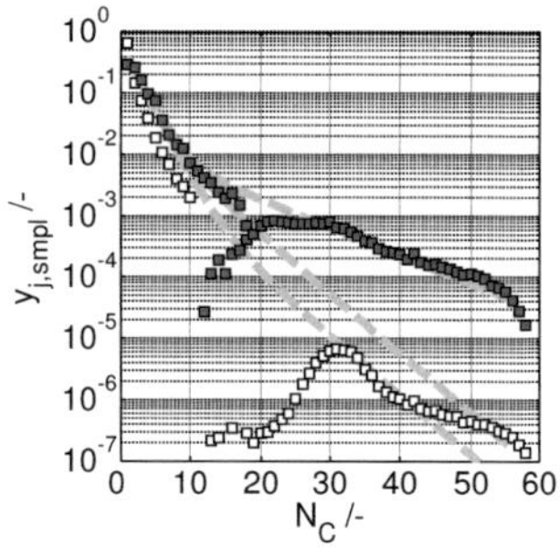

Figure 12.10. Combined product analysis of gas, liquid and waxy hydrocarbons (wax samples) of the Fe-GTLX catalyst for (■) $H_2/CO = 2$ and (□) $H_2/CO_2 = 3$ syngas.
* Fe-GTLX, T = 270 °C, P = 1 MPa, $\tau_{mod,n}$ = 2000 kg s/m³.
Estimated ASF parameters: $H_2/CO = 2$: $\alpha_1 = 0.6$, $\alpha_{2,C9+} = 0.9$; $H_2/CO_2 = 3$: $\alpha_1 = 0.6$, $\alpha_{2,C15+} = 0.8$

increasing yield of long chain hydrocarbons, as the ampoule method and the gravimetric wax sampling are prone to error. About 12% of the carbon is missing in the case of the Co-GTL4 catalyst. The specific production rates were as follows under the given conditions:

- Fe-GTLX $H_2/CO = 2$: 51 g gaseous and liquid hydrocarbons/(h kg catalyst), 10 g wax/ (h kg catalyst)
- Co-GTL4 $H_2/CO = 2$: 87 g gaseous and liquid hydrocarbons/(h kg catalyst), 50 g wax/ (h kg catalyst)

The analysis of Fe-GTLX wax samples (Figure 12.10) confirmed that that the wax contains hydrocarbons with more than 55 carbon atoms (C_{55+}).

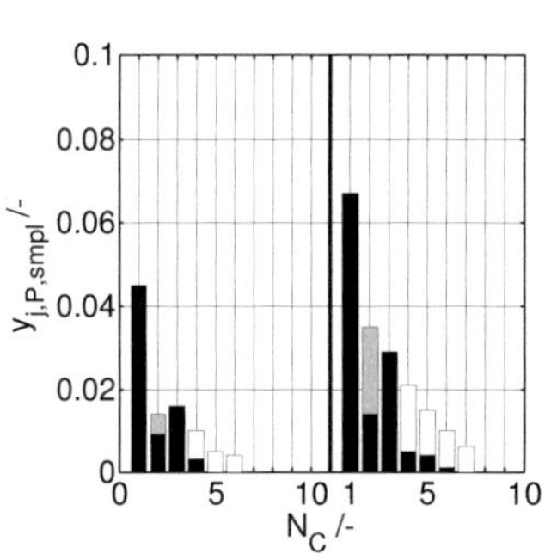

Figure 12.11. Fraction of the total distribution of volatile hydrocarbons (carbon selectivity) found on the permeate side of the CSP2 membrane for Φ = 0.7 (left) and Φ = 0.15 (right).
■: n-alkanes, □: n-alkenes, : oxygenates.
Note: samples analyzed by online-GC (not able to separate $C_2/C_{2=}$ and $C_3/C_{3=}$).
* Fe-GTLX/ CSP2, T = 270 °C, P = 1 MPa, $\tau_{mod,n}$ = 2000 kg s/m³, Ψ = 3.3, $(H_2)_S$

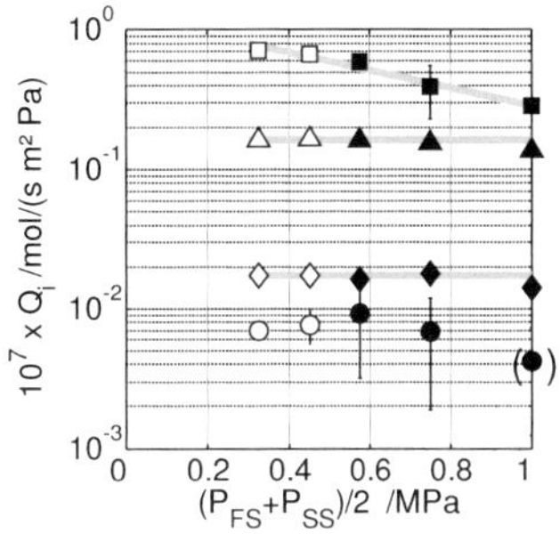

Figure 12.12. CSP2 (P84®) membrane: Permeances Q_i of (□■) H_2O, (△▲) H_2, (◇◆) CO_2 and (○●) CO as function of the mean pressure P_{mean}. Variation of the feed P_{FS} pressure (□△◇○) at constant sweep pressure P_{SS}; variation of the sweep pressure P_{SS} (■▲◆●) at constant feed pressure P_{FS}. Lines: trends.
* CSP2/dry gas mixture: $H_2/CO/CO_2$ (50/30/20 vol%), 25 vol% H_2O, T = 250 °C, Ψ = 2, $(Ar)_S$

12.4 Mass transfer in hydrophilic membranes

12.4.1 Mass transfer limitations

During experiments with gas mixtures in the gas permeation cell, the pressures on the sweep and feed side were altered. By doing this, the total pressure difference and therefore the mean pressure between feed and sweep side was varied. An increasing permeance with increasing mean pressure would indicate viscous flow. The results in Figure 12.12 are in agreement with the single gas experiments and confirm that the permselective polymer layer and carbon gaskets are defect free and viscous flow does not occur. The permeance of H_2 and CO_2 are independent of the mean pressure. (The permeance of CO scatters strongly as the transmembrane flux of CO is very small making an accurate determination of the permeance difficult.) The drop of H_2O permeance with increasing mean pressure represents an exception. A possible explanation could be a pressure dependence of the solubility or diffusion coefficient. Or mass transfer limitations become prevailing at isobaric conditions ($P_{FS} = P_{SS}$), resulting in a lowered, overall permeance.

External mass transfer limitations

As H_2O is the fastest permeating species, its overall permeance would be affected first if a mass transfer limited regime is approached. The transport resistances within the membrane and within the boundary layer are put into relation to each other, yielding an expression analogous to the Biot number used in heat transfer problems:

$$Bi = \frac{\Delta P_{i,m}}{\Delta P_{i,ext}} = \frac{\beta_{ext}}{Q_i RT} \tag{12.3}$$

with $\Delta P_{i,m}$ as partial pressure difference across the membrane and $\Delta P_{i,ext}$ as partial pressure difference in the external boundary layer on the fluid side and β_{ext} as mass transfer coefficient in the boundary layer on the fluid side. Large

Biot numbers indicate that mass transfer limitations are within the membrane. β_{ext} is calculated using correlations for heat transfer in annular gaps (VDI Wärmeatlas 1994, Gd1). β_{ext} is estimated with $1.5 \cdot 10^{-2}$ m/s for the feed side and $7 \cdot 10^{-3}$ m/s for the sweep side. For the feed and the sweep side, Biot numbers of 45 and 15 are calculated, respectively, assuming a permeance Q_i of $1 \cdot 10^{-7}$ mol/(s m^2 Pa). Hence, the external mass transfer is not limiting, even at low gas velocities.

Internal mass transfer limitations

The isobaric diffusion experiment ($P_{FS} = P_{SS}$, $\Phi = 1$) represents a special case, as Knudsen and molecular diffusion are the determining mass transfer mechanisms in the support layers. The transport resistances within the functional polymer layer and within the porous support layers are put into relation to each other, yielding the following expression:

$$\frac{\Delta P_{i,fxn}}{\Delta P_{i,spprt}} = \frac{\varepsilon}{\tau} \frac{\delta}{Q_i RT \Delta s} \tag{12.4}$$

with $\Delta P_{i,fxn}$ as partial pressure difference across the functional polymer layer, $\Delta P_{i,spprt}$ as partial pressure difference across the support layers, δ as binary diffusion coefficient ($7 \cdot 10^{-6}$ m^2/s, H_2O in Ar at 250 °C and 1 MPa) and Δs (3000 μm) as thickness of the tubular membrane. Assuming a permeance Q_i of $1 \cdot 10^{-7}$ mol/(s m^2 Pa), expression (12.4) indicates that transport resistances in the polymer and support layers are in the same magnitude. Therefore, internal mass transfer limitations may play a role when isobaric conditions ($\Phi \rightarrow 1$) are approached.

12.4.2 Membrane unit (permeation cell) performance

The following figures give an overview on the performance of the gas permeation cell equipped with a CSP2 membrane. The gas permeation cell was operated in co-current sweeping mode; the objective was to remove H_2O from synthesis gas stream with an acceptable loss of H_2, CO_2 and CO.

In Figure 12.13, the partial pressures on the feed and sweep side are plotted along the axial coordinate of the gas permeation cell which is operated at 250 °C, low sweep pressure (Φ = 0.15) and high sweep ratio (Ψ = 3.3). The curves are calculated on basis of the mathematical model for the permeation cell assuming an ideal solution-diffusion membrane. The permeances are constant and taken from the experiments (permeances, refer to Table 5.3; activation energies for the permeation step, refer to Table 12.19).

The H_2O partial pressure declines quickly. Due to the low pressure and the high gas flow on the sweep side, H_2O accumulation on the sweep can be

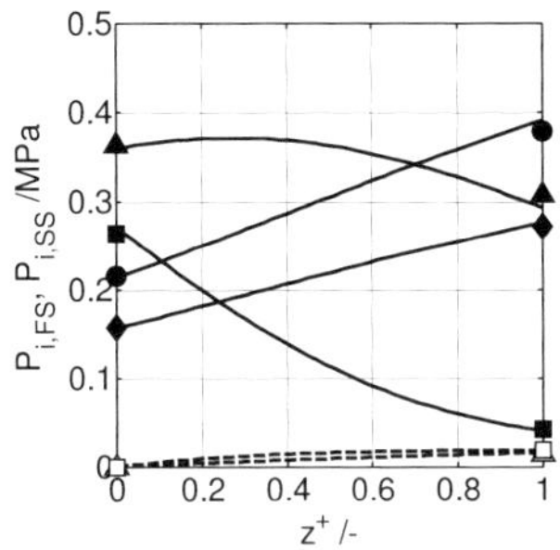

Figure 12.13. Partial pressures of H_2O (■□), H_2 (▲△), CO_2 (◆) and CO (●) on feed (■▲◆●) and sweep (□△) side along axial coordinate of the gas permeation cell.

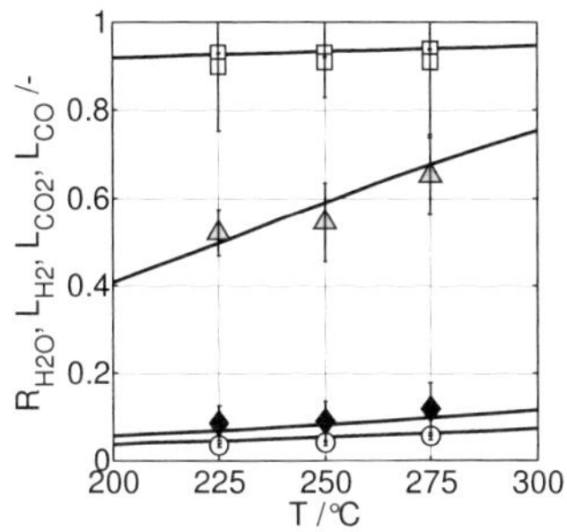

Figure 12.14. Removal of the H_2O (□) and loss of H_2 (▲), CO_2 (◆) and CO (○) to the permeate side as function of temperature.

Curves: calculated with model for permeation cell based on experimentally determined permeances. * Reference case: CSP2/ dry gas mixture: $H_2/CO/CO_2$ (50/30/20 vol%), 25 vol% H_2O, T = 250 ℃, P_F = 1 MPa, Ψ = 2, Φ = 0.15, $(Ar)_S$

limited, and more than 90% of H_2O fed to the permeation cell is removed from the gas stream (Figure 12.14). The slower permeating species are retained on the feed side, resulting in increasing or only slowly dropping partial pressures. This is an important aspect with regard to the membrane reactor application.

At low sweep pressure, about 60% H_2 is lost to the sweep side. The loss of CO_2 is below 10%, the loss of CO below 5%. Due to the temperature independent H_2O permeance, higher temperatures are not beneficial for the overall performance of the permeation cell. Increasing H_2O concentrations in the feed accelerate specifically the H_2O transport kinetics resulting in a high degree of H_2O removal (R_{H2O} > 0.9). However, the permeation of the other species is not or only slightly suppressed (Figure 12.15, a).

The significant loss of H_2 is not acceptable, and therefore, the CSP2 membrane is not suitable for dehydration of H_2 rich gas streams. H_2 loss can be reduced by adaptation of the operating parameters of the gas permeation cell, i.e. reduction of the feed side pressure (Figure 12.15, b), increase of the sweep side pressure (Figure 12.15, c and d) or using H_2 (or H_2-rich gas) as sweep gas (Figure 12.15, d). All proposed variations lead to a reduced loss of H_2, but also in a reduced degree of H_2O removal.

With regard to the FT experiments in the membrane reactor, the deficiencies in membrane permselectivity can be reduced if H_2 is applied as sweep gas at a reduced pressure ratio (Figure 12.15, d).

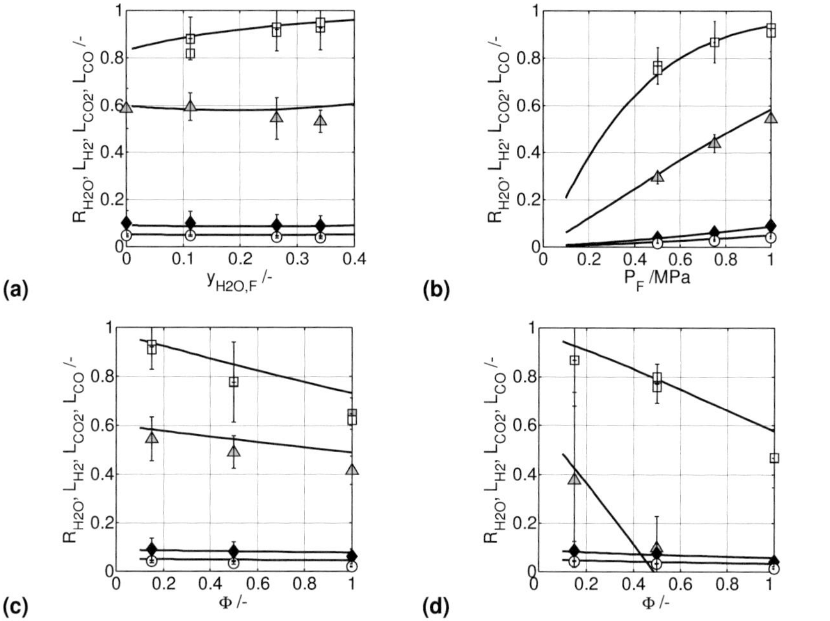

Figure 12.15. Removal of the H_2O (□) and loss of H_2 (▲), CO_2 (◆) and CO (○) to the permeate side as function of (a) molar fraction of H_2O co-fed (H_2O permeance correlated with data in Table 12.21), (b) pressure of the feed side, (c) pressure ratio Φ with Ar as sweep gas, (d) pressure ratio Φ with H_2 as sweep gas.

Curves: calculated with model for permeation cell based on experimentally determined permeances. * Reference case: CSP2/ dry gas mixture: H_2/CO/CO_2 (50/30/20 vol%), 25 vol% H_2O, T = 250 °C, P_F = 1 MPa, Ψ = 2, Φ = 0.15, $(Ar)_S$

12.4.3 Dynamic permeation experiments

Figure 12.16 shows the pressure response of a dynamic permeation experiment (see configuration in Figure 3.4, c). Initially, the membrane was swept with Ar on the feed side. The permeate side was blocked-in and filled with Ar at the same pressure level. Due to a sudden change from Ar to H_2 as feed gas, H_2 started permeating from the feed side to the permeate side, while Ar started permeating in the opposite direction. As H_2 permeates faster than Ar, the pressure on the permeate side runs through a maximum. The system required more than 60 minutes to equilibrate. A step change from H_2 to Ar as feed gas yielded an inverted pressure response.

A simple model assuming fully back-mixed compartments on the feed and permeate side, neglecting transient diffusion effects in the membrane and applying constant permeances derived from the previous permeation

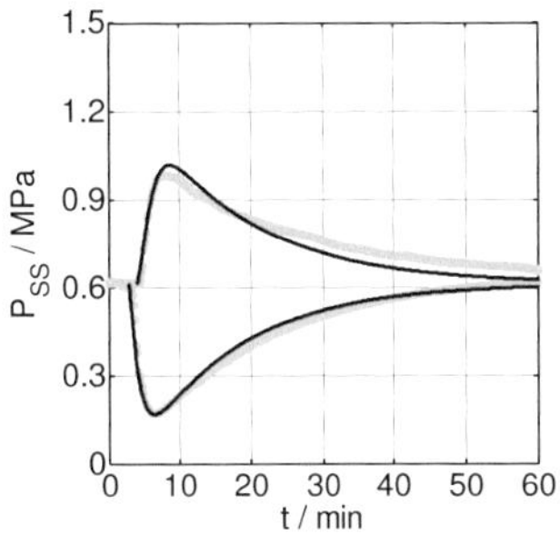

Figure 12.16. Dynamic permeation experiment: pressure response within the blocked-in sweep side as result of a step-change in the feed gas composition: Ar → H_2 (top), H_2 → Ar (bottom).
Curves: experimental data (grey), calculated with simple model with fixed permeances (black).
* CSP2: P_F = 0.6 MPa, $P_{SS,0}$ = 0.6 MPa, T = 250 °C, V_{FS} = 28.4 cm³, V_{SS} = 25 cm³, Q_{H2} = $0.16 \cdot 10^{-7}$ mol/(s m² Pa), Q_{Ar} = $0.01 \cdot 10^{-7}$ mol/(s m² Pa)

experiments (see also Table 5.3) was able to describe the pressure response of the experiment with Ar/H_2 (Figure 12.16) and of experiments with gas mixtures without H_2O (Ar → H_2/CO/CO_2) and with H_2O (H_2/CO/CO_2 → H_2O/H_2/CO/CO_2).

A lag time between the step change and the pressure response was observed, which should be used here to estimate the diffusion coefficient. Generally time-lag experiments are carried out in dedicated apparatus and the membrane compartments are evacuated for several hours before gas is introduced on the feed side (Koros & Chern 1987). From time-lag experiments, the diffusion coefficient is determined as follows:

$$D_i = \Delta s^2 / (6 t_{lag}) \tag{12.5}$$

Considering the non-ideal experimental set-up, the following numbers can give only a rough estimation. For H_2 in the Ar/H_2 experiment and for H_2O in the H_2O/H_2/CO/CO_2 experiment, time-lags of about 3.5 minutes and 2.2 minutes were observed, respectively. With a membrane thickness of 1 µm, one estimates a diffusion coefficient for H_2 > $8 \cdot 10^{-12}$ cm²/s (0.6 MPa, 250 °C) and for H_2O > $1.3 \cdot 10^{-11}$ cm²/s (1 MPa, 250 °C). A H_2 solubility of < 0.5 cm³/(cm³ Pa) corresponds to a H_2 permeance of Q_{H2} = $0.16 \cdot 10^{-7}$ mol/(s m² Pa); a H_2O solubility of < 1.1 cm³/(cm³ Pa) corresponds to a H_2O permeance of Q_{H2O} = $0.58 \cdot 10^{-7}$ mol/(s m² Pa).

As the polymer should be in a glassy state ($T < T_g$), low diffusion coefficients are expected, however, these diffusion coefficients appear several order of magnitudes too low (in glassy polymers diffusion coefficients around 10^{-10} cm²/s are observed). But the numbers show qualitatively, that H_2O dissolves preferably in the hydrophilic polymer matrix. The higher solubility in combination with an apparently faster diffusion transport results in a higher permeance for H_2O than for H_2.

Table 12.19. Permeance 10^7 x Qi in mol/(s Pa m²) for various components determined by steady state and transient pressure drop measurements and by steady-state permeation experiments with gas mixtures. Dry gas mixture: $H_2/CO/CO_2$ (50/30/20 vol%); P_F = 1 MPa, P_S = 0.15 MPa (Ar), dry feed flow rate: 200 ml/min (NTP), sweep flow rate: 400 ml/min (NTP).

gas	method	$y_{H2O,F}$ /-	T /°C							$E_{A,P}$
			100	150	200	225	250	275	300	/kJ/mol
H_2O	multi/ steady	0.26				0.663	0.588	0.670		0.318
H_2O [a]	multi/ steady	0.3			0.76		0.6		0.9	(3.3)
H_2	single/ trans.	-	0.041	0.073	0.118		0.166			15.2
							0.166 [b]			
							0.174 [c]			
							0.175 [d]			
H_2	single/ steady	-	0.039	0.049	0.124		0.176			16.3
					0.121					
H_2	multi/ steady	0.0					0.153	0.192		
H_2	multi/ steady	0.26				0.134	0.164	0.206		19.4
H_2 [a]	multi/ steady	0.3			0.11		0.16		0.37	(26.9)
CO_2	single/ trans.	-	0.003	0.005	0.009		0.012			9.5
CO_2	multi/ steady	0.0					0.019	0.021		
CO_2	multi/ steady	0.26				0.013	0.016	0.017		13.3
CO_2 [a]	multi/ steady	0.3			0.029		0.035		0.066	(18.4)
CO	single/ trans.	-	0.004	0.006	0.009		0.010			10.7
CO	single/ steady	-	0.004	0.006	0.008		0.011			9.6
CO	multi/ steady	0.0					0.010	0.010		
CO	multi/ steady	0.26				0.009	0.009	0.012		12.0
Ar	single/ trans.		0.003	0.005	0.009		0.012			
CH_4 [a]	single/ steady	-	0.005		<0.03					
CH_4 [a]	multi/ steady	0.3			0.012		0.021		0.039	(26.5)

[a] results from ECN (Vente 2006): 47% H_2, 30% H_2O, 20% CO_2, 3% CH_4; pressure drop 9 bar
[b] reverse flow, [c] higher sweep pressure, [d] repro

Table 12.20. Permeance 10^7 x Q_i in mol/(s Pa m²) for various components determined by steady state permeation experiments with gas mixtures with methanol and ethanol co-feeding. Dry gas mixture: $H_2/CO/CO_2$ (50/30/20 vol%); P_F = 1 MPa, P_S = 0.15 MPa (Ar), dry feed flow rate: 200 ml/min (NTP), sweep flow rate: 400 ml/min (NTP).

gas	method	$y_{MeOH,F}$ /- $y_{EtOH,F}$ /-	T /°C			
			200	250	270	290
MeOH	multi/ steady	0.25	0.026	0.028	0.034	0.047
H_2			0.10	0.149	0.186	0.244
CO_2			0.011	0.025	0.017	0.024
CO			0.003	0.014	0.007	0.009
EtOH	multi/ steady	0.26	0.002	0.002	0.003	0.005
H_2			0.123	0.153	0.223	0.299
CO_2			0.0171	0.026	0.024	0.030
CO			0.003	0.014	0.005	0.006

Table 12.21. Permeance 10^7 x Q_i in mol/(s Pa m²) for various components determined by steady state by steady-state permeation experiments with gas mixtures with varying contents of H_2O in the feed gas. Dry gas mixture: $H_2/CO/CO_2$ (50/30/20 vol%); P_F = 1 MPa, P_S = 0.15 MPa (Ar), dry feed flow rate: 200 ml/min (NTP), sweep flow rate: 400 ml/min (NTP).

gas	method	T /°C	$y_{H2O,F}$ /-						
			0	0.1	0.11	0.2	0.26	0.30	0.34
H_2O	multi/ steady	250	-		0.438		0.588		0.801
H_2	multi/ steady		0.152		0.157		0.164		0.167
CO_2	multi/ steady		0.017		0.019		0.016		0.019
CO	multi/ steady		0.008		0.009		0.009		0.008
H_2O [a]	multi/ steady	250		0.49		0.72		0.73	
H_2 [a]	multi/ steady			0.16		0.17		0.17	
CO_2 [a]	multi/ steady			0.032		0.036		0.028	
H_2O [a]	multi/ steady	200		0.56		0.69		0.75	
H_2 [a]	multi/ steady			0.11		0.11		0.11	
CO_2 [a]	multi/ steady			0.026		0.030		0.029	

[a] results from ECN (Vente 2006): 47% H_2, 30% H_2O, 20% CO_2, 3% CH_4; pressure drop 9 bar

12.5 Kinetics and selectivity

12.5.1 Fischer-Tropsch synthesis

12.5.1.1 Activation behaviour of Fe- and Co-based FT catalysts

The H_2 consumption during temperature programmed reduction (TPR) of the Fe- and Co-based FT catalysts and of the low temperature shift catalyst is given in Figure 12.17.

Figure 12.18 (a-d) shows the activation behaviour of the Fe-5K, Fe-GTL3, Fe-GTLX and Co-GTL4 FT catalysts. After the reduction of the catalyst, the reactor is cooled down to the activation temperature under a steady Ar flow and pressurized to 1 MPa. At t=0, the feed stream is switched from Ar to H_2/CO syngas by turning a 4-way valve. From this point on, the formation of the FT regime starts; this is a process that runs through several transient kinetic regimes (episodes) until the state of the highest FT activity is reached. Schulz et al. (1999) and Schulz, Nie & Usmanov (2002) studied these transient kinetic regimes in detail for K-promoted Fe-based catalyst and various Co-based catalysts, respectively.

One distinguishes for K-promoted Fe-based catalyst – see e.g. Figure 12.18 (a) – the following episodes: (I) apparent high CO conversion due to CO retention and CO adsorption, (II-III) carbiding of the catalyst, (IV) strong increase in FT and CO/CO_2 shift activity and shift of the product selectivity, (V) established FT regime with maximum FT activity, followed by (VI) catalyst deactivation. Strong catalyst deactivation was observed for the Fe-5K catalyst

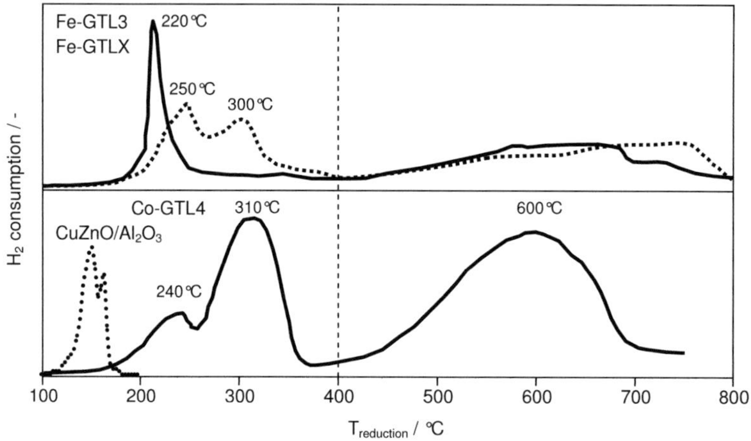

Figure 12.17. H_2 consumption during temperature programmed reduction (TPR) of the applied Fe-based FT catalysts Fe-GTL3 (—) and Fe-GTLX (----) and the Co-based catalyst GTL4 (—) and low-temperature shift catalyst $CuZnO/Al_2O_3$ (----). TPR conditions (carried out by Quantachrome GmbH, Odelzhausen, Germany): H_2/Ar (1/3), 20 $cm^3/(min\ g)$, 0.5 g catalyst, 25 °C – 2 K/min – 900 °C

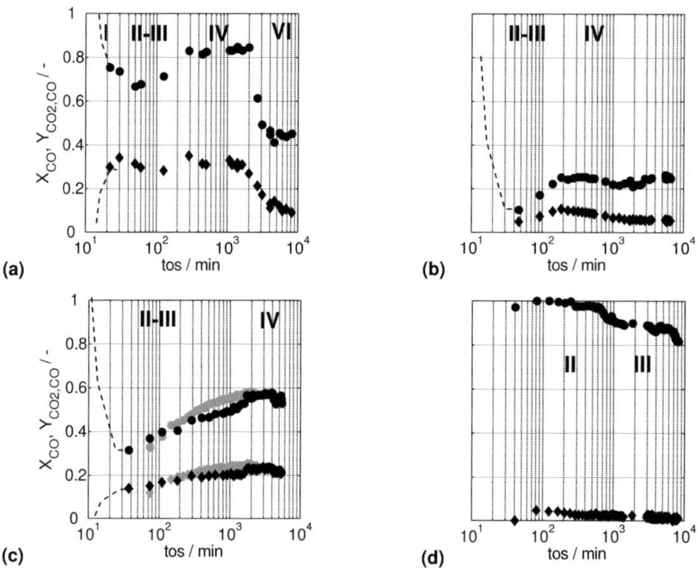

Figure 12.18. CO conversion and CO_2 yield versus time on stream (tos) during activation and stabilisation of the following catalysts:

(a) Fe-5K: T = 250 °C, H_2/CO = 2, P = 1 MPa, $\tau_{mod,n}$ = 4000 kg s/m^3, offline GC (ampoule)

(b) Fe-GTL3: T = 225 °C, H_2/CO = 2, P = 1 MPa, $\tau_{mod,n}$ = 5000 kg s/m^3, online GC

(c) Fe-GTLX: T = 250 °C, H_2/CO = 2, P = 1 MPa, $\tau_{mod,n}$ = 2000 kg s/m^3, online GC

(d) CO-GTL4 (#2b): T = 230 °C, H_2/CO = 2, P = 1 MPa, $\tau_{mod,n}$ = 4000 kg s/m^3, online GC

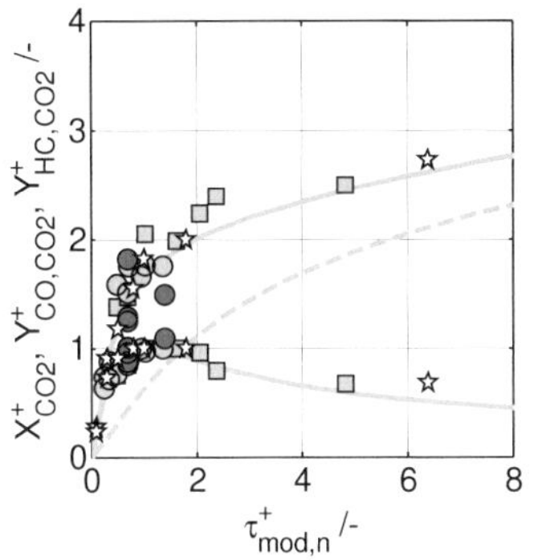

Figure 12.19. Dimensionless plot of CO_2 conversion through the intermediate CO to hydrocarbons in isothermal fixed-bed reactors with Fe-based catalysts (H_2/CO_2 = 3). Conversion and yield are scaled to the maximum CO yield obtained; residence time is scaled to the characteristic residence time at which maximum CO yield is reached. Curves: calculated.

● Fe-GTLX (Rohde)
● Fe-GTLX (Mena 2009)
■ Fe-5K (Unruh 2006)
☆ Fe (Riedel 2003, 300 °C)

due to its high degree of alkalization (only for H_2/CO syngases); a pronounced deactivation regime was not found for the other tested catalysts.

Schulz, Nie & Usmanov (2002) distinguish three episodes for Co-based catalysts: (I) apparent high CO conversion due to CO retention and CO adsorption, (II) built-up of FT activity by solid-state transformations ('in-situ construction' of the FT catalyst) and shift in the product selectivity to much higher chain growth probability and increasing olefinicity, (III) established FT regime with maximum FT activity. These regimes cannot be recognized in Figure 12.18 (d), as the Co-GTL4 was activated under high conversion conditions (up to 100% CO conversion). It took around 3000 min (50h) until the FT regime was established and a stable operating point was achieved. Due to its strong deactivation, the Fe-5K catalyst was discarded.

12.5.1.2 Benchmarking of Fe- and Co-based FT catalysts

CO_2 hydrogenation

The conversion of CO_2 to hydrocarbons by CO/CO_2-shift active Fe-based catalyst takes place via CO as intermediate. The dimensionless plot in Figure 12.19 merges data from various CO_2 hydrogenation experiments reported in literature and highlights the typical behaviour of CO as intermediate. The CO yield $Y_{CO,CO2}$ runs through a maximum with increasing residence time.

Comparison of catalyst performance

Figure 12.20 compares the Fe-GTLX catalyst (Rohde) to the Fe-5K catalyst (Unruh 2006) on basis of reaction rates of FT and CO_2/CO shift reaction and on basis of the differential selectivity ${}^{d}S$ (12.6) and stability of the intermediate λ_{CO} (12.7). The differential selectivity and the stability of the intermediate are key parameters in reaction engineering to discuss parallel and consecutive reactions.

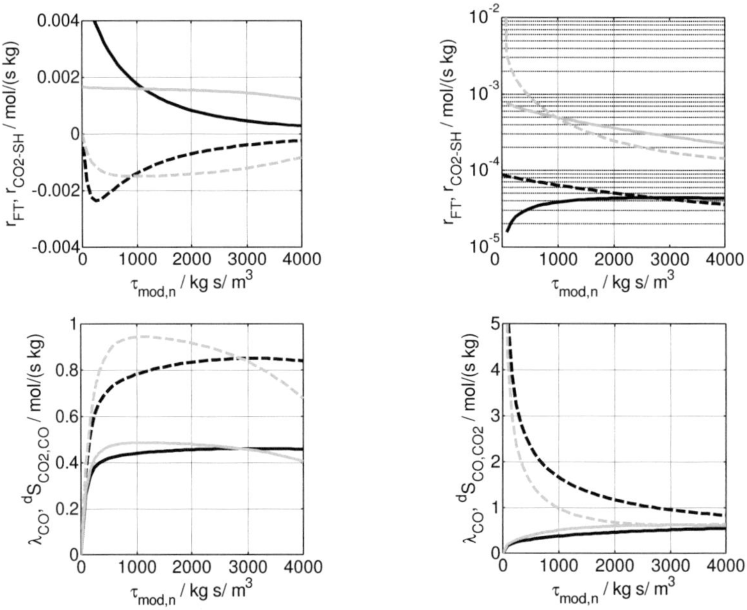

Figure 12.20. Top row: Reaction rates of the FT (solid line) and CO_2/CO shift (broken line) reaction, calculated according to Fe-GTLX (black) and Fe-5K (grey, Unruh 2006) kinetics for H_2/CO (left) and H_2/CO_2 (right) syngases.
Bottom row: Differential selectivity ^dS (solid line) and stability of the intermediate product λ_{CO} (broken line), calculated according to Fe-GTLX (black) and Fe-5K (grey, Unruh 2006) kinetics for H_2/CO (left) and H_2/CO_2 (right) syngases.
* Fe-GTLX: T = 270℃, P = 1 MPa, Fe-5K: T = 250℃, P = 1 MPa

$$^dS_{CO,CO2} = \frac{r_{CO2-SH}}{r_{FT} + r_{CO2-SH}} \qquad \left[\frac{r_{CO2\rightarrow CO}}{r_{CO\rightarrow HC} + r_{CO2\rightarrow CO}}\right] \tag{12.6}$$

$$\lambda_{CO} = \left|\frac{r_{CO2-SH}}{r_{FT}}\right| \qquad \left[\frac{r_{CO2\rightarrow CO}}{r_{CO\rightarrow H2}}\right] \tag{12.7}$$

From Figure 12.20 (left), the following conclusions can be drawn for H_2/CO syngases (solely based on the reactor model and applied kinetic rate laws):

- The rate of the reverse CO_2/CO shift reaction is negative, as additional CO_2 is formed. The rate initially accelerates due to the high driving force of the thermodynamic equilibrium and due to H_2O formed by the FT reaction. At very high residence times, the reverse CO_2/CO shift reaction will switch signs, indicating that FT reaction is drawing CO out of the equilibrium.
- The selectivity towards CO_2 – expressed by $^dS_{CO2,CO}$ – is limited to 50% due to stoichiometric limitations (dry syngas).

Table 12.22. Results of benchmarking tests of Fe- and Co-based catalyst under conditions of FTS and CO_2 hydrogenation in lab-scale fixed-bed reactors. Additional data given from literature references [1] Unruh (2006) and [2] Mena (2009). All catalyst activated according to procedure given in Table 4.2.

FT synthesis: $H_2/CO = 2$							
catalyst	T /°C	P /MPa	$\tau_{mod,n}$ / kg s/m³	X_{CO} /-	$Y_{CO2,CO}$ /-	$S_{CO2,CO}$ /-	WTY_{C1+} / g/(h kg)
1 Fe-5K	250	1	4000	0.83	0.31	0.37	97
2 Fe-5K [1]	250	1	4000	0.67	0.28	0.42	73
3 Fe-GTLX	250	1	4000	0.76	0.28	0.37	89
4 Fe-GTLX[2]	250	1	4000	0.70	0.29	0.41	77
5 Fe-GTL3	235	1	5625	0.35	0.08	0.22	36
6 Co-GTL4	230	1	2000	0.58	0.02	0.03	207
- Co-GTL1	205	1	2670	0.74	0	0	206
CO_2 hydrogenation: $H_2/CO_2 = 3$							
catalyst	T / °C	P /MPa	$\tau_{mod,n}$ / kg s/m³	X_{CO2} /-	$Y_{HC,CO2}$ /-	$S_{HC,CO2}$ /-	WTY_{C1+} / g/(h kg)
1 Fe-5K	250	1	4000	0.28	0.22	0.79	22
2 Fe-5K[1]	250	1	4000	0.21	0.15	0.73	21
3 Fe-GTLX	270	1	4000	0.13	0.03	0.27	5
4 Fe-GTLX[2]	250	1	4000	0.10	0.04	0.39	5
5 Fe-GTL3	255	1	6900	0.06	0.01	0.18	1
6 Co-GTL4	-	-	-	-	-	-	-
- Co-GTL1	205	1	2670	0.74	0	0	206

- Within this temperature range (250-270°C), the rate of the FT reaction is higher than the rate of the CO/CO_2 shift reaction ($\lambda_{CO} < 1$).

From Figure 12.20 (right), the following conclusions can be drawn for H_2/CO_2 syngases (solely based on the reactor model and applied kinetic rate laws):

- The rate of the CO_2/CO shift reaction is significantly higher than the rate of FT reaction ($\lambda_{CO} > 1$), indicating a high initial stability of the intermediate CO.
- The maximum in CO yield $Y_{CO,CO2}$ ($\lambda_{CO} = 1$) indicates that FT reaction consumes more CO.

For H_2/CO syngases, both catalysts show a similar performance as the average FT rates are similar. However, for H_2/CO_2 syngases, Fe-5K exhibits a much higher FT and CO_2/CO shift activity, leading to a pronounced maximum in CO yield and a significant higher hydrocarbon yield and production.

Table 12.23. Kinetic parameter values of the Fischer-Tropsch and CO_2/CO-shift reaction on the K-promoted Fe-GTL3 catalyst. Kinetic rate parameters according to the rate equations of Zimmerman & Bukur (1990), refer to Table 4.4 (FTS1/CO2-SH1).
* Fe-GTL3 , P = 1 MPa, T = (235-)275°C, $\tau_{mod,n}$ = 0- 15000 kg s/m³, $z_{H2,F}$ =1, 0 < $z_{CO2,C,F}$ < 1

Fe-GTL3		Fischer-Tropsch	CO_2/CO-shift
$k_{235°C}$	/mol / (s·kg·MPa)	$3.1 \cdot 10^{-13}$	$2.8 \cdot 10^{-14}$
$k_{255°C}$	/mol / (s·kg·MPa)	$1.5 \cdot 10^{-12}$	$1.2 \cdot 10^{-13}$
$k_{275°C}$	/mol / (s·kg·MPa)	$2.9 \cdot 10^{-12}$	$4.0 \cdot 10^{-13}$
a_{CO}	/-	1	1
b_{H2O}	/-	$1.2 \cdot 10^{-3}$	$2.6 \cdot 10^{-2}$
c_{CO2}	/-	$6.1 \cdot 10^{-3}$	$1 \cdot 10^{-12}$

12.5.1.3 Kinetics of FT catalysts

Kinetics of Fe-GTLX catalyst (additional figures)

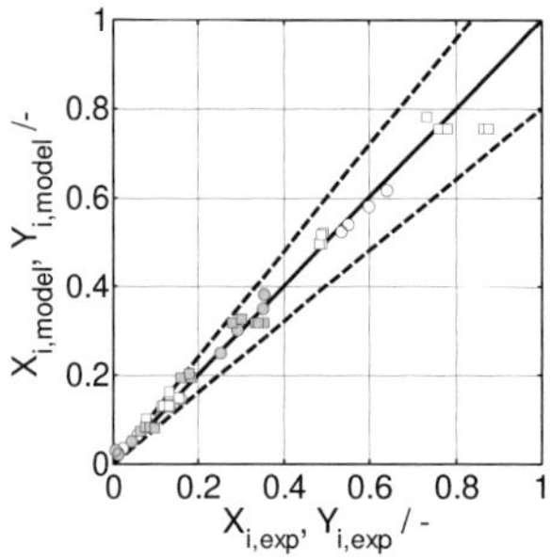

Figure 12.21. Parity plot between the experimental and model values (FTS2/CO2-SH2) of conversion data (□○) and yield data (■●); experiments: variation of residence time (□■), H_2O co-feeding (○●); broken lines: error ±20%.
* Fe-GTLX, T = 270℃, $P_{H2,F} + P_{CO,F} + P_{CO2,F}$ = 1 MPa

Kinetics of Fe-GTL3 catalyst

During fixed-bed experiments with the Fe-GTL3 catalyst without membrane, the experimental conditions as residence time and temperature were varied systematically for H_2/CO and H_2/CO_2 (Figure 12.22) syngases. Kinetic rate parameters were determined by simultaneous regression analysis of conversion-residence time and yield-residence time data for H_2/CO and H_2/CO_2 syngases. The kinetic rate parameters are summarized in Table 12.23. For H_2/CO syngases, the catalyst exhibits a very high FT activity with a very low CO/CO_2 shift activity. Fe-GTL3 shows low activity in converting CO_2 to hydrocarbons. With the shift from H_2/CO to H_2/CO_2 syngas, the total hydrocarbon yield drops sharply (Figure 12.23).

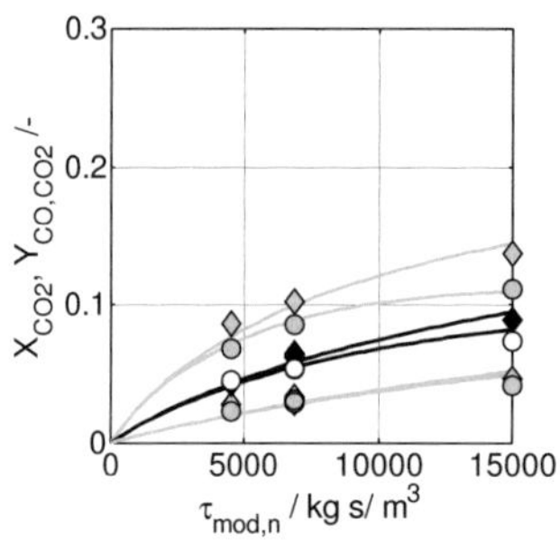

Figure 12.22. Fe-GTL3: Measured conversions (◆◆◇) and CO yields (○●●) for H_2/CO_2 = 3/1 as function of the modified residence time at 255℃ (◆○) and at 235℃ and 275℃ (◆●), respectively. Curves: calculated with kinetic model.
* Fe-GTL3, P = 1 MPa

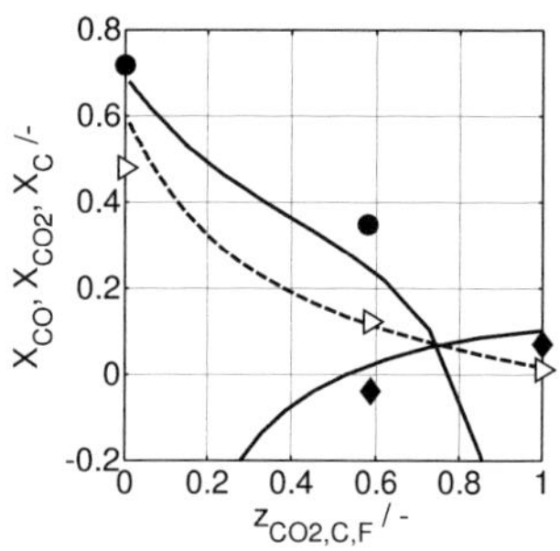

Figure 12.23. CO (●), CO_2 (◆) and total carbon conversion (▷) as function of the stoichiometric syngas composition; lines: calculated with kinetic model, broken line: total carbon conversion.
* Fe-GTL3, T = 275℃, P = 1 MPa, $\tau_{mod,n}$ = 6900 kg s/m^3

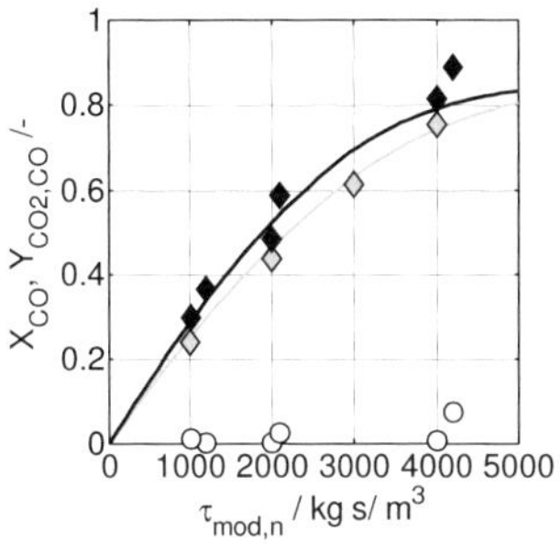

Figure 12.24. Co-GTL4 (#2a/b): Measured conversions (◆◆) and CO_2 yields (○) for H_2/CO = 2/1 as function of the modified residence time. X_{CO} curves: calculated with kinetic model; (◆—): data from Mena (2009) as reference.
* Co-GTL4 (#2a/b), T = 230 °C, P = 1 MPa

Kinetics of Co-GTL4 catalyst

The Co-GTL4 catalyst was tested in three independent experimental runs. The first run (#1) focussed on the effect of CO_2 containing syngases on catalyst performance and selectivity, while run two and three (#2a/b) were related to H_2O co-feeding experiments. Though the catalyst for each run was taken out of the same batch, the activity found in the first run (#1) was significant lower than in runs (#2a) and (#2b). The apparent rate constant in run (#1) is about 50% of the rate constant determined for the runs (#2a/b) (see Table 12.24). These findings can possibly be attributed to different holding times under inert gas conditions during activation.

Mena (2008) confirmed the higher activity of runs (#2a/b) by independent experiments in a different experimental set-up and reported an activation energy of 120 kJ/mol. The kinetic rate parameters of the two catalyst batches are summarized in Table 12.24; residence time variations are shown for Co-GTL4 (#1) in Figure 4.8 and for Co-GTL4 (#2) in Figure 12.24.

H_2O co-feeding experiments with Co-GTL4 catalyst

Figure 12.25 shows the results related to H_2O co-feeding experiments (Table 12.25). During these experiments, the total pressure was increased so that the partial pressure of the reactants $P_{H2,F}$ and $P_{CO,F}$ were kept constant in the feed.

Table 12.24. Kinetic parameter values of the Co-GTL4 (batch #1, #2) catalyst. Kinetic rate parameters of FT reaction according to the rate equation of Yates & Satterfield (1991).
* P = 1 MPa, T = 230 °C, $\tau_{mod,n}$ = 0 - 4000 kg s/m^3, $z_{H2,F}$ = 1, 0 < $z_{CO2,C,F}$ < 1
* Co-GTL4 (#1): E_A = 219 kJ/mol, k_0 = 7.4 $\cdot 10^9$ mol/(s·kg·Pa^2)
* Co-GTL4 (#2): E_A = 125 kJ/mol, k_0 = 3.13 mol/(s·kg·Pa^2)

		Co-GTL4 (batch #1)	Co-GTL4 (batch #2)
$k_{220°C}$	/mol / (s·kg·Pa^2)	$5.1 \cdot 10^{-14}$	$4.9 \cdot 10^{-14}$
$k_{230°C}$	/mol / (s·kg·Pa^2)	$1.46 \cdot 10^{-13}$	$1.3 \cdot 10^{-13}$
$k_{240°C}$	/mol / (s·kg·Pa^2)	$4.1 \cdot 10^{-13}$	$3.3 \cdot 10^{-13}$
a_{CO}	/ Pa^{-1}	$9.2 \cdot 10^{-6}$	$9.2 \cdot 10^{-6}$

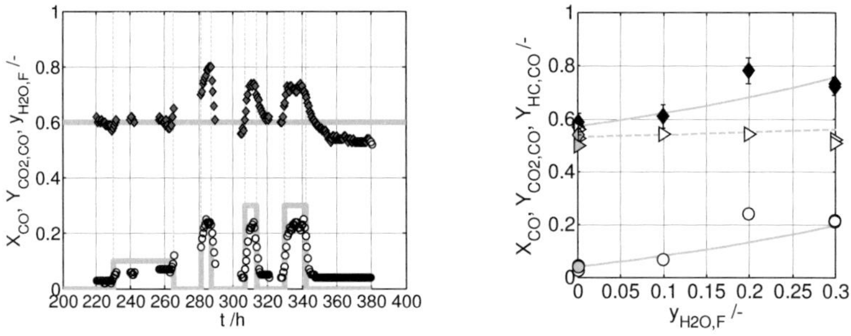

Figure 12.25. Measured CO conversion (◆), CO_2 yield (○) and hydrocarbon yield (▷) for a H_2/CO = 2/1 syngas as function of the H_2O fraction (⊓) co-fed in the fresh feed. (◆●▶) repro after H_2O co-feeding experiments. Curves: (grey) calculated with FT kinetics considering CO/CO_2 shift reaction.
* Co-GTL4 (#2b), T = 230 °C, $P_{H2,F} + P_{CO,F}$ = 1 MPa, $\tau_{mod,n}$ = 2000 kg s/m^3

- With increasing H_2O fraction in the feed stream, the CO conversion increases due to an increased formation of CO_2. The Co-GTL4 exhibits a noticeable CO/CO_2 shift activity under these conditions. This is not agreement with general observations in literature. Overall, the hydrocarbon yield remains constant.
- The co-feeding experiments were carried out for about 140 h (6d). At the end, the CO conversion of the reference point without co-feeding dropped from 58% to 53% (Figure 12.25). The Co-GTL4 catalyst deactivated though the critical H_2O/H_2 ratio (~0.7) and the critical H_2O partial pressure (~0.6 MPa) were not exceeded considerably (Table 12.25).
- Further experiments are needed to investigate H_2O-driven deactivation of Co-based catalysts, in particular at lower temperatures. The mean deactivation rate during the co-feeding experiments was about -0.04% CO conversion per hour. E.g. Davis (2003) and Storsaeter et al. (2005a) report deactivation rates of -0.03%/h for a Co/SiO_2 catalyst and -0.2%/h for a Co/Al_2O_3-Re catalyst.

12.5.1.4 Selectivity of Fe- and Co-based FT catalysts

FT product spectra are often displayed as ASF-plots (see Figure 2.5). The molar fraction $y_{j,HC}$ of hydrocarbons molecules containing j carbon atoms is

Table 12.25. H_2O co-feeding experiments with Co-GTL4 (#2b) catalyst, see Figure 12.25.

$y_{H2O,F}$	X_{CO}	H_2O/H_2	P_{H2O} [b] / MPa
0	0.58 (0.53)[a]	0.25	0.28
0.1	0.61	0.44	0.40
0.2	0.78	0.63	0.48
0.3	0.73	1	0.65

[a] repro after end of H_2O co-feeding; [b] at 230 °C: $P_{H2O,sat}$ = 2.8 MPa

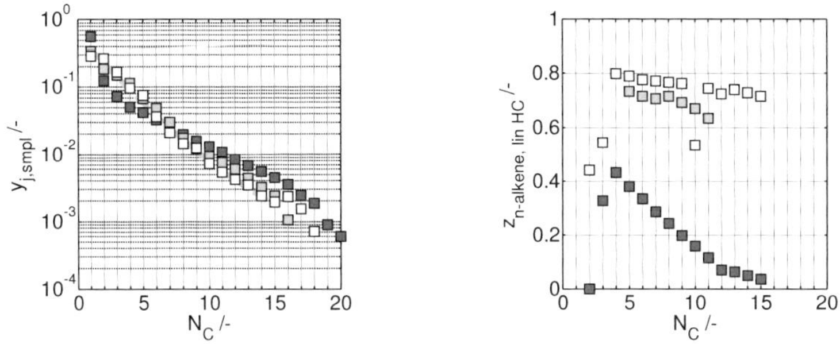

Figure 12.26. ASF distribution of the volatile hydrocarbons (left) and molar alkene fraction in the fraction of linear hydrocarbons as function of the carbon number for the following catalyst: Co-GTL4 (■), Fe-GTL3 (▒), Fe-GTLX (□).
* Co-GTL4 (#2b), $H_2/CO = 2$, $T = 230°C$, $P = 1$ MPa, $\tau_{mod,n} = 2100$ kg s/m^3
* Fe-GTL3, $H_2/CO = 2$, $T = 225°C$, $P = 1$ MPa, $\tau_{mod,n} = 5000$ kg s/m^3
* Fe-GTLX, $H_2/CO = 2$, $T = 270°C$, $P = 1$ MPa, $\tau_{mod,n} = 2000$ kg s/m^3

calculated on basis of the total number of organic molecules detected in the FID chromatogram of the GC. The molar fractions depend on the sampling method and the quality of the analytical method and therefore, the experimentally derived molar fractions $y_{j,HC}$ are denoted as $y_{j,smpl}$. The applied on-line GC covers only the hydrocarbon range C_1-C_{10}, while the off-line GC (in combination with the ampoule method) covers C_1-C_{20}.

$$y_{j,smpl} = \dot{N}_j / \sum_{j=1}^{N_{smpl}} \dot{N}_j \tag{12.8}$$

The distribution between linear alkenes (n-alkenes) and alkanes (n-paraffins) is a good indicator for the extent of secondary reactions during FT synthesis (12.9).

$$z_{n-alkene,lin\ HC} = \frac{y_{j,n-alkene}}{y_{j,n-alkene} + y_{j,n-alkane}} \tag{12.9}$$

The next figures address the effect of the operating conditions on the product selectivity of Co- and Fe-based FT catalysts.

Comparison of Fe- and Co-based FT catalysts

Figure 12.26 compares the hydrocarbon distributions of the two Fe-based catalysts (Fe-GTLX and Fe-GTL3) and one Co-based catalyst (Co-GTL4). The overall product of the Co-based catalyst is heavier compared to Fe-based catalysts, though it exhibits a significantly higher CH_4 selectivity. The primary products of the Fischer-Tropsch synthesis are linear 1-alkenes, with a primary selectivity of about 80 mol% (Schulz 1977, Riedel 2003). Compared to Co-based catalysts, secondary reactions of alkenes play an inferior role for Fe-

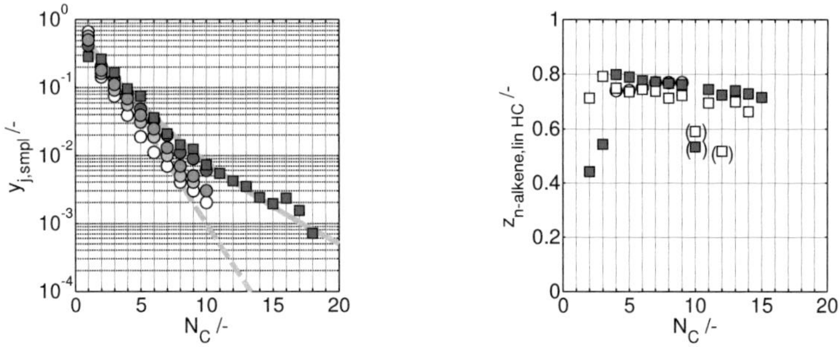

Figure 12.27. ASF distribution of the volatile hydrocarbons (left) and molar alkene fraction in the fraction of linear hydrocarbons as function of the carbon number for the following syngases with increasing CO_2 content: $z_{CO2,C,F}$ ■●: 0 ■●: 0.54 ■●: 0.78 □○: 1. Square symbols: ampoule method/ offline GC, round symbols: online GC.
ASF parameters: $H_2/CO = 2$ (solid line): $\alpha_1 = 0.6$, $\alpha_{2,C9+} = 0.8$; $H_2/CO_2 = 3$ (broken line): $\alpha = 0.5$
* Fe-GTLX, T = 270 °C, P = 1 MPa, $\tau_{mod,n}$ = 2000 kg s/m^3

based catalysts, indicated by an only slowly dropping alkene fraction with increasing carbon number.

Effect of synthesis gas composition – $z_{CO2,C,F}$

Figure 12.27 - Figure 12.28 compare the effect of syngas composition $H_2/CO/CO_2$ – expressed by $z_{CO2,C,F}$ – on the product selectivity for Fe- GTLX and Co-GTL4 catalysts.

Figure 12.27 (left) is an ASF plot of the C_1-C_{17} molar product distribution (carbon number distribution) of the Fe-GTLX catalyst for various synthesis gases with increasing CO_2 fraction in the feed. With increasing $z_{CO2,C,F}$, the chain growth probability decreases, whereas the methane selectivity increases. For the H_2/CO syngas, the carbon number distribution can be

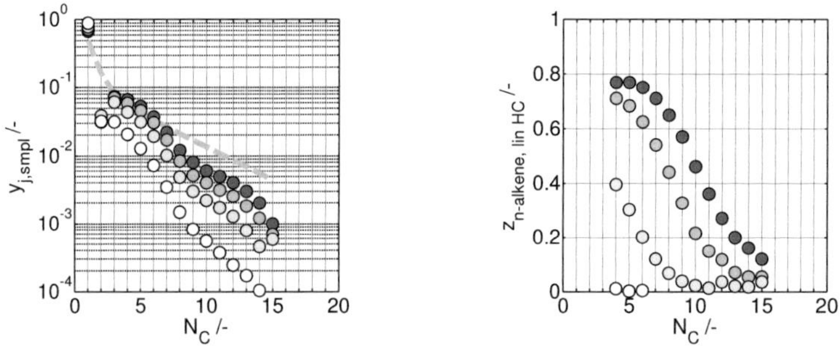

Figure 12.28. ASF distribution of the volatile hydrocarbons (left) and molar alkene fraction in the fraction of linear hydrocarbons as function of the carbon number for the following syngases with increasing CO_2 content: $z_{CO2,C,F}$: ●: 0, ●: 0.3, ●: 0.5, ○: 0.7.
* Co-GTL4 (#1), $H_2/CO = 2$, T = 230 °C, P_{CO} = 1 MPa, $\tau_{mod,n}$ = 2000 kg s/m^3

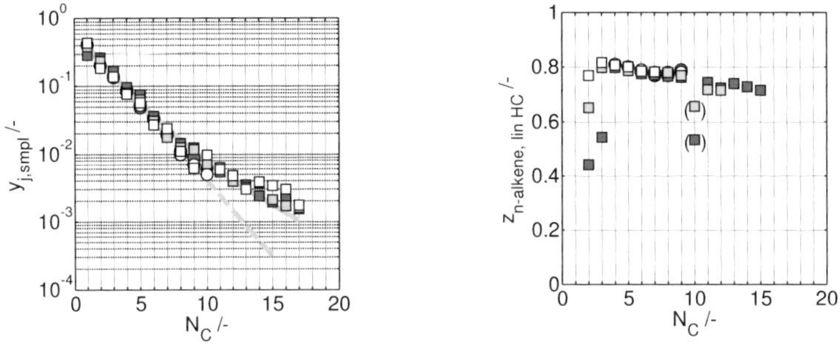

Figure 12.29. ASF distribution of the volatile hydrocarbons (left) and molar alkene fraction in the fraction of linear hydrocarbons as function of the carbon number for the various amounts of H2O co-fed to the feed stream: ■●: 0, ■●: 8, □○: 17 mol-% H_2O. Square symbols: ampoule method/ offline GC, round symbols: online GC.
ASF parameters: H_2/CO = 2 (solid line): $\alpha_1 = 0.6$, $\alpha_{2,C9+} = 0.8$; reference (broken line): $\alpha = 0.6$
* Fe-GTLX, H_2/CO =2, T = 270 °C, $P_{H2,F} + P_{CO,F}$ = 1 MPa, $\tau_{mod,n}$ = 2000 kg s/m^3

described by a simple two-α ASF model; for higher carbon numbers, a chain growth probability of 0.8 is found, yielding an average chain length of 2.9. In the case of the H_2/CO_2 syngas, α drops down to 0.5, yielding an average chain length of 2.5. With respect to the applied reactor model, the choice of C_3H_6 as pseudo-component, representing the complex product distribution, was justified.

In the case of the Co-GTL4 catalyst (Figure 12.28), an increase in $z_{CO2,C,F}$ leads to a steep rise of the methane selectivity S_{CH4} and a steep decay of the selectivity towards long-chain hydrocarbons. This indicates that the FT regime cannot be maintained under the H_2/CO_2 syngas, and the main reaction is the methanation of CO_2 to CH_4. The steep increase in CH_4 selectivity (or the strong decay in C_{5+} selectivity) can be seen as indicator that the FT regime cannot be maintained at low CO partial pressures or high H_2/CO partial pressure ratios contrary to Fe-based catalysts (Riedel 2003, Unruh 2006).

Effect of synthesis gas composition – $y_{H2O,F}$

Figure 12.29 - Figure 12.30 compare the effect of H_2O addition on the product selectivity for the Fe- GTLX and Co-GTL4 catalysts. During the H_2O co-feeding experiments, the total pressure was increased in such a way, that the partial pressure of the reactants $P_{H2,F}$ and $P_{CO,F}$ and the overall residence time were kept constant.

Figure 12.29 is an ASF plot of the C_1-C_{17} molar product distributions of the Fe-GTLX catalyst (carbon number distribution) for the reference case without H_2O co-feeding, and for 8 mol-% and 17 mol-% H_2O in the feed (H_2/CO =2). The effect of H_2O on the overall chain growth probability is not significant. Ampoule

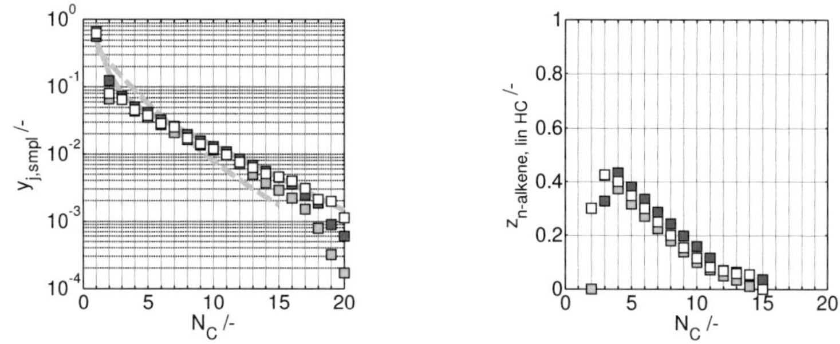

Figure 12.30. ASF distribution of the volatile hydrocarbons (left) and molar alkene fraction in the fraction of linear hydrocarbons as function of the carbon number for the various amounts of H_2O co-fed to the feed stream: ■: 0, ■: 20, □: 30 mol-% H_2O. Square symbols: ampoule method/ offline GC.
ASF parameters: H_2/CO = 2 (solid line): $\alpha_1 = 0.2$, $\alpha_{2,C2+} = 0.8$; reference to Fe-GTLX (broken line): $\alpha_1 = 0.6$, $\alpha_{2,C9+} = 0.8$
* Co-GTL4 (#2b), H_2/CO =2, T = 230°C, $P_{H2,F} + P_{CO,F}$ = 1 MPa, $\tau_{mod,n}$ = 2000 kg s/m^3

analyses suggest that the methane selectivity increases. Experiments from Satterfield et al. (1986) showed that at H_2O fractions > 20 mol-%, the methane selectivity starts dropping. H_2O addition leads to an increase in ethane and propane fraction, while the fractions of higher alkenes remain unaffected. This finding is in agreement with Satterfield (1986) and Dry (1981).

For the Co-GTL4 catalyst, the effect of H_2O co-feeding at constant inlet partial pressure of the reactants is small. Figure 12.30 is an ASF plot of the C_1-C_{20} molar product distributions for the reference case without H_2O co-feeding, and for 20 mol% and 30 mol% H_2O in the feed (H_2/CO =2). A significant decrease in S_{CH4} and increase in S_{C5+}, as reported in literature by various authors (van Steen 1993, Iglesia 1997, Stoersaeter et al. 2005) cannot be extracted from the ASF plot. H_2O addition leads rather to an increase in methane formation, a decrease in $S_{C5+,HC}$, and increased formation of CO_2 via the CO/CO_2 shift reaction.

Effect of synthesis gas composition – $z_{H2,F}$

For the Co-GTL4 catalyst, the C_{5+} selectivity and olefinicity of the product increase with decreasing H_2/CO ratio (Figure 12.31). Therefore, it is beneficial to operate at H_2/CO ratios below 2, if a high selectivity towards long chain hydrocarbons is desired.

Effect of other operating parameters

In the case of the Co-based catalyst Co-GTL4, the temperature was varied over the range 200-240°C. The effect on $S_{C5+,HC}$ and $S_{CH4,HC}$ was relatively small; the secondary reactions as hydrogenation of alkenes were accelerated,

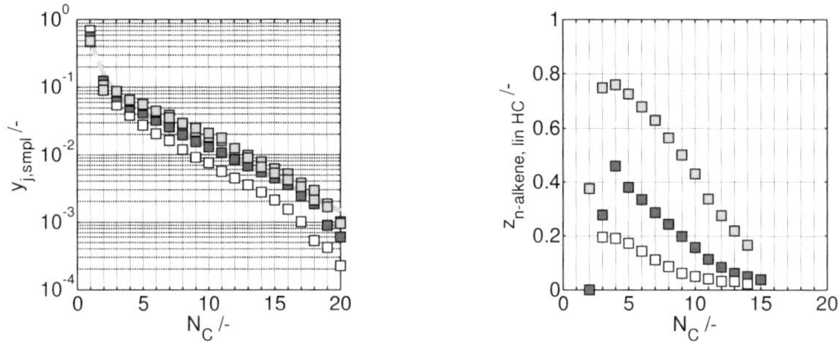

Figure 12.31. ASF distribution of the volatile hydrocarbons (left) and molar alkene fraction in the fraction of linear hydrocarbons as function of the carbon number for H_2/CO = 1 (■), 2 (■) and 4 (□).
* Co-GTL4 (#2b), H_2/CO = var., T = 230 °C, P_{CO} = 0.33 MPa, $\tau_{mod,n}$ = 2000 kg s/m^3

resulting into lower alkene/alkane ratios. Variations in residence time affect in particular the distribution of Co-based catalyst. The hydrocarbon distribution is much more sensitive to changes in H_2/CO ratio compared to Fe-based catalyst. If the feed ratio differs strongly from the usage ratio of the FT reaction, the product distribution will change with increasing residence time. For $H_2/CO > 2$, the products will become lighter as the average H_2/CO will increase with residence time, for $H_2/CO < 2$, the products will become heavier as the average H_2/CO will drop with residence time.
For balanced synthesis gases, the effect of residence time variations on chain lengths was minor; however, there was a clear relationship between decreasing olefinicity with increasing residence time.

12.5.2 Low temperature CO/CO_2 shift reaction

A mixture of a Co-based FT catalyst and a low-temperature shift catalyst was used in experiments for *in-situ* H_2O removal by chemical reaction (Chapter 5.4). The alumina supported CuZnO catalyst (V1578) is a commercial low-temperature shift catalyst provided by BASF, Ludwigshafen, Germany.

The catalyst pellets were crushed to powder. 0.5 g or 1 g of $CuO/ZnO/Al_2O_3$ catalyst ($100 < d_P < 160$ μm) are mixed together with 9.5 or 9 g of calcined SiO_2 and filled into the annular gap of the fixed-bed glass reactor (Figure 4.1). The catalyst bed sits on 50 mm high SiO_2 ($200 < d_P < 400$ μm) packing, which is hold in position by a porous glass disk. The oxidic catalyst pre-cursor of the low temperature shift catalyst is reduced and activated. The reduction procedure is given in Table 4.2. The reduction temperature is limited to 250 °C to avoid the formation of α-brass.

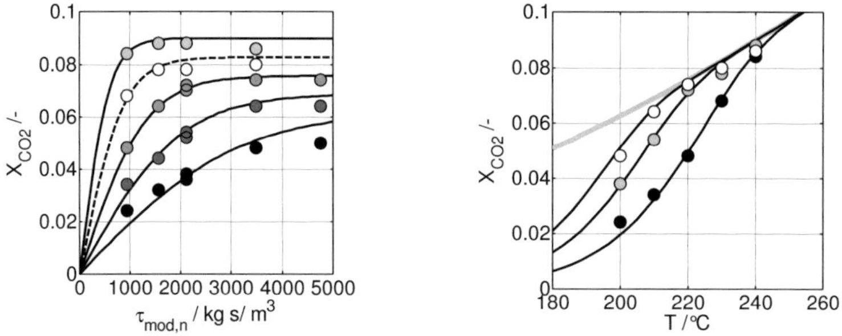

Figure 12.32. CO_2/CO shift reaction (RWGS): Measured CO_2 conversion X_{CO2} as function of the modified residence time $\tau_{mod,n}$ (left) for various temperatures: (●) 200, (●) 210, (●) 220, (○) 230, (●) 240 °C and as function of the temperature (right) for various modified residence time $\tau_{mod,n}$: (●) 950, (●) 2100, (○) 3500 kg s/m^3. Curves: calculated with kinetic model, (—) thermodynamic equilibrium conversion $X_{CO2,eq}$.
* Cu/ZnO, P = 0.1 MPa, $(H_2/CO_2)_F = 1$

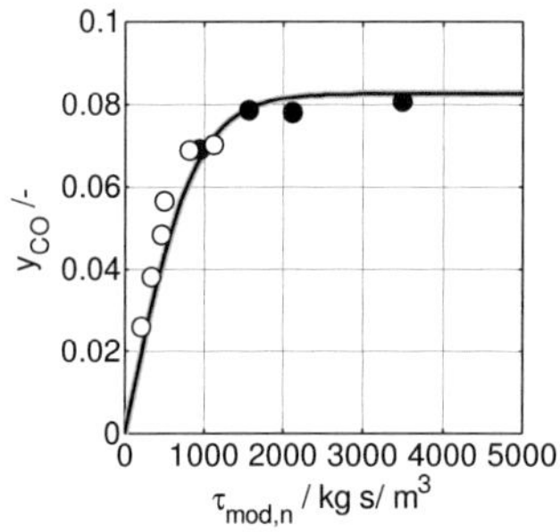

Figure 12.33. CO_2/CO shift reaction (RWGS): Measured molar fraction of CO as function of the modified residence time $\tau_{mod,n}$ at two pressure levels: (●) 0.1, (○) 1 MPa. Curves: calculated with kinetic model at 0.1 MPa (—) and 1 MPa (—).
* CuZnO, P = var., $(H_2/CO_2)_F = 1$, T = 230 °C

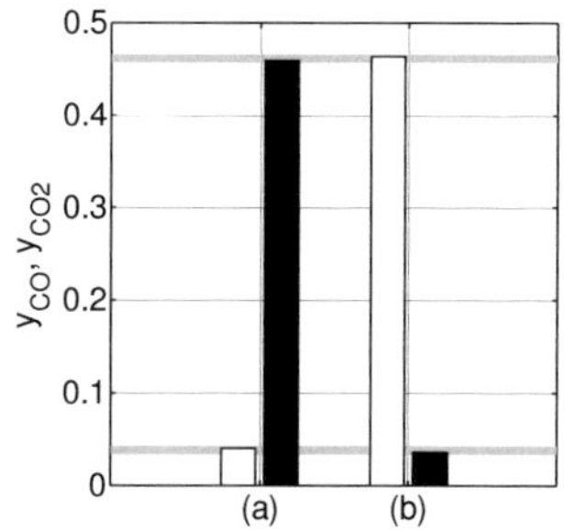

Figure 12.34. CO/CO_2 (a) and CO_2/CO (b) shift reaction: Measured molar fraction of CO_2 (■) and CO (□) at T = 220 °C and a modified residence time $\tau_{mod,n}$ = 2100 kg s/m^3, (—) indicates thermodynamic equilibrium composition.
* (a) CuZnO, P = 0.1 MPa, $(H_2O/CO)_F = 1$
* (b) CuZnO, P = 0.1 MPa, $(H_2/CO_2)_F = 1$

The results of the kinetic experiments are shown in Figure 12.32 - Figure 12.34. The operating window of the CO_2/CO shift reaction is between 210 and 240 °C and overlaps with temperature window of the Co-based FTS (190-240 °C). Above 220 °C, the equilibrium conversion is attained already at low residence times ($\tau_{mod,n}$ > 2000 kg s/m^3). Van Herwijnen & de Jong (1980) observed that the forward shift reaction is about 40 times faster than the reverse shift reaction. This explains that the equilibrium conversions were reached in all experiments starting with CO/H_2O feed gas under similar conditions, even for low residence times (see Figure 12.34).

Table 12.26. Kinetic parameter values of the CO_2/CO shift reaction on $CuZnO/Al_2O_3$ catalyst.
* CuZnO, P = 0.1-1 MPa, T = 200-240°C, $\tau_{mod,n}$ = 1000 - 4750 kg s/m^3, H_2/CO_2 = 1-2.2
* CuZnO: E_A = 103.1 kJ/mol, $k_0 = 1.07\cdot10^{10}$ mol/(s·kg·Pa^2)

		CO_2/CO-shift (rev)
$k_{200°C}$	/mol / (s·kg·Pa)	0.044
$k_{210°C}$	/mol / (s·kg·Pa)	0.075
$k_{220°C}$	/mol / (s·kg·Pa)	0.13
$k_{230°C}$	/mol / (s·kg·Pa)	0.21
$k_{240°C}$	/mol / (s·kg·Pa)	0.34
a_{CO}	/ 1/Pa	1
b_{H2}	/ 1/Pa	10
		$r_{CO2-SH} = k' \frac{P_{CO2}P_{H2} - P_{H2O}P_{CO}K_{P,CO\cdot SH}}{(1 + aP_{CO} + bP_{H2})^2}$

Experimental results at different pressure levels (Figure 12.33) indicate that the rate of the CO_2/CO shift reaction depends only weakly on total pressure. This suggests a rate equation with a squared denominator or a denominator with an inhibition term using the product of two partial pressures (as proposed by van Herwijnen & de Jong 1980).

Kinetic rate parameters of CO_2/CO shift reaction were determined by non-linear regression of the conversion-residence time data for H_2/CO_2 syngases. The kinetic rate parameters are summarized in Table 12.26. The activation energy of 103 kJ/mol is concordant with the values found by Campbell (1970, 117 kJ/mol) and by Spencer (1995, 110 kJ/mol) for low temperature shift catalysts in the non mass-transport limited regime. During the experiments, no deactivation of the catalyst was observed.

12.5.3 DME/DEE synthesis

The dehydration of methanol and ethanol to dimethyl and diethyl ether was used as an example reaction for H_2O removal by membranes in a chemical reactor (Chapter 5.3). As catalyst, a commercial available acidic γ-Al_2O_3 catalyst (Merck) was applied, which is used for dehydration of methanol to dimethyl ether (DME). The reaction kinetics was determined in independent experiments.

8 g γ-Al_2O_3 catalyst ($100 < d_P < 160$ µm) were diluted with calcined SiC and filled into the annular gap of the membrane reactor equipped with a membrane replica (Figure 4.1). The catalyst bed sits on a coarse SiC ($200 < d_P < 250$ µm) packing. A porous glass disk supports the catalyst /SiC mixture and the inert bed.

A simplified kinetics of methanol dehydration to DME over γ-Al_2O_3 catalyst is determined by adjusting the reference kinetics of Bercic & Levec (1992) to a

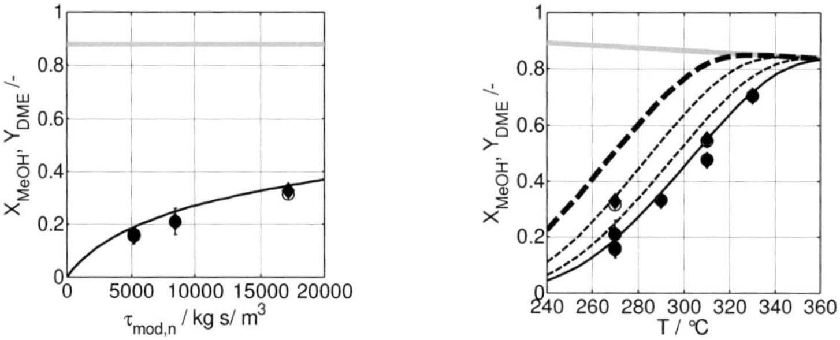

Figure 12.35. Measured methanol conversion (♦) and DME yield (○) as function of the modified residence time $\tau_{mod,n,MeOH}$ (left) at T = 270 °C and as function of the temperature (right) for $\tau_{mod,n,MeOH}$ = 5200 (—), 8500 (--) and 17000 (--) kg s/m^3. Curves: calculated with own kinetic model (–, --), with reference kinetics (---) taken from literature (Bercic 1992), (—) thermodynamic equilibrium conversion $X_{MeOH,eq}$.
* γ-Al_2O_3, $P_{MeOH,F}$ = 0.67 MPa, $P_{Ar,F}$ = 0.33 MPa (carrier gas)

data set, collected in own experiments, varying residence time and temperature (Figure 12.35). The kinetic parameters values are summarized in Table 12.27. As Bercic & Levec (1992) determined the kinetics at higher temperatures (320-360 °C) and lower pressures (0.14 MPa), the kinetic parameters had to be adjusted slightly, otherwise the kinetics predicts significant higher methanol conversions (Figure 12.35). The same approach was applied to derive a simple kinetics for the dehydration of ethanol to DEE.

Table 12.27. Kinetic parameter values of the DME synthesis from methanol and DEE synthesis from ethanol on γ-Al_2O_3 catalysts, parameter values given as factors of the parameter values of the reference DME kinetics determined by Bercic & Levec (1992).
* γ-Al_2O_3/ DME, T = 270-330 °C, $P_{MeOH,F}$ = 0.67 MPa, $P_{Ar,F}$ = 0.33 MPa, $\tau_{mod,n,MeOH}$ = 5200-17000 kg s/m^3,* γ-Al_2O_3/ DEE, T = 270-330 °C, $P_{EtOH,F}$ = 0.70 MPa, $P_{Ar,F}$ = 0.30 MPa, $\tau_{mod,n,EtOH}$ = 7700 kg s/m^3

		DME	DEE	Bercic et al. (1992)
E_A	/ kJ/mol	1.07 x	1.085 x	143.47
$\Delta H_{MeOH/EtOH}$	/ kJ/mol	1 x	1 x	70.51
ΔH_{H2O}	/ kJ/mol	0.98 x	1 x	42.13
k_0	/ mol / (s·kg)	1.06 x	1.06 x	$1.43 \cdot 10^{13}$
$a_{0,MeOH/EtOH}$	/ m^3/mol	1 x	1 x	$0.54 \cdot 10^{-6}$
$b_{0,H2O}$	/ m^3/mol	1.2 x	1.2 x	$8.44 \cdot 10^{-5}$
$K_{P,DME/DEE}$	/ -	$\exp\left(\frac{2611}{T} - 2.22\right)$	$\exp\left(\frac{2421}{T} - 2.47\right)$	$\exp\left(\frac{2611}{T} - 2.22\right)$

$$k = k_0 \exp\left(\frac{-E_A}{RT}\right) \qquad a_{MeOH/EtOH} = a_{0,MeOH/EtOH} \exp\left(\frac{\Delta H_{MeOH/EtOH}}{RT}\right) \qquad b_{H2O} = b_{0,H2O} \exp\left(\frac{\Delta H_{H2O}}{RT}\right)$$

$$r = k \frac{\left(c^2_{MeOH/EtOH} - c_{H2O} c_{DME/DEE} / K_{P,DME/DEE}\right)}{\left(1 + 2\sqrt{a_{MeOH/EtOH} c_{MeOH/EtOH}} + b_{H2O} c_{H2O}\right)^4}$$

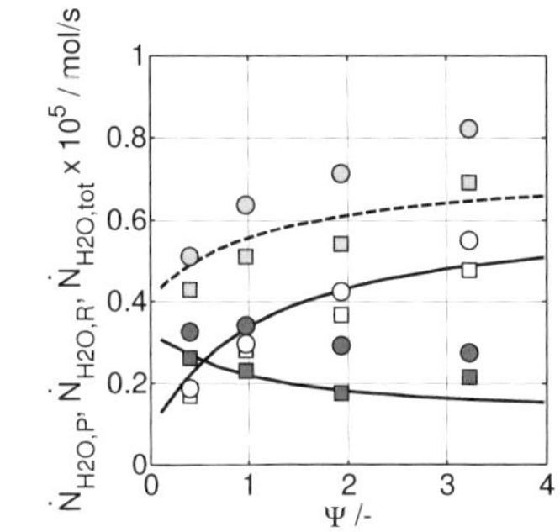

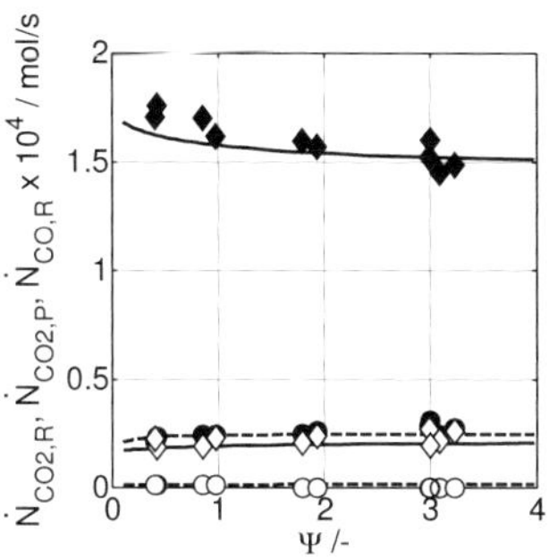

Figure 12.36. Molar flow rates of H_2O on the retentate (●■) and permeate side (○□), and the total flow rate (●■), determined gravimetrically (squares), by GC and oxygen balance (circles) as function of the sweep ratio.

Figure 12.37. Molar flow rates of CO_2 (◆) and CO (●) on the retentate side and CO_2 (◇) and CO (○) on the permeate side as function of the sweep ratio.

Curves: calculated with membrane reactor model.

* Fe-GTLX/ CSP2: T= 270 °C, P_F = 1 MPa, $\tau_{mod,n}$ = 2000 kg s/m^3, $(H_2/CO_2)_F$ = 3; Ψ = var., Φ = 0.7, $(H_2)_S$

12.6 *In-situ* H_2O removal experiments

12.6.1 *In-situ* H_2O removal during CO_2 hydrogenation

Additional figures with regard to variations in sweep and pressure ratio are listed below.

Variation of sweep ratio

Figure 12.36 plots the H_2O flow rates on the permeate and retentate side as function of the sweep ratio. With increasing sweep ratio, the amount of H_2O formed by the reaction increases, indicating that H_2O removal leads to a higher reaction rate (Figure 12.39) and conversion level. This is an agreement with Figure 12.37, which shows a dropping CO_2 flow rate in the retentate and a constant CO_2 flow rate in the permeate. An increased conversion level is therefore linked to an increasing sweep ratio and degree of H_2O removal.

Reduced H_2O partial pressures result in an overall faster reaction rate due to diminished inhibition terms. Figure 12.39 plots the relative reaction rates at a given modified residence time as function of the H_2O recovery; the steep increase of reciprocal denominator reflects the positive effect on the inhibition terms for FT and CO_2/CO-shift reaction.

H_2O recovery was varied by variation of the sweep ratio at a constant, reduced pressure ratio. At a sweep ratio close to zero, about 20% of the H_2O is removed due to the low pressure on the sweep side. At very high sweep

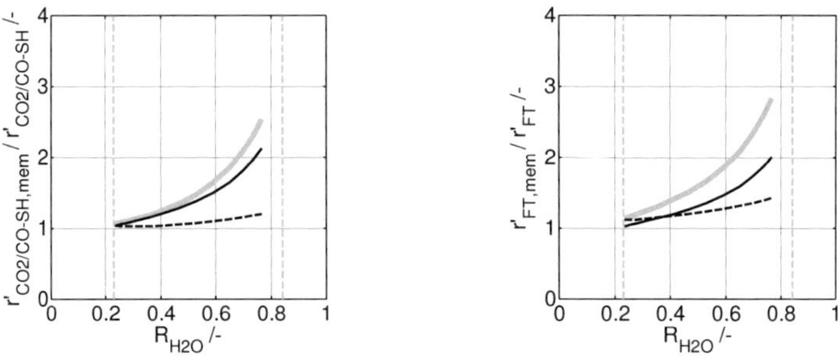

Figure 12.39. Relative reaction rates at $\tau_{mod,n}$ = 2000 kg s/m³ of the CO_2/CO shift (left) and FT reaction (right) as function of the degree of H_2O removal R_{H2O}. Reaction rates related to reaction rates without *in-situ* H_2O removal at $\tau_{mod,n}$ = 2000 kg s/m³.
Reaction rate (—), reciprocal denominator (inhibition terms) of reaction rate (—), nominator of reaction rate (---). Curves: calculated with membrane reactor model and kinetic model (Fe-GTLX).
* Fe-GTLX/ CSP2: T= 270 °C, P_F = 1 MPa, $\tau_{mod,n}$ = 2000 kg s/m³, $(H_2/CO_2)_F$ = 3; Ψ = var., Φ = 0.7, $(H_2)_S$

ratios, the H_2O recovery levels out at 84%. Higher sweep ratios will not bring any improvement as the driving force cannot be increased further. The permeation characteristics and the H_2O permeance are limiting.

Variation of pressure ratio

Membrane deficiencies in permselectivity can be balanced by the proper choice of sweep gas and pressure level on the sweep side (Figure 12.40). As the H_2O/H_2 permselectivity (<5) of the CSP2 membrane is still too low to avoid a significant loss of H_2, H_2 is applied as sweep gas. A pressure ratio of 0.7 guarantees that the loss of H_2 is reduced too a minimum and that no H_2 is co-fed to the reaction zone. By doing so, the membrane has apparently a much

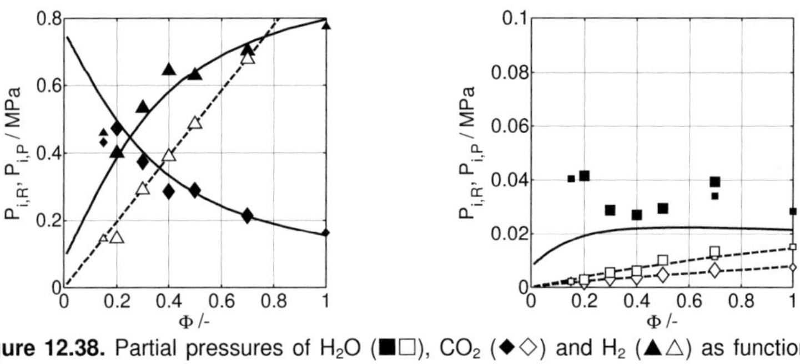

Figure 12.38. Partial pressures of H_2O (■□), CO_2 (◆◇) and H_2 (▲△) as function of the pressure ratio Φ on the retentate (■◆▲) and permeate side (□◇△). Curves: calculated with membrane reactor model.
* Fe-GTLX/ CSP2: T= 270 °C, P_F = 1 MPa, $\tau_{mod,n}$ = 2000 kg s/m³, $(H_2/CO_2)_F$ = 3; Ψ = 3.3, Φ = var., $(H_2)_S$

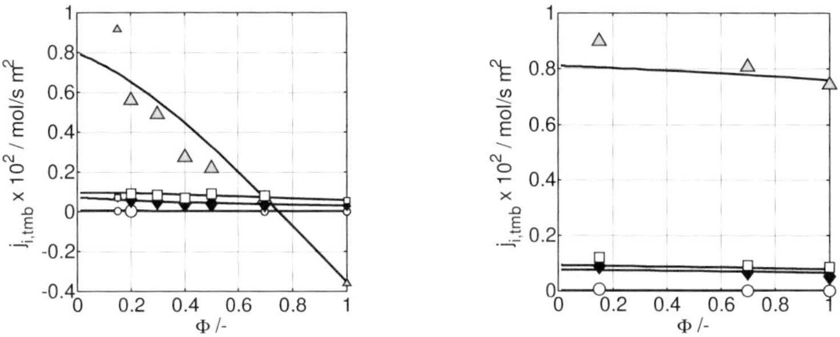

Figure 12.40. Transmembrane molar fluxes of (▲) H_2, (□) H_2O, (◆) CO_2 and (○) CO as function of the pressure ratio Φ for H_2 (left) and Ar (right) as sweep gas. Curves: calculated with membrane reactor model.
* Fe-GTLX/ CSP2: T= 270°C, P_F = 1 MPa, $\tau_{mod,n}$ = 2000 kg s/m^3, $(H_2/CO_2)_F$ = 3; Ψ = 3.3, Φ = var.

higher H_2O/H_2 permselectivity.

12.6.2 *In-situ* H_2O removal during DME/DEE synthesis

Figure 12.41 compares the partial pressure profiles of reaction of methanol to DME with and without *in-situ* H_2O removal. H_2O and methanol partial pressure are expected to be much lower, while the DME partial pressure will increase. The reduced H_2O and methanol partial pressures result in an overall faster reaction rate due to diminished inhibition terms, resulting in an increased conversion level. Figure 12.42 plots the relative reaction rates at a given modified residence time as function of the H_2O recovery; the steep increase of reciprocal denominator reflects the effect on the inhibition terms.
A series of *in-situ* H_2O removal experiments was carried out with ethanol instead of methanol (Figure 12.43). H_2O removal led here as well to increased ethanol conversion to the diether. However, these experiments were characterized by fast membrane decay, clearly demonstrating that the CSP2 membrane cannot be operated under these conditions.

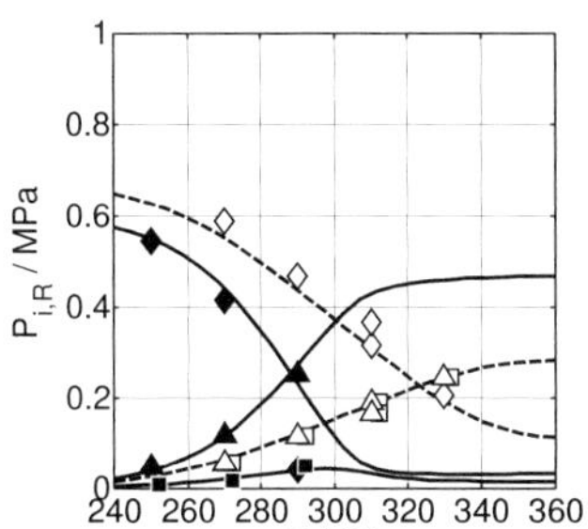

Figure 12.41. Partial pressures of methanol (◇◆), H_2O (□■, estimated) and DME (△▲) as function of the temperature, with CSP2 membrane (◆■▲), without membrane (◇□△).
Curves: calculated with reactor model with (–) and without (--) integrated hydrophilic membrane.
* γ-Al_2O_3/ CSP2, T = var., $P_{MeOH,F}$ = 0.67 MPa, $P_{Ar,F}$ = 0.33 MPa (carrier gas), $\tau_{mod,n,MeOH}$ = 5200 kg s/ m^3, Ψ = 3.3, Φ = 0.15, $(Ar)_S$

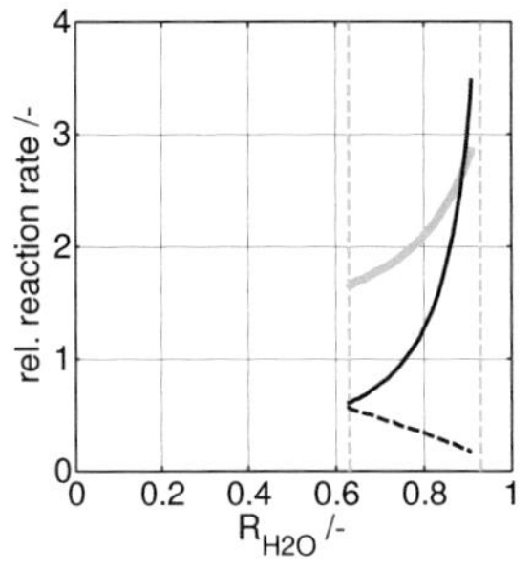

Figure 12.42. Reaction rate of the methanol etherification to DME with *in-situ* H_2O removal relative to the reaction rate without *in-situ* H_2O removal as function of the degree of H_2O removal R_{H2O}. Relative rate of reaction (—), and reaction rate broken down into relative reciprocal denominator (inhibition terms) of the rate equation (–, scaled down by a factor of 5), relative nominator of the rate equation (---).
Curves: calculated with membrane reactor and kinetic model.
* γ-Al_2O_3/ CSP2, T = 290 °C, $P_{MeOH,F}$ = 0.67 MPa, $P_{Ar,F}$ = 0.33 MPa (carrier gas), $\tau_{mod,n,MeOH}$ = 5200 kg s/ m³, Ψ = var., Φ = 0.15, $(Ar)_S$

The permeances of DME and DEE are below the methanol permeances, but still larger than the ethanol permeance, though molecules are more complex indicating that solubility plays a key role (Figure 12.44). The permeances of oxygenates increase quickly with temperature. The presence of the alcohols and diethers at elevated temperature led to fast decay of selective membrane properties.

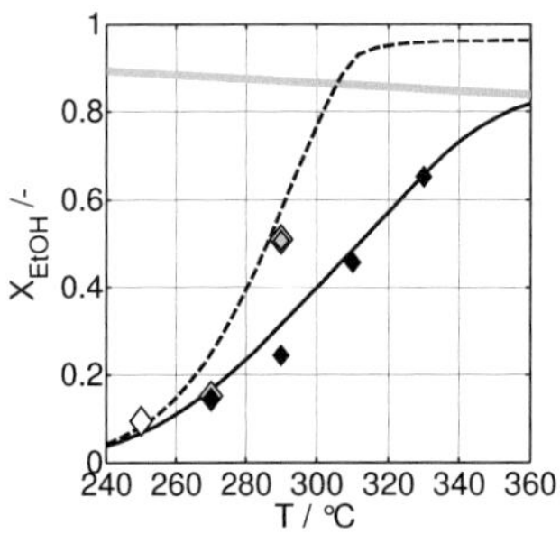

Figure 12.43 Measured ethanol conversion X_{EtOH} as function of the temperature for a fixed bed reactor without (◆—) and with integrated hydrophilic membrane (◇--), (◆) repro with fresh membrane.
Curves: calculated with membrane reactor model, (—) thermodynamic equilibrium conversion $X_{EtOH,eq}$.
* γ-Al_2O_3/ CSP2, T = var., $P_{MeOH,F}$ = 0.7 MPa, $P_{Ar,F}$ = 0.3 MPa (carrier gas), $\tau_{mod,n,EtOH}$ = 7700 kg s/ m³, Ψ = 3.3., Φ = 0.15, $(Ar)_S$

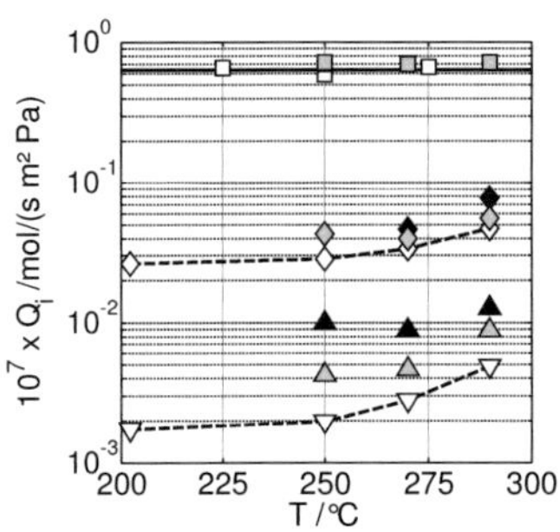

Figure 12.44. Permeances Q_i of H_2O (■), methanol (◆ ◆) and DME (▲▲) determined under reactive conditions of DME synthesis. For reference, H_2O (□), methanol (◇) and ethanol (▽) permeances measured in non-reactive experiments (membrane test cell) are indicated.
* γ-Al_2O_3/ CSP2, T = var., $P_{MeOH,F}$ = 0.67 MPa, $P_{Ar,F}$ = 0.33 MPa (carrier gas), $\tau_{mod,n,MeOH}$ = 5200 kg s/ m³, Ψ = 3.3, Φ = 0.15, 0.7, $(Ar)_S$

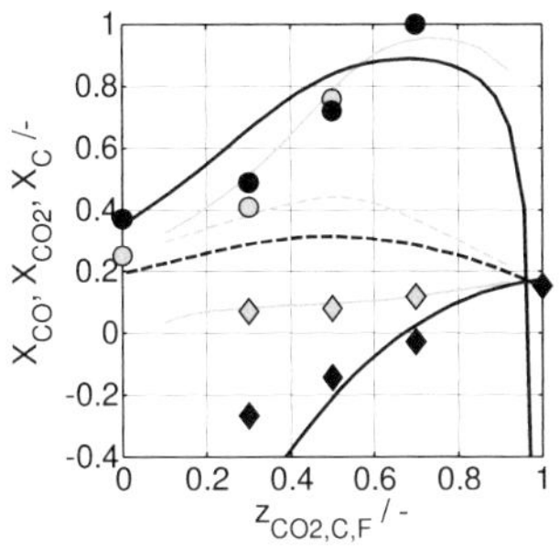

Figure 12.45. CO (● ●), CO_2 (◆ ◆) and total carbon conversion (---) as function of the stoichiometric syngas composition $z_{CO2,C,F}$ for the Co-GTL4 catalyst only (●–, ◆–) and for the Co-GTL4(#1)/ CuZnO combination (●–, ◆–).
Curves: calculated with FT, CO/CO_2 shift & simple CO_2 methanation kinetics, Co-GTL4 catalyst activity deactivated by 30%.
* 2g Co-GTL4(#1), 2g Co-GTL4(#1) / 2g CuZnO, T = 230 °C, P = 1 MPa, $\tau_{mod,n}$ = 1900 kg s/m^3 based on Co-GTL4 catalyst mass only

12.6.3 *In-situ* H_2O removal during FT synthesis

In additional experiments, the physical mixture of the Co-based catalyst and the low-temperature shift catalyst was exposed to CO_2 containing synthesis gases. The objective was to determine if the shift-active Co-based catalyst system is able to convert CO_2 to hydrocarbons.

Figure 12.45 compares the response of the Co-GTL4 and the Co-GTL4/CuZnO combination on increasing CO_2 content and raising H_2/CO ratios. The Co-GTL4 catalyst is able to convert small amounts CO_2 (<20%), which is mainly converted to methane. CO is converted at $z_{CO2,C,F} > 0.6$ more or less completely to methane and short-chain molecules (see also Figure 12.28).

The Co-GTL4/CuZnO combination exhibits the behaviour of a Fe-based catalyst (Figure 12.45): at low to medium CO_2 contents, the CO_2 conversion is negative as additional CO_2 is formed via the CO/CO_2 shift reaction (internal shift and removal of H_2O). At high CO_2 contents, the CO_2 conversion is positive, i.e. CO_2 is converted via CO as intermediate to hydrocarbons. However, the CO_2 conversion found with integrated shift function is similar to the one measured for Co-GTL4 alone. This suggests that the methanation reaction is still the determining reaction, and indeed, the FT regime cannot be maintained under these conditions. The CO partial pressures are too low, and therefore, methane and short-chain hydrocarbons are the main products. In terms of product selectivity, a functionalized Co-based catalyst cannot mimic a Fe-based catalyst under these conditions.

12.7 Reactor and reactor models

12.7.1 (Membrane) reactor model – Additional calculations

12.7.1.1 Material balance

Equation (12.10) represents the general, differential material balance for the gas phase of a fixed-bed reactor. This one-dimensional material balance takes into account accumulation, dispersive flux, convective flux and a reaction network as source/ sink.

$$\frac{\partial c_{iG}}{\partial t} = D_{i,ax}\frac{\partial^2 c_{iG}}{\partial z^2} - \frac{\partial c_{iG}u_G}{\partial z} + \rho_{cat}\frac{(1-\varepsilon)}{\varepsilon}\sum_j \nu_{ij}r_j \qquad \text{with } u_{sf} = u_G\varepsilon \tag{12.10}$$

In this thesis, only steady-state processes are considered (12.11) and accumulation term is set to zero. The Bodenstein number Bo_{ax} indicates if dispersive fluxes can be neglected.

$$\frac{\partial c_{iG}u_G}{\partial z} = D_{i,ax}\frac{\partial^2 c_{iG}}{\partial z^2} + \rho_{cat}\frac{(1-\varepsilon)}{\varepsilon}\sum_j \nu_{ij}r_j \qquad \text{with } Bo_{ax} = \frac{\varepsilon u_G L}{D_{i,ax}} \tag{12.11}$$

If this is the case, the simplified material balance yields the ideal design-equation of a packed-bed reactor (Fogler 1999) expressed in molar flow rates:

$$\frac{d\dot{N}_i}{dz^+} = m_{cat}\sum_j \nu_{ij}r_j \qquad \text{with } \dot{N}_i(z^+ = 0) = \dot{N}_{i,F} \tag{12.12}$$

Equation (12.12) is applied as mathematical model of the integral fixed-bed lab-scale reactor in the process of kinetic rate parameter estimation. The model of the lab-scale fixed-bed reactor is based on the following assumptions:

- Plug-flow model: Calculations and RTD measurements show that axial dispersion, channelling and bypassing effects can be neglected.
- No impulse balance: Experiments show the pressure drop is less than 2% of the total pressure and therefore negligible.
- Pseudo-homogenous model: Calculations and data from literature indicate that no internal and external mass- and heat transfer limitations occur, i.e. gas, liquid and solid phase can be treated as a single phase.
- No energy balance: Due to the high dilution of the catalysts bed by inert material, temperature effects are negligible, experiments show that an isothermal temperature profile ±1 °C is obtained.

These assumptions are discussed in detail in the paragraph below, where a large part is dedicated to the effect of back-mixing and residence-time

Table 12.28. Criteria to determine mass and heat transfer limitations. Taken from Fogler (1999) and Baerns, Hofmann & Renken (1992).

	internal (catalyst particle)	external (fluid-to-particle)
mass transfer	TR1: $\frac{(-\nu_{ij}r_j)\rho_b\, d_P^2}{4\, D_{eff}\, c_{i,G}} < 1$	TR3: $\frac{(-\nu_{ij}r_j)\rho_b\, d_P}{2\beta\, c_{i,G}} < 0.15$
heat transfer	TR2: $\frac{(-\Delta H_R)(-\nu_{ij}r_j)\rho_b\, d_P^2}{4\lambda_{eff}\, T} < \frac{RT}{E_A}$	TR4: $\frac{(-\Delta H_R)(-\nu_{ij}r_j)\rho_b\, d_P E_A}{2\alpha T^2 R} < 0.15$

distribution and to the results of a sensitivity analysis on model parameters. The relative accuracy of the model on conversion and yield data should be within ± 5%.

12.7.1.2 Mass- and heat transfer limitations

The derivation of the rate parameters from the fixed-bed reactor experiments and the analysis of the membrane reactor experiments are based on a *pseudo-homogenous* lab-scale fixed-bed reactor model. It can be easily assessed according to criteria published in textbooks, if indeed internal and external mass- and heat transfer limitations are not present (Table 12.28).

A conservative assessment was carried out for Fe- and Co-based catalysts in the membrane and kinetic reactor set-up under typical reaction conditions (i.e. base points). The results are summarized in Table 12.29. The initial rate of reaction – taken from the kinetic rate equations of the Fe-GTLX and Co-GTL4 – was chosen as rate of reaction $(-\nu_{ij}r_j)_{init}$. The effective diffusion coefficient of CO or CO_2 was calculated on the assumption of liquid filled catalyst pores, also taking into account the catalyst porosity and tortuosity; the diffusion coefficient was taken from van der Laan (1999).

The heat- and mass transfer coefficients α and β can be calculated according to the correlation of Gnielinski (VDI-Wärmeatlas 1994, see Nu_1 and Sh_1 in Table 12.29). However, as Martin (1978) highlighted, at low Péclet numbers most of the experimentally obtained particle-to-fluid heat and mass transfer coefficients in packed beds were found to be several orders of magnitude below the values predicted for a single sphere in cross flow. Martin was able to explain this discrepancy by a simple model accounting for a non-uniform distribution of the void fraction, i.e. by taking into account bypassing or channelling.

Due to very small particles and the low superficial velocity in the lab-scale reactors, the Reynolds and Péclet numbers are very low. Channelling and bypassing cannot be excluded and therefore, the Nusselt and Sherwood numbers determined with the Gnielinski correlation were decreased by more than two magnitudes (see Nu_2 and Sh_2 in Table 12.29).

Table 12.29. Estimation of the significance of internal and external mass- and heat transfer limitations in the membrane and kinetic reactor, on basis of literature correlations and the criteria TR1-TR4 listed in Table 12.28.

		membrane reactor			kinetic reactor		
		Fe-GTLX H_2/CO	Fe-GTLX H_2/CO_2	Co-GTL4 H_2/CO	Fe-GTLX H_2/CO	Fe-GTLX H_2/CO_2	Co-GTL4 H_2/CO
conditions							
T	°C	270	270	230	270	270	230
P	MPa		1			1	
$\dot{V}_{F,n}$	ml/min		120			60	
reactor							
A_{cs}	m^2		$1.71\cdot10^{-4}$			$3.77\cdot10^{-5}$	
L	m		0.15			0.18	
u_{sf}	m/s		0.0023			0.0053	
reaction							
$(-\nu_{ij}r_j)_{init}$	mol/s kg	$5.2\cdot10^{-3}$	$8.2\cdot10^{-5}$	$7.3\cdot10^{-2}$	$5.2\cdot10^{-3}$	$8.2\cdot10^{-5}$	$7.3\cdot10^{-2}$
E_A	kJ/mol		86.4			86.4	
ΔH_R	kJ/mol	-160	-120	-160	-160	-120	-160
ρ_b	kg/m^3		150...1000			150...1000	
catalyst							
d_P	m		$< 200\cdot10^{-6}$			$< 200\cdot10^{-6}$	
D_{eff}	m^2/s		$7.5\cdot10^{-10}$			$7.5\cdot10^{-10}$	
λ_{eff}	W/m K		0.25...1			0.25...1	
fluid-to-particle							
$c_{i,G}$	mol/m^3	220	220	240	220	220	240
ρ_G	kg/m^3	2.3	2.8	2.5	2.3	2.8	2.5
η_G	Pa s	$1.9\cdot10^{-5}$	$1.7\cdot10^{-5}$	$1.6\cdot10^{-5}$	$1.9\cdot10^{-5}$	$1.7\cdot10^{-5}$	$1.6\cdot10^{-5}$
ν_G	m^2/s	$8\cdot10^{-5}$	$6\cdot10^{-5}$	$7\cdot10^{-5}$	$8\cdot10^{-5}$	$6\cdot10^{-5}$	$7\cdot10^{-5}$
λ_G	W/m K	0.21	0.22	0.18	0.21	0.22	0.18
$c_{P,G}$	kJ/kg K	10	11	10	10	11	10
$\delta_{H2,CO2}$	m^2/s	$2\cdot10^{-5}$	$2\cdot10^{-5}$	$2\cdot10^{-5}$	$2\cdot10^{-5}$	$2\cdot10^{-5}$	$2\cdot10^{-5}$
a_G	m^2/s	$9\cdot10^{-5}$	$10\cdot10^{-5}$	$8\cdot10^{-5}$	$9\cdot10^{-5}$	$10\cdot10^{-5}$	$8\cdot10^{-5}$
characteristic numbers							
Re_P	-	0.06	0.07	0.07	0.13	0.17	0.15
Re_ε	-	0.15	0.19	0.17	0.33	0.43	0.38
Pr	-	0.09	0.06	0.08	0.09	0.06	0.08
Sc	-	0.38	0.35	0.31	0.38	0.35	0.31
$Pe = Re_\varepsilon Pr$	-	0.01	0.01	0.01	0.03	0.03	0.03
Nu_1	-	4	4	4	4.1	4.1	4.1
Nu_2	-	0.05	0.05	0.05	0.05	0.05	0.05
Sh_1	-	2.2	2.2	2.2	2.3	2.3	2.3
Sh_2	-	0.01	0.01	0.01	0.01	0.01	0.01
α (Nu_2)	W/m^2 K	51	56	313	51	56	313
β (Sh_2)	m/s	$1\cdot10^{-3}$	$9\cdot10^{-4}$	$1\cdot10^{-3}$	$1\cdot10^{-3}$	$9\cdot10^{-4}$	$1\cdot10^{-3}$
transport resistances		(see Table 12.28)					
mass transfer (internal)	TR1	0.31 < 1	0.005 < 1	4.1 > 1	0.31 < 1	0.005 < 1	4.1 > 1
heat transfer (internal)	TR2	$6\cdot10^{-5}$ $< 5.2\cdot10^{-2}$	$7\cdot10^{-5}$ $< 5.2\cdot10^{-2}$	$1\cdot10^{-3}$ $< 4.8\cdot10^{-2}$	$6\cdot10^{-5}$ $< 5.2\cdot10^{-2}$	$7\cdot10^{-5}$ $< 5.2\cdot10^{-2}$	$1\cdot10^{-3}$ $< 4.8\cdot10^{-2}$
mass transfer (external)	TR3	$2\cdot10^{-3}$ < 0.15	$4\cdot10^{-5}$ < 0.15	$3\cdot10^{-3}$ < 0.15	$2\cdot10^{-3}$ < 0.15	$4\cdot10^{-5}$ < 0.15	$3\cdot10^{-3}$ < 0.15
heat transfer (external)	TR4	$6\cdot10^{-2}$ < 0.15	$6\cdot10^{-4}$ < 0.15	1.1 > 0.15	$6\cdot10^{-2}$ < 0.15	$6\cdot10^{-4}$ < 0.15	1.1 > 0.15

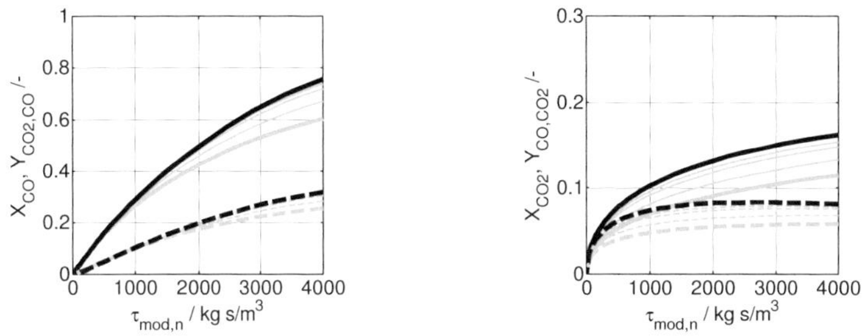

Figure 12.46. Effect of back-mixing on reactor performance: Calculated conversion (—, —) and yield (- -, - -) profiles of a CO/CO_2-shift active Fe catalyst for H_2/CO = 2/1 (left) and H_2/CO_2 = 3/1 (right) syngases as function of the modified residence time for various degrees of back-mixing: (—,- -) N=∞, ideal plug-flow reactor (PFR); (—,- -) N=1, perfectly mixed reactor (CSTR); (,) N=2, N=5, N=10, various degrees of back-mixing.
Curves: calculated with kinetic model (Fe-GTLX, T = 270°C, P = 1 MPa) and CSTR-in-series model, see eq. (12.13).

On basis of this conservative approach one can conclude from Table 12.29 that *internal heat transfer* (TR2) and *external mass transfer* (TR3) limitations can be neglected. For the Fe-based catalyst, *internal mass transfer* (TR1) and *external heat transfer* (TR4) do not play a role, in particular for H_2/CO_2 syngases. However, if particle size d_P is increased above 400 μm, mass and heat transfer may become limiting for H_2/CO case. This is in accordance with the results from Claeys (1997), who tested K-promoted Fe catalyst of various particle sizes.

The criteria indicate that *internal mass transfer* (TR1) and *external heat transfer* (TR4) limitations may occur for the Co-based catalyst due its higher activity. However, the assessment is based on conservative assumptions.

12.7.1.3 Axial back-mixing (axial dispersion)

Axial dispersion can have a significant effect on reactor performance, as clarified in Figure 12.46 for the FT synthesis on a CO/CO_2-shift active Fe catalyst with H_2/CO and H_2/CO_2 syngases. Here, a cascade of CSTRs is applied to account for the nonideality of the reactor, where the number of CSTRs N defines the degree of back-mixing:

$$m_{cat,k} \sum \nu_{ij} r_j = \dot{N}_{i,k-1} - \dot{N}_{i,k} \qquad \text{with } k = 1 \ldots N \qquad (12.13)$$

The results for ideal PFR (N=∞) differ significantly from the results of the fully back-mixed CSTR (N=1). However, the performance of a cascade of ten CSTRs is already similar to that of the PFR (i.e. within the experimental error).

The derivation of the rate parameters from the fixed-bed reactor experiments and the analysis of the membrane reactor experiments are based on the assumption that axial dispersion or back-mixing is negligible. The removal of the axial dispersion term from the reactor equation simplifies the model to an ideal plug-flow reactor, reducing the complexity of the model and solution time significantly.

This assumption has to be justified. The degree of back-mixing can be estimated by the Bodenstein number Bo_{ax}, which puts the convective flux into relation to the dispersive flux:

$$Bo_{ax} = \frac{u_{sf} L}{D_{ax}} = PE_{ax} \frac{L}{d_P} \qquad (12.14)$$

For $Bo_{ax} > 50$, axial dispersion can be neglected. The VDI-Wärmeatlas (1994, Mh4) recommends calculating the coefficient D_{ax} for axial dispersion in random packings of equal-sized sphere as follows:

$$\frac{1}{PE_{ax}} = \frac{\delta_{bed}/\delta}{Pe_0} + \frac{1}{K_{ax}} \qquad K_{ax} = 2 \qquad (12.15)$$

$$PE_{ax} = \frac{u_{sf} d_P}{D_{ax}} \qquad Pe_0 = \frac{u_{sf} d_P}{\delta} \qquad \delta_{bed}/\delta = 1 - \sqrt{1-\varepsilon} \qquad (12.16)$$

with PE_{ax} as the "effective" Péclet number and Pe_0 as the "molecular" Péclet number. δ and δ_{bed} represent the diffusion coefficients in the solid-free space and within the packing, respectively. Baerns, Hofmann & Renken (1992) give the following correlation to calculate D_{ax}:

$$PE_{ax} = \left(\frac{0.3}{Re_P\, Sc} + \frac{0.5}{1 + 3.8/(Re_P\, Sc)} \right)^{-1} \qquad (12.17)$$

Both correlations yield similar axial dispersion coefficients between $4 \cdot 10^{-6}$ and $6.3 \cdot 10^{-6}$ m^2/s. The resulting Bodenstein numbers are for the kinetic reactor > 100, for the membrane reactor > 50 (Table 12.30). The criterion of Finlayson (Fogler 1999) confirms that the axial dispersion can be neglected with respect to forced axial convection (Table 12.30):

$$\left| \frac{(-\nu_{ij} r_j)\rho_b\, d_P}{u_{sf}\, c_{i,G}} \right| << \left| \frac{u_{sf} d_P}{D_{ax}} \right| \qquad (12.18)$$

Table 12.30. Estimation of the significance of axial back-mixing in the membrane and kinetic reactor, based on literature correlations, based on data in Table 12.29.

		membrane reactor			kinetic reactor		
		Fe-GTLX H_2/CO	Fe-GTLX H_2/CO_2	Co-GTL4 H_2/CO	Fe-GTLX H_2/CO	Fe-GTLX H_2/CO_2	Co-GTL4 H_2/CO
VDI-Wärmeatlas (1994)							
Pe_0	-	0.02	0.03	0.02	0.05	0.06	0.05
δ_{bed}/δ	-			0.23			0.23
PE_{ax}	-	0.09	0.11	0.09	0.2	0.23	0.19
D_{ax}	m^2/s	$5\cdot10^{-6}$	$4\cdot10^{-6}$	$5\cdot10^{-6}$	$5.3\cdot10^{-6}$	$4.6\cdot10^{-6}$	$5.2\cdot10^{-6}$
Bo_{ax}	-	70	81	65	181	207	168
Baerns et al. (1992)							
PE_{ax}	-	0.07	0.09	0.07	0.17	0.2	0.16
D_{ax}	m^2/s	$6.3\cdot10^{-6}$	$5.4\cdot10^{-6}$	$6.3\cdot10^{-6}$	$6.3\cdot10^{-6}$	$5.4\cdot10^{-6}$	$6.3\cdot10^{-6}$
Bo_{ax}	-	55	65	51	151	176	140
Fogler (1999)							
criterion		$2\cdot10^{-3}$	$3\cdot10^{-5}$	$3\cdot10^{-2}$	$9\cdot10^{-4}$	$1\cdot10^{-5}$	$1\cdot10^{-2}$
Finlayson		$\ll 0.09$	$\ll 0.09$	< 0.09	$\ll 0.21$	$\ll 0.21$	< 0.20

In addition, the residence-time distribution (RTD) of the membrane reactor was determined in pulse-tracer experiments for various flow rates to determine the degree of back-mixing experimentally and to identify channelling or dead zones. Significant deviations from near plug-flow behaviour can be expected as the feed stream to the membrane reactor is diverted by 90 degrees into the annular channel around the membrane (Figure 4.1).

The membrane reactor was equipped with a non-permeable membrane replica and filled with a catalyst/ inert particle mixture. N_2 flowed through the reactor with a constant flow rate (100-900 ml/min), and at t=0, an Ar tracer pulse (1 ml) was injected in the feed stream (pulse input). The outlet concentration of Ar in the exit stream (pulse response) was measured and registered by a thermal conductivity detector (TCD). The detailed experimental set-up and the applied methods are described elsewhere in detail (Zürcher, Pabst & Schaub 2008). On basis of the pulse response, the residence-time distribution function or exit-age distribution function E(t) was calculated:

$$E(t) = \frac{c(t)}{\int_0^\infty c(t)dt} \quad (12.19)$$

The obtained RTDs are analysed by a tank-in-series model, where N CSTRs are put in series, with the following material balance for each CSTR in the cascade:

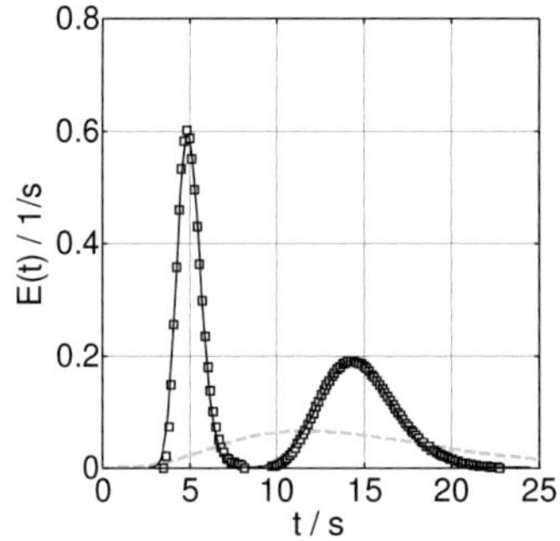

Figure 12.47. Residence-time distribution function E(t) of the pulse response, determined (□) for the membrane reactor (equipped with a non-permeable membrane replica) at 25°C and 1 atm for two different flow rates (100 and 300 Nml/min).
RTD curves calculated by N tanks-in-series model (short residence time: N = 55, long residence time: N = 50), broken curve: example RTD curve for N = 5.

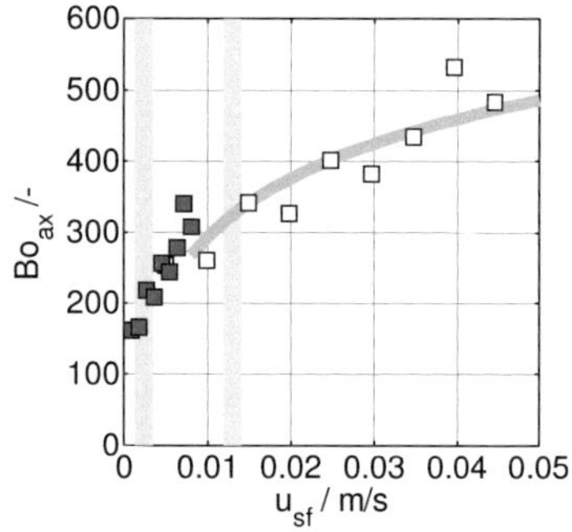

Figure 12.48. Degree of back-mixing: Bodenstein number Bo_{ax} determined as function of the volume flow rate for the membrane reactor (with non-permeable insert/ dummy membrane), measured (□) at 25°C and 1 atm, extrapolated (■) to 270°C and 1 MPa; grey bars: superficial velocity of base point ($\tau_{mod,n}$ = 2000 kg s/m^3) at standard and reaction conditions.

$$\varepsilon_{tot} V_{R,k} \frac{dc_k}{dt} = \dot{V}c_{k-1} - \dot{V}c_k \qquad V_{R,k} = V_R / N \tag{12.20}$$

Here, ε_{tot} takes into account as well the porosity of the packed bed ε as the porosity of the catalyst particles ε_P.

Figure 12.47 shows the RTD functions obtained from two experiments, one with a long and one with a short residence time. The pulse responses are fitted by the RTD function E(t) of a cascade of N CSTRs, which is derived as follows (Fogler 1999):

$$E(t) = \frac{t^{N-1}}{(N-1)!\,\tau_k^N} e^{-t/\tau_k} \qquad \tau_k = \frac{\varepsilon_{tot} V_{R,k}}{\dot{V}} \tag{12.21}$$

The tank-in-series model is able to represent the measured RTD functions very well, with N=50 CSTRs for long residence time experiment (100 ml/min N_2) and N=55 CSTRs for the short residence time experiment (300 ml/min N_2). There are no significant deviations of the experimental data from the calculated curves at the tail end, and therefore no pronounced dead zones.

The large number of cells confirms that axial dispersion is negligible under these conditions (see Figure 12.46). On basis of the spread of the RTD function, expressed by the square of the standard deviation or variance σ^2, and the mean residence time t_{mean}, one can estimate the Bodenstein number Bo_{ax} (Fogler 1999):

$$\frac{\Delta\sigma^2}{t_{mean}^2} = \frac{1}{N} = \frac{2}{Bo_{ax}} \tag{12.22}$$

For ambient conditions, very high $Bo_{ax} > 100$ are found (Figure 12.48). The relevant Bo_{ax} under reaction conditions (1 MPa, 270 °C) are calculated on basis of the ideal gas law (u_{sf} scales with P and T^{-1}) and the correlation of Fuller (Reid, Prausnitz & Poling 1987) for binary diffusion coefficients (δ scales with P^{-1} and $T^{1.75}$). Figure 12.48 indicates that axial dispersion effects will be negligible even under reaction conditions, confirming the results derived on basis of the literature correlations (Table 12.30).

12.7.1.4 Radial concentration profiles (due to membrane transport)

Radial concentration profiles will develop if the dispersive and convective fluxes in radial direction limit the transport across the membrane. The analysis of the membrane reactor experiments is based on the assumption that no radial concentration profiles are present. This occurs if the transmembranal flux is the limiting step as expressed in the following criterion:

$$Q_i \Delta P_i << \frac{D_r \Delta c_{i,G}}{\Delta r} \qquad \frac{Q_i P_i \Delta r}{D_r c_{i,G}} << 1 \tag{12.23}$$

The VDI-Wärmeatlas (1994, Mh6) recommends calculating the coefficient D_r for radial dispersion in random packings of equal-sized sphere as follows:

$$\frac{1}{PE_r} = \frac{\delta_{bed}/\delta}{Pe_0} + \frac{1}{K_r} \qquad K_r = 8\left(2 - \left(1 - \frac{2}{d_R/d_P}\right)^2\right) \tag{12.24}$$

$$PE_r = \frac{u_{sf} d_P}{D_r} \qquad Pe_0 = \frac{u_{sf} d_P}{\delta} \qquad \delta_{bed}/\delta = 1 - \sqrt{1-\varepsilon} \tag{12.25}$$

with PE_r as the *effective* Péclet number and Pe_0 as the *molecular* Péclet number. δ and δ_{bed} represent the diffusion coefficients in the solid-free space and within the packing, respectively. The diffusion coefficient δ_{bed} within the packing is - assuming a porosity ε of 0.4 - only 22.5% of the diffusion coefficient in the solid-free space δ. Due to the small particles and the low superficial velocity in the lab-scale membrane reactor, the dispersion coefficient is determined by the diffusion coefficient δ_{bed}, the contribution of dissipative radial mixing is negligible:

$$D_r = \delta_{bed} + \frac{u_{sf} d_P}{K_r} \delta \tag{12.26}$$

The data in Table 12.31 confirms that the criterion (12.23) is fulfilled in the case for the membrane test cell and the membrane reactor, even for a highly permeable membrane with a permeance of $Q_i = 1 \cdot 10^{-7}$ mol/ (s m^2 Pa).

Table 12.31. Estimation of the significance of radial concentration profiles in the membrane reactor and membrane test cell, on basis of literature correlations on data in Table 12.29.

		membrane reactor		membrane test cell	
δ	m^2/s	$2.5\cdot10^{-5}$	$2.5\cdot10^{-5}$	$2.5\cdot10^{-5}$	$2.5\cdot10^{-5}$
Δr	m	0.0033	0.0033	0.0033	0.0033
Q_i	mol/s Pa m^2	$\mathbf{1\cdot10^{-7}}$	$\mathbf{1\cdot10^{-8}}$	$\mathbf{1\cdot10^{-7}}$	$\mathbf{1\cdot10^{-8}}$
VDI-Wärmeatlas (1994)					
Pe_0	-	0.018	0.018	0.03	0.03
δ_{bed}/δ	-	0.23	0.23	1	1
PE_r	-	0.08	0.08	-	-
D_r	m^2/s	$5.7\cdot10^{-6}$	$5.7\cdot10^{-6}$	$2.5\cdot10^{-5}$	$2.5\cdot10^{-5}$
see equation (12.23)					
criterion		0.26 < 1	0.026 << 1	0.06 << 1	0.006 << 1

δ: binary diffusion coefficient, H_2O in H_2 at 250 °C and 1 MPa; Δr: radial diffusion distance; Q_i: membrane permeance

12.7.1.5 Sensitivity analysis

A sensitivity analysis on the *process parameters* for a fixed-bed reactor (lab-scale) with a Fe-GTLX catalyst yields the following results (Figure 12.49):

- Conversion increases linearly with increasing pressure P and catalyst amount m_{cat} and decreasing feed flow rate.
- Conversion is very sensitive to slight temperature variations. The calculations are based on the following activation energies determined for Fe-GTLX catalyst: $E_{A,FT}$ = 92 kJ/mol and $E_{A,FT}$ = 154 kJ/mol (Mena 2009). Measured variations in the axial temperature profile were found in the range of $\Delta x_{rel} = \pm 0.75\%$.
- An increase in H_2/CO and in particular in H_2/CO_2 ratio (expressed by $z_{H2,F}$) leads to an increase in conversion. Therefore, co-feeding of H_2 during

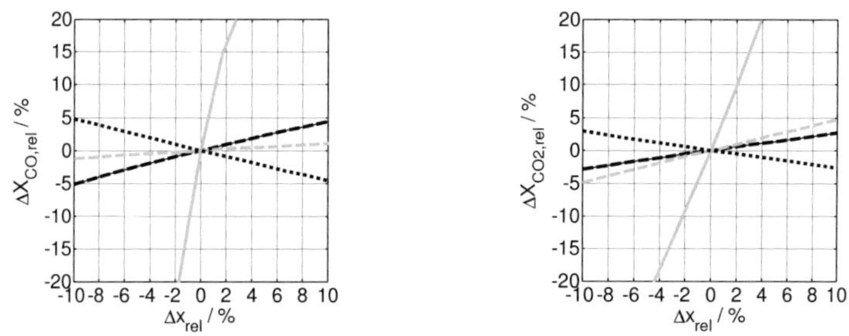

Figure 12.49. Effect of a change Δx_{rel} of the process parameter x on CO conversion (left, feed gas composition: $H_2/CO = 2$) or CO_2 conversion (right, feed gas composition: H_2/CO_2 = 3), with x either (—) T, (—) P, (- -) m_{cat}, (···) $\dot{V}_{F,n}$ or (- -) $z_{H2,F}$.

* Fe-GTLX, T = 270 °C, P = 1 MPa, $\tau_{mod,n}$ = 4000 kg s/m^3

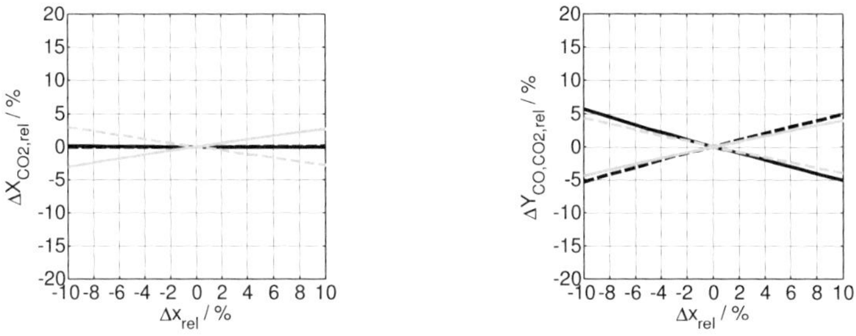

Figure 12.50. Effect of a change Δx_{rel} of the kinetic rate parameter x on CO_2 conversion (left) or CO yield (right), with x either (—) k_{FT}, (- -) a_{FT}, (—) k_{SH} or (- -) a_{SH}.
* Fe-GTLX, T = 270 °C, P = 1 MPa, $\tau_{mod,n}$ = 4000 kg s/m^3, $(H_2/CO_2)_F$ = 3

membrane experiments (with H_2 as sweep gas) needs to be avoided and the pressure ratio has to be chosen carefully.

The results of a sensitivity analysis on the *kinetic rate parameters* determined for Fe-GTLX catalyst are shown in Figure 12.50.

A sensitivity analysis on the *membrane transport parameters* yields the following results (Figure 12.51):

- CO_2 conversion increases linearly with increasing H_2O permeance Q_{H2O} and with increasing overall membrane permeance Q_{ref}.
- CO_2 conversion is not affected by changes in H_2 permeance Q_{H2}.
- CO_2 conversion decreases slightly with increasing CO_2 permeance Q_{CO2}. An increased loss of CO_2 across the membrane will not lead to an apparent increase in CO_2 conversion as per definition.
- In increase of the overall membrane permeance Q_{ref} (e.g. due to different permeabilities of new membranes) at constant permselectivities leads to an increase in CO_2 conversion.

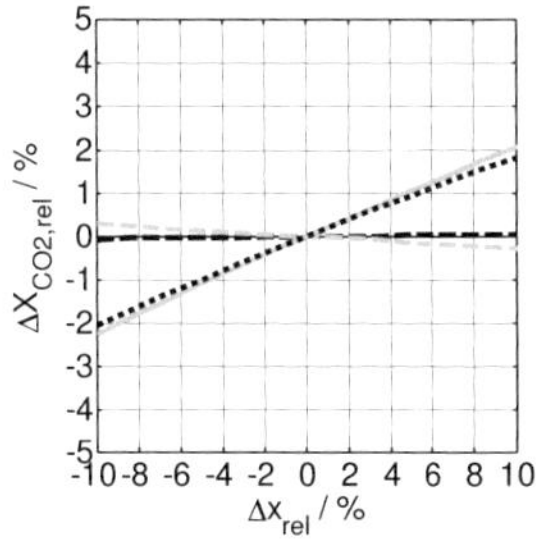

Figure 12.51. Effect of a change Δx_{rel} of the membrane permeance x on CO_2 conversion, with x either (—) Q_{H2}, (- -) Q_{CO}, (—) Q_{H2O}, (- -) Q_{CO2} or (···) Q_{ref}.
* Fe-GTLX, H_2/CO_2 = 3, T = 270 °C, P = 1 MPa, $\tau_{mod,n}$ = 4000 kg s/m^3; CSP2 membrane: Ψ = 3.3, Φ = 0.7, sweep gas: H_2

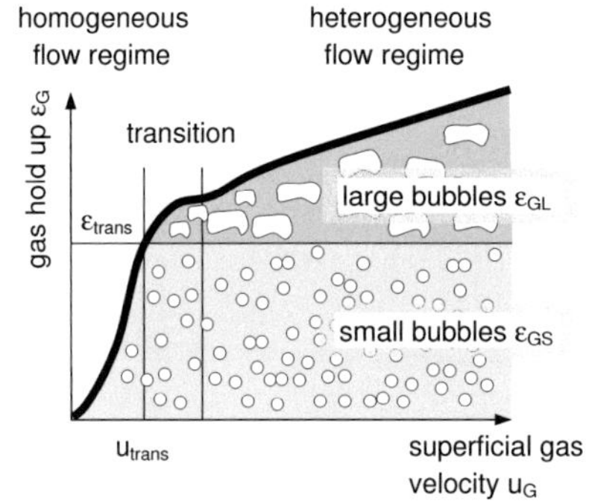

Figure 12.52. Gas hold up of small and large bubble as function of the superficial gas velocity, (according to Sie & Krishna 1999).

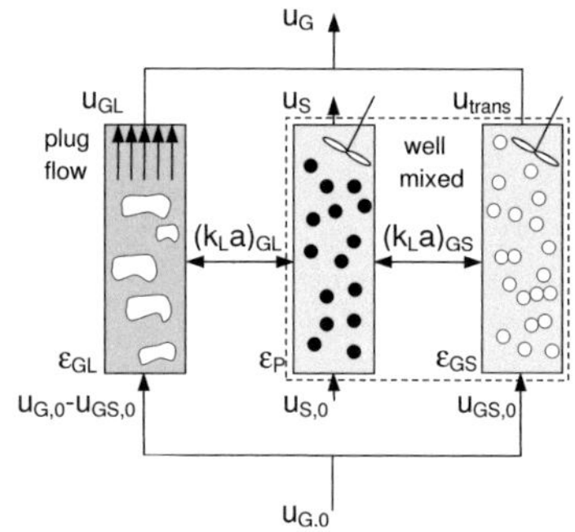

Figure 12.53. Schematic of the "generalized two phase model": (a) "dilute phase" represented by large bubble class in plug-flow, (b) "dense phase" presented by completely mixed liquid phase (with suspended catalyst particles) and completely mixed small gas bubble class (according to Krishna et al. 1996).

12.7.2 Three-phase slurry bubble column reactor

The reaction engineering model for a three-phase slurry bubble column reactor was adopted from van der Laan (1999), which was based on the generalized two-phase model (Krishna, Ellenberger & Sie 1996, de Swart 1996).
The slurry bubble column is operated in the heterogeneous flow regime which is characterized by the presence of two distinct bubble classes: large bubbles and small bubbles (Figure 12.52). The large bubble class is considered in the reactor model as dilute phase in plug-flow; the small bubble class and the liquid phase with the suspended catalyst are considered as completely mixed phases (Figure 12.53). The model equations are summarized in Table 12.32. The hydrodynamic parameters and the physical properties and distribution coefficients are given in Table 12.33 and Table 12.34.

There are two differences with regard to the set of the formulas applied by van der Laan (1999). First, the gas velocity of the small bubbles u_{GS} was set to u_{trans} instead of $u_{GS,0}$; the inlet velocity of the small bubble class $u_{GS,0}$ was calculated by the species balance (12.29), resulting in $u_{GS,0} > u_{GS} = u_{trans}$. This has an effect on gas distribution between dilute and dense phase; furthermore, deviations in mass balance can be avoided.

Second, an exchange term representing the hydrophilic membrane was included in equation (12.30) for the completely mixed liquid phase. This is definitely a simplified approach, but due to the fast mass transfer between the liquid and the gas phase, it was regarded as satisfactory. More complex models considering the mass transfer contributions of the dense and dilute phases are presented in Deshmukh et al. (2007).

The detailed membrane integration such as horizontal or vertical membrane modules or co-current, counter-current or cross-flow operation was not considered as this would have defeated the purpose of a case study.

Table 12.32. Model equations for an isothermal slurry bubble column reactor (equations based on van der Laan 1999).

Species balance

Large bubbles GL, plug-flow:

$$\frac{d\left(u_{GL} c_{GL,i}\right)}{dz} = -(k_L a)_{GL,i}\left(\frac{c_{GL,i}}{m_i} - c_{i,L}\right) \tag{12.27}$$

Small bubbles GS, completely mixed:

$$\frac{1}{H}\left(u_{GS,0} c_{GS,0,i} - u_{GS} c_{GS,i}\right) = (k_L a)_{GS,i}\left(\frac{c_{GS,i}}{m_i} - c_{i,L}\right) \tag{12.28}$$

$$u_{GS,0} = u_{GS} + \frac{H}{c_G}\sum_{i=1}^{N}(k_L a)_{GS,i}\left(\frac{c_{GS,i}}{m_i} - c_{i,L}\right) \quad \text{with} \quad u_{GS} = u_{trans} \tag{12.29}$$

Liquid phase L, completely mixed:

$$\frac{1}{H}\int_0^H (k_L a)_{GL,i}\left(\frac{c_{GL,i}}{m_i} - c_{i,L}\right)dz + (k_L a)_{GS,i}\left(\frac{c_{GS,i}}{m_i} - c_{i,L}\right) + \frac{1}{H}\left(u_{S,0} c_{L,0,i} - u_S c_{L,i}\right) + \varepsilon_L \varepsilon_P \rho_P \sum_{j=1}^{J} \nu_{ij} r'_j + \varepsilon_L \varepsilon_P \rho_P a_m Q_{ref} S_{i,ref}\left(c_{L,i} m_i RT - P_{S,i}\right) \tag{12.30}$$

Table 12.33. Hydrodynamic parameters for the isothermal slurry bubble column reactor model (van der Laan 1999).

$u_{GS} = u_{trans} = \varepsilon_{GS} u_{GS,ref}\left(1 + 0.8\varepsilon_P / u_{GS,ref}\right)$	with $u_{GS,ref} = 0.095$ m/s
$\varepsilon_{GS} = \varepsilon_{GS,ref}\left(\rho_G / \rho_{G,ref}\right)^{0.48}\left(1 - 0.7\varepsilon_P / \varepsilon_{GS,ref}\right)$	with $\varepsilon_{GS,ref} = 0.27$ and $\rho_{ref} = 1.3$ kg/m^3
$\varepsilon_{GL} = 0.3\left(\bar{u}_G - u_{GS}\right)^{0.58}\left(\rho_G / \rho_{G,ref}\right)^{0.5}$	
$\varepsilon_G = \varepsilon_{GL} + \varepsilon_{GS}\left(1 - \varepsilon_{GL}\right)$	
$(k_L a)_{GL,i} = 0.5\varepsilon_{GL}\sqrt{D_i / D_{ref}}$	with $D_{ref} = 2 \cdot 10^{-9}$ m^2/s
$(k_L a)_{GS,i} = 1.0\varepsilon_{GS}\sqrt{D_i / D_{ref}}$	with $D_{ref} = 2 \cdot 10^{-9}$ m^2/s

Table 12.34. Physical properties and distribution coefficients of key components applied in the isothermal slurry bubble column reactor model (calculated according to Marano & Holder 1997, van der Laan 1999).

			H_2	CO	CO_2	H_2O	N_2	CH_4
230°C	$m_i = c_{G,i} / c_{L,i}$	/-	6.47	5.23	2.38	0.82	6.14	2.94
250°C	$m_i = c_{G,i} / c_{L,i}$	/-	5.83	4.86	2.38	0.9	5.66	2.82
270°C	$m_i = c_{G,i} / c_{L,i}$	/-	5.27	4.51	2.37	0.98	5.22	2.70
250°C	D_i	/ 10^{-8} m^2/s	3.85	1.53	1.26	1.89	1.54	